Carl Wehner

FOUNDATIONS OF DISCRETE MATHEMATICS

FOUNDATIONS OF DISCRETE MATHEMATICS

Albert D. Polimeni
H. Joseph Straight
State University College
Fredonia, New York

Brooks/Cole Publishing Company
Monterey, California

Brooks/Cole Publishing Company
A Division of Wadsworth, Inc.

Printed in the United States of America

10 9 8 7 6 5 4 3 2 1

Library of Congress Cataloging in Publication Data

Polimeni, Albert D. [date]
Foundations of discrete mathematics.

Bibliography: p.
Includes index.
1. Mathematics—1961– . I. Straight, H. Joseph,
[date]. II. Title.
QA39.2.P67 1985 510 84-23309
ISBN 0-534-03612-0

Sponsoring Editor: Craig Barth
Production Editors: Penelope Sky and Cece Munson
Manuscript Editor: Susan Gerstein
Interior and Cover Design: Victoria A. Van Deventer
Cover Photo: Jeffrey R. Helwig
Art Coordinator: Judith Macdonald
Interior Illustration: John Foster
Typesetting: Bi-Comp, Inc.
Cover Printing: Phoenix Color Corporation
Printing and Binding: R. R. Donnelley & Sons Company

Preface

This book is an outgrowth of notes used for a sophomore-level course at S.U.N.Y. College at Fredonia. It is intended to be a first encounter with abstract mathematics and, at the same time, an introduction to those topics in mathematics basic to the study of computer science. As such, we feel that the subject matter covered forms the basis of a required course for both mathematics and computer science majors. Students with one year of college-level mathematics, including a semester of calculus and an introductory programming course using a high-level language, are adequately prepared for such a course.

Recent CUPM guidelines recommend two courses in the area of discrete mathematics. One is a course in discrete methods and included topics from graph theory and combinatorics; the other is a course in abstract algebra with applications. This book contains prerequisite material for both of these courses. In addition, such material provides a treatment of those concepts needed by computer science students for upper-level coursework in data structures, algorithms, and the theory of computation.

We have attempted throughout the text to keep the level of discussion within easy reach of the student. At the same time care has been taken to respect rigor in the mathematics. It is essential at this level that the student learn to read mathematical proofs. This is initially dealt with in Chapter 1, where the formal language of mathematics and various proof techniques are covered. In the remaining chapters, theorems are clearly displayed and stated. If the proof of a theorem is given and the method used is other than direct, then the particular technique of proof is specified, and we have strived to make the presentation both precise and readable.

In many instances we have chosen to illustrate mathematical principles and theorems in the form of algorithms, usually presented as program segments written in the Pascal language. In the exercises, students are often asked to create an iterative or recursive algorithm. Chapter 3, besides providing much needed practice with the elementary properties of divisibility and prime factorization, introduces the student to some important arithmetic algorithms.

Since there is no universal agreement as to what topics should be included in a text of this sort, we have endeavored to cover those areas from which a course at the sophomore level can be reasonably designed. In our one-semester, three-hour course, most of the material in Chapters 1 through 5 is covered, with a few additional topics selected from Chapters 6 and 7 as time permits. For this reason, the last two chapters are designed so as to provide flexibility. Chapter 6 is arranged so that if an instructor wishes to emphasize the material on Boolean algebras and switching networks, this can be conveniently done by covering Sections 6.1, 6.4, and 6.5. Likewise, the first six sections of Chapter 7 provide a moderate introduction to graph theory, with some indications of its applications. The remaining sections address several of the better-known graph-theoretic algorithms, and these can be treated selectively. For example, the course could proceed directly to Section 7.8, Minimum Spanning Trees, after covering Section 7.4, Trees. Whatever outline the instructor chooses to follow, our hope is that the exposition will aid in preparing students for additional study in both mathematics and computer science.

There are exercises at the end of each section as well as problems at the end of each chapter. The exercises focus on the material covered in that section, providing the student with needed practice in applying those concepts most recently encountered. Chapter problems, on the other hand, are intended not only for review but to deepen the student's exposure to key ideas. Many of the problems ask the student to prove something, and some problems invite the student to explore some aspect of a topic not covered in detail in the text. The more challenging exercises are indicated with a star (★). Answers are provided in Appendix A for most of the odd-numbered exercises and problems.

We wish to express our sincere thanks to Nancy Boynton, Fred Byham, Gary Chartrand, Bruce Chilton, and Eugene Rozycki for class-testing this manuscript and for providing us with many helpful suggestions and corrections. We would also like to express our sincere appreciation to the reviewers, who supplied many valuable ideas for additional topics, exercises, and the general layout of the book. They are Richard Brualdi, University of Wisconsin; David Buchthal, University of Akron; John Buoni, Youngstown State University; Kent Harris, Western Illinois University; Larry Heath, University of Texas; Frieda Holley, Metropolitan State College; David Jonah, Wayne State University; S. Lakshmivarahan, University of Oklahoma at Norman; Ron Loser, Adams State College; James Maxwell, Oklahoma State University; Wayne Powell, Oklahoma State University; James Simpson, University of Kentucky; and

Thomas Upson, Rochester Institute of Technology. Finally, we are especially grateful to Craig Barth, Mathematics Editor at Brooks/Cole, for his valuable assistance, his patience with us, and his encouragement throughout the preparation of this text.

Albert D. Polimeni
H. Joseph Straight

Contents

FOUNDATIONS OF DISCRETE MATHEMATICS

One Logic

1.1 INTRODUCTION TO LOGIC

The theorems of any mathematical system and the algorithms that make up a program system are developed and put together using the rules of logic. Thus, in order to understand mathematical proofs and the development of program systems, it is important to be comfortable with these rules.

In this chapter, emphasis will be placed on certain common forms of exposition and reasoning, independent of any particular application. If these are learned well, then the student will find it easier to follow the line of reasoning in upper-level mathematics and computer science courses. As a result, the student can focus on the content of a course and not be distracted by the logical forms being used.

The logic of mathematics furnishes a set of ground rules governing the validity and development of a mathematical proof. Logic is the one aspect of proof that can be learned; the rest is intuition, creativity, imagination, and instinct.

We ask, then, "What is the logic of an argument?" Crudely put, we could say that the logic of an argument is what is left over when the particular meaning of the argument has been removed. In other words, the logic of an argument is its form or syntax. A similar distinction can be made about the statements in a programming language such as Pascal. The statements must obey certain syntactic rules, apart from their actual semantic interpretations. For example, the Pascal assignment statement

```
SUM := SUM + NEXT
```

is syntactically correct; however, when taken out of the context of a particular program, its meaning is lost.

An example will help to clarify the preceding comments.

Example 1.1 Consider the following argument about Randy, who is a student in a particular course.

> If Randy has the ability and works hard, then Randy will be successful in this course. Therefore, if Randy is not successful in this course, then Randy does not have the ability or Randy does not work hard.

This argument can be symbolized by letting p, q, and r represent the following statements:

p: Randy has the ability.
q: Randy works hard.
r: Randy will be (is) successful in this course.

The logical form of the initial argument may then be given as follows:

If p and q, then r. Therefore, if not r, then not p or not q. ■

Note that the abstracted form of an argument such as the one in the preceding example is not bound to any particular content. Logic provides a means for validating such an abstract form in a way that is independent of the truth or falsity of its constituent statements. This is done through the use of "symbolic logic" and the construction of "truth tables."

The term "statement" has already been used several times in the foregoing discussion, and it is important to understand what is meant by this term. Sentences such as

Everyone in this class will get an A.

or

The number 72 is positive.

are examples of "declarative sentences"—each is an assertion. On the other hand, the question

What time is it?

is not a declarative sentence.

For purposes of mathematical logic, we consider declarative sentences to which a *truth-value* can be assigned; in particular, a declarative sentence that is true will have truth-value "T," while a false declarative sentence will have truth-value "F." A *statement* (or *proposition*) is a declarative sentence for which the truth-value is definitely known or can be validly determined. For instance, each of the sentences

If 6 is divided by 3, then the quotient is an odd integer.

and

The product of 117, 118, and 119 is divisible by 6.

are statements. In fact, the first is false and the second is true. However, each of the declarative sentences

The number x is positive.

and

He plays baseball.

is not a statement, because offhand, its truth-value cannot be determined. If, in the first sentence, "x" is replaced by "-3" to obtain

The number -3 is positive.

then the sentence becomes a (false) statement. Similarly, we might replace "He" in the second sentence by "Reggie Jackson," forming the (true) statement

Reggie Jackson plays baseball.

Symbols like "x," as seen in the above example, are called *variables,* because they are used to represent any one of a number of possible values. We shall have more to say later about sentences that become statements when the variables in the sentence are given particular values.

Lest the student be misled, it should be pointed out that certain sentences that contain variables are also statements. For instance, the sentence

If $x = 2$, then $x^3 = 8$.

is a statement, since it is true for any value of x, as will be seen in the next section.

In what follows, we shall use lower-case letters, such as p, q, or r, and letters with subscripts, such as p_1, q_2, or r_3, to denote statements; accordingly, these are called *propositional variables*.

Exercises 1.1

Indicate which of the following are declarative sentences and which are statements.

a. The integer 24 is even.
b. Is the integer $3^{15} - 1$ even?
c. The product of 2 and 3 is 7.
d. The sum of x and y is 3.
e. If the integer x is odd, is x^2 odd?
f. It is not possible for the integer x to be both even and odd.
g. The product of the integers x^2 and x^3 is x^6.
h. The integer $2^{524287} - 1$ is prime. (Recall that an integer $n \geq 2$ is prime provided the only positive factors of n are 1 and n.)

1.2 THE LOGICAL CONNECTIVES

In research, in the classroom, or at a roadside pub, mathematicians (and possibly others) are frequently interested in determining the truth-value of a given mathematical statement. Most mathematical statements are combinations of simpler statements, formed through some choice of the words *or, and,* and *not,* or the phrases *if—then—,* and *if and only if.* These are called the *logical connectives* (or simply the *connectives*) and are defined next. Many mathematical statements can be symbolized by a single propositional variable or by some meaningful combination of propositional variables and logical connectives. A propositional variable will be referred to as a *primitive* (or simple) *statement,* while any combination of propositional variables and logical connectives will be called a *compound statement*.

DEFINITION 1.1

Let p and q be propositional variables.

1. The *disjunction* of p and q is the compound statement

$$p \text{ or } q$$

and is true provided at least one of p or q is true; otherwise, it is false. We denote the disjunction of p and q by $p \vee q$.

2. The *conjunction* of p and q is the compound statement

$$p \text{ and } q$$

and is true provided both p and q are true. We denote the conjunction of p and q by $p \wedge q$.

3. The *negation* of p is the compound statement

$$\text{not } p$$

and has truth-value opposite that of p. It is denoted by $\sim p$.

Example 1.2 Which integers x satisfy the following condition?

$$x > -2 \quad \text{and} \quad x < 3$$

Solution The integers $-1, 0, 1, 2, 3, 4$, etc., satisfy the condition $x > -2$. The integers $2, 1, 0, -1, -2, -3$, etc., satisfy the condition $x < 3$. We want those integers that satisfy both conditions, namely, those that are both greater than -2 and less than 3. There are four such integers: $-1, 0, 1$, and 2. ■

Example 1.3 The Boston Red Sox, the Cleveland Indians, the Milwaukee Brewers, and the New York Yankees are professional baseball teams. Suppose on a given night that the Red Sox play the Yankees and the Indians play the Brewers. Both games are completed, and the next morning someone makes the statement

The Brewers or the Red Sox won last night.

Find the negation of this compound statement.

Solution The given compound statement is of the form $p \vee q$, where p represents the statement

The Brewers won last night.

and q represents the statement

The Red Sox won last night.

By definition, $p \vee q$ is true provided at least one of p or q is true. Thus $p \vee q$ is false, and $\sim(p \vee q)$ is true, only when *both* p and q are false. It follows that the negation of the given compound statement is

Both the Brewers and the Red Sox lost last night.

(In the next section we will show that the negation of any compound statement of the form $p \vee q$ is the compound statement $\sim p \wedge \sim q$.) ■

Example 1.4 Determine the output of the following Pascal program segment.

```
X := 1;
WHILE NOT (X > 50) DO
  BEGIN
    WRITELN(X);
    X := 2 * X
  END;
```

Solution The value of the variable X is initialized to 1. Then, as long as the value of X is not greater than 50, the value is output and doubled. Thus the values 1, 2, 4, 8, 16, and 32 will be output. When the value of X reaches 64, the condition NOT (X $>$ 50) is false, which completes the program segment. ■

These examples, though somewhat elementary, demonstrate the use of the logical connectives *or, and,* and *not* in both mathematical and programming settings. Indeed, these connectives are used in a notable way—as "Boolean" operators—in many programming languages; their use in Pascal is typical.

Pascal provides a built-in Boolean data type. A variable that is declared to be of this type has exactly one of two possible values: FALSE or TRUE. For instance, suppose a person is writing a program to keep score in a volleyball game. It might be desirable to have variables A and B of type Integer for the scores of the two teams, and a Boolean variable AWINS, which will have value TRUE if team A wins the game and value FALSE otherwise. For team A to win the game, they must score at least 15 points and be at least 2 points ahead of team B. It would thus make sense and be syntactically correct for such a program to contain the assignment statement

```
AWINS := (A >= 15) AND (A - B >= 2)
```

The expression "(A >= 15) AND (A − B >= 2)" is called a "Boolean expression" because its value is either TRUE or FALSE, rather than an integer, or real, or value of some other type. (Note that each of the conditions "A >= 15" and "A − B >= 2" is either true or false.) Hence it is perfectly valid to assign the value of the expression to the Boolean variable AWINS.

Pascal provides the Boolean operators NOT, AND, and OR, whose operands are Boolean expressions. Specifically, the syntax is

NOT ⟨Boolean expression⟩
⟨Boolean expression$_1$⟩ AND ⟨Boolean expression$_2$⟩
⟨Boolean expression$_1$⟩ OR ⟨Boolean expression$_2$⟩

In addition, Pascal uses Boolean expressions in conditional statements and in statements used to program loops. Consider, for example, the syntax for the two-branch conditional

IF ⟨Boolean expression⟩ THEN ⟨statement$_1$⟩ ELSE ⟨statement$_2$⟩

or that for the WHILE loop

WHILE ⟨Boolean expression⟩ DO ⟨statement⟩

Example 1.5 A data structure composed of a fixed number of elements of the same type, organized as a simple linear sequence, is called a *vector*. The elements of the vector are normally *subscripted* using an integer subrange, or by some enumeration. A vector is also termed a *one-dimensional* or *linear array*.

A standard special case is that of a vector that is subscripted using the positive integers 1, 2, . . . , MAX, where MAX is a constant. In Pascal we could declare a type VECTOR, with 50 elements of type REAL, as follows:

```
CONST MAX = 50;
TYPE SUBSCRIPT = 1..MAX;
     ELEMENTTYPE = REAL;
     VECTOR = ARRAY [SUBSCRIPT] OF ELEMENTTYPE;
```

(Note that the subscript range or the type of elements for the type VECTOR may be modified by changing the constant MAX or the type ELEMENTTYPE.)

An important problem concerning vectors is that of searching a given vector V for a given value KEY. Suppose we wish to know whether one of the elements V[1], V[2], . . . , V[LAST] is equal to KEY, and if so, then we would also like to know a subscript POSITION such that V[POSITION] = KEY.

A simple and general method for answering these questions is called *linear searching*; it simply compares KEY with V[1], V[2], and so on, in turn. How should such a search be controlled in Pascal? Well, suppose V[CURRENT] is the current value with which KEY is being compared, and that the Boolean variable FOUND indicates whether a value equal to KEY has been found. Initially, then, CURRENT = 1 and FOUND = FALSE. The search should continue as long as FOUND = FALSE and CURRENT <= LAST. This suggests using a WHILE loop to control the search, as seen in the following program segment:

```
(*begin search*)
CURRENT := 1;
FOUND := FALSE;
WHILE (NOT FOUND) AND (CURRENT <= LAST) DO
  IF V[CURRENT] = KEY THEN
    BEGIN
      FOUND := TRUE;
      POSITION := CURRENT
    END
  ELSE CURRENT := CURRENT + 1;
(*end search*)
```
■

The truth-value of a given compound statement can be determined from the truth-values of the primitive statements of which it is composed. We now demonstrate how a *truth table* can be used to examine a given compound statement. The table will contain a column for each propositional variable in the statement and a column for the whole statement. If the statement is particularly complex, other columns may be used for certain substatements. The table contains a row for each possible combination of truth-values of the propositional variables involved.

In Figure 1.1 we give the truth tables for *or, and,* and *not*. It should be noted that there are four possible combinations for the truth-values of two propositional variables p and q.

Consider next the compound statement

$$\text{If } p, \text{ then } q.$$

p	q	$p \vee q$
T	T	T
T	F	T
F	T	T
F	F	F

p	q	$p \wedge q$
T	T	T
T	F	F
F	T	F
F	F	F

p	$\sim p$
T	F
F	T

FIGURE 1.1 Truth tables for $p \vee q$, $p \wedge q$, and $\sim p$

How is the truth-value of this statement determined from the truth-values of the propositional variables p and q? An example will help to clarify the situation.

Example 1.6 Consider the compound statement

> If you score 90 or more on the final exam, then you get an A for this course.

which an instructor might make to a student. Letting p and q denote the statements

p: You score 90 or more on the final exam.
q: You get an A for this course.

then symbolically the instructor's statement is represented

If p, then q.

Let us analyze this statement for each of the four possible truth-value combinations. (Assume that the semester is over and that the student's final exam score and course grade are known.)

Case 1. Both p and q are true. In this case the student scored 90 or more on the final and got an A, just as the instructor promised. So the given statement is true.

Case 2. p is true and q is false. The student scored at least 90 on the final, but for some reason did not get an A. Perhaps the instructor made an error, but based on the evidence we must conclude that the instructor's statement is false.

Case 3. p is false and q is true. The student scored less than 90 on the final but got an A anyway. Perhaps the student got an 89 on the final and the instructor, being in a good mood, decided to give the student the A. The point is that this does not contradict the instructor's statement; we must take the statement to be true.

Case 4. Both p and q are false. The student scored less than 90 on the final and did not get an A. This is much like the previous case. In a sense, the validity of the instructor's statement has not been tested, so we must take it to be true.

Thus we see in this example that the only case in which the statement

If p, then q.

is false is when p is true and q is false. Examples like this one motivate the next definition. ■

DEFINITION 1.2

Let p and q be propositional variables. The compound statement

If p, then q.

is called the *implication of q by p*. It is false when p is true and q is false, and it is true otherwise. We symbolize the statement by $p \to q$. It is common to refer to the compound statement $p \to q$ as a *conditional statement* or simply as an *implication*. We often read $p \to q$ as "p implies q," and call p the *hypothesis* and q the *conclusion*.

The truth table for the implication is shown in Figure 1.2.

p	q	$p \to q$
T	T	T
T	F	F
F	T	T
F	F	T

FIGURE 1.2 Truth table for $p \to q$

Example 1.7 Consider the following Pascal program segment:

```
A := 1;
IF (X < -1) OR (Y > 2) THEN A := 2;
WRITELN(A);
```

What will be output by this program segment if X and Y have, respectively, each of the following values?

(a) −2, 1 (b) 1, 4 (c) −2, 4 (d) 1, 1

Solution (a) 2 (b) 2 (c) 2 (d) 1

(Notice that A will be assigned the value 2 unless both of the conditions $X < -1$ and $Y > 2$ are false.) ■

Next consider the statements

(a) If $x = 1$, then $x^2 = 1$.
(b) If $x^2 = 1$, then $x = 1$.

which concern a real number x. It is clear that these statements are different yet related. Here (a) is true and (b) is false. On the other hand, the statements

(a) If $x = 1$ or $x = -1$, then $x^2 = 1$.
(b) If $x^2 = 1$, then $x = 1$ or $x = -1$.

though related in the same way, are both true. In each case the statements, taken together, are of the type

(a) If p, then q.
(b) If q, then p.

DEFINITION 1.3

Let p and q be propositional variables. The conditional statement

If q, then p.

is called the *converse* of the conditional statement

If p, then q.

It occurs frequently in mathematics that a compound statement with the logical form

$$(p \rightarrow q) \wedge (q \rightarrow p)$$

is true; that is, that both the statement $p \rightarrow q$ and its converse $q \rightarrow p$ are true. Hence there is good reason to make the next definition.

DEFINITION 1.4

Let p and q be propositional variables. The compound statement

$$(p \rightarrow q) \wedge (q \rightarrow p)$$

is called the *biconditional* or *equivalence*. The statement is read "p if and only if q" and is denoted by $p \leftrightarrow q$. A common shorthand form of "p if and only if q" is "p iff q."

The truth table for the biconditional $p \leftrightarrow q$, shown in Figure 1.3, is not difficult to complete when the truth tables for the conditional and the conjunction are kept in mind. Notice that $p \leftrightarrow q$ is true only when p and q have the same truth-value. Let us emphasize once again the meaning of the statement $p \leftrightarrow q$, namely, that in order to show that $p \leftrightarrow q$ is true it suffices to show that both $p \rightarrow q$ and $q \rightarrow p$ are true.

p	q	$p \rightarrow q$	$q \rightarrow p$	$p \leftrightarrow q$
T	T	T	T	T
T	F	F	T	F
F	T	T	F	F
F	F	T	T	T

FIGURE 1.3 Truth table for $p \leftrightarrow q$

This is a good spot to discuss the use of parentheses in a compound statement. For example, consider the statement $\sim p \vee q$. Does it mean $(\sim p) \vee q$ or does it mean $\sim(p \vee q)$? In fact, convention has it that $(\sim p) \vee q$ is the correct interpretation. Questions of this sort can always be resolved by employing parentheses, but their overuse can be quite tedious and can at times make a statement somewhat difficult to read. Hence we shall adopt a basic rule that allows us to omit certain parentheses in compound statements. This rule is based on the idea of "precedence" of the logical connectives.

Precedence rule for the logical connectives

In a parenthesis-free statement, the logical connectives are applied in the following order:

Connective	*Precedence*
$\sim$	first
$\wedge$	second
$\vee$	third
$\rightarrow$	fourth
$\leftrightarrow$	fourth

We say that *not* precedes *and, and* precedes *or,* and so on. (These precedence rules for *not, and,* and *or* agree with the rules for evaluating Boolean expressions in Pascal and FORTRAN77.) Note that "$\rightarrow$" and "$\leftrightarrow$" are at the same precedence level. If a compound statement involves these two connectives, we shall agree to use parentheses to make it clear how they are to be applied.

Example 1.8 Here are several compound statements and their correct interpretations.

	Statement	*Interpretation*
(a)	$\sim p \vee q$	$(\sim p) \vee q$
(b)	$p \vee \sim q \wedge r$	$p \vee ((\sim q) \wedge r)$
(c)	$\sim q \rightarrow \sim p$	$(\sim q) \rightarrow (\sim p)$
(d)	$(p \rightarrow q) \leftrightarrow \sim p \vee q$	$(p \rightarrow q) \leftrightarrow ((\sim p) \vee q)$
(e)	$p \wedge q \rightarrow (p \leftrightarrow q)$	$(p \wedge q) \rightarrow (p \leftrightarrow q)$

■

Exercises 1.2

1. Let p, q, and r denote the following statements:

p: Ralph reads The New York Times.
q: Ralph watches the MacNeil–Lehrer Report.
r: Ralph jogs 3 miles.

Write each of the following compound statements symbolically.

a. Ralph reads The New York Times and watches the MacNeil–Lehrer Report.
b. Ralph reads The New York Times or jogs 3 miles.
c. If Ralph reads The New York Times, then he doesn't watch the MacNeil–Lehrer Report.
d. Ralph reads The New York Times if and only if he jogs 3 miles.
e. It is not the case that if Ralph jogs 3 miles then he reads The New York Times.
f. Ralph watches the MacNeil–Lehrer Report or jogs 3 miles, but not both.

2. Let the propositional variables p, q, and r be defined as in Exercise 1. Express each of the following compound statements in words.

a. $p \wedge r$ **b.** $q \vee r$
c. $(p \wedge q) \vee r$ **d.** $\sim p \vee \sim q$
e. $p \rightarrow q$ **f.** $q \leftrightarrow r$

3. Consider the following statements:

p: Dodger pitcher Fernando Valenzuela had a sore arm in 1981.
q: The Los Angeles Dodgers won the 1981 World Series.

The statement p is false; q is true. Express each of the following compound statements symbolically and determine its truth-value.

a. Dodger pitcher Fernando Valenzuela had a sore arm in 1981 or the Dodgers won the 1981 World Series.
b. Dodger pitcher Fernando Valenzuela had a sore arm in 1981 and the Dodgers won the 1981 World Series.
c. If Dodger pitcher Fernando Valenzuela had a sore arm in 1981, then the Dodgers won the 1981 World Series.

d. If the Los Angeles Dodgers did not win the 1981 World Series, then Dodger pitcher Fernando Valenzuela had a sore arm that year.

e. Dodger pitcher Fernando Valenzuela had a sore arm in 1981 if and only if the Dodgers won the World Series that year.

f. The Dodgers won the 1981 World Series if and only if pitcher Fernando Valenzuela did not have a sore arm that year.

4. For each of the following compound statements, first identify the primitive statements p, q, r, etc., of which it is composed. Then express the statement symbolically.

a. If n is an integer, then either n is even or n is odd.

b. If an integer is prime and is greater than 2, then the integer is odd.

c. If the number x is not negative and its square is less than 4, then either x is zero or x is positive and less than 2.

d. If an integer is even and greater than 2, then it is not prime.

5. Express each of the following compound statements symbolically.

a. If triangle ABC is equilateral, then it is isosceles.

b. The number $x = 3$ if and only if $2x - 3 = 3$.

c. If $\pi^{\sqrt{2}}$ is a real number, then either $\pi^{\sqrt{2}}$ is rational or $\pi^{\sqrt{2}}$ is irrational.

d. The product $xy = 0$ if and only if $x = 0$ or $y = 0$.

e. If the integer n is greater than 200, then, if n is prime, n is greater than 210.

f. If line k is perpendicular to line m and line m is parallel to line n, then line k is perpendicular to line n.

g. If n is an integer, then either n is positive or n is negative or $n = 0$.

h. If a and b are integers and $b \neq 0$, then a/b is a rational number.

6. a. Rewrite the program segment for the linear search (Example 1.5) using a REPEAT-UNTIL loop instead of a WHILE loop.

b. How is the Boolean expression that controls the REPEAT-UNTIL loop related to the Boolean expression that controls the WHILE loop?

7. We mentioned the precedence rule for the Boolean operators NOT, AND, and OR; namely, NOT precedes AND and AND precedes OR. Pascal, of course, also has arithmetic operators, such as "+" and "*," and relational operators, such as "=" and "<."

a. How do the Boolean operators fit into the overall precedence scheme?

b. Is the expression "A >= 15 AND A − B >= 2" valid? Explain.

c. Is the expression "NOT FOUND AND (CURRENT <= LAST)" valid? Explain.

★8. A *list* is a linear data structure composed of a variable number of elements of possibly different types. However, if the elements all have the same type, then one may define a data type LIST in Pascal by using a record with two fields: a field ELEMENT (of type VECTOR) to hold the elements of the list, and a field LENGTH (of type 0..MAX) to hold the current length of the list. The type LIST, and a variable L of type LIST, could be declared as follows:

```
CONST MAX = 50;   (*or any other positive integer*)
TYPE  SUBSCRIPT = 1..MAX;
      ELEMENTTYPE = REAL;   (*or whatever type you
      wish*)
      VECTOR = ARRAY [SUBSCRIPT] OF ELEMENTTYPE;
      LIST = RECORD
        ELEMENT : VECTOR;
        LENGTH : 0..MAX
      END;
VAR   L : LIST;
```

Then L.LENGTH is the length of the list L, and if L.LENGTH is positive, then L.ELEMENT[J] is the Jth element of L, $1 \leq J \leq$ L.LENGTH.

a. Write a Pascal procedure SEARCH to perform a linear search of the list L for the value KEY. Return FOUND and POSITION, as defined in Example 1.5.

b. Assume that the elements of the list L have been sorted so that L.ELEMENT[1] ≤ L.ELEMENT[2] ≤ . . . ≤ L.ELEMENT[L.LENGTH]. Write a procedure COMPACT that removes any duplicate values from the list L; that is, COMPACT will return a list of distinct values.

9. In volleyball it is important to know which team is serving, for a team scores a point only if that team is serving and wins a volley. If the serving team loses the volley, then the other team gets to serve. Thus in a program to keep score in a volleyball game (between teams A and B), it may be useful to have Boolean variables ASERVES and AWINSVOLLEY. (ASERVES is true if team A is serving, false otherwise; AWINSVOLLEY is true if team A wins the current volley, false if team B wins it.)

a. Write a Boolean expression that is true if team A scores a point, false otherwise.

b. Write a Boolean expression that is true if team B scores a point, false otherwise.

c. Write a Boolean expression that is true if the serving team loses the current volley, false otherwise.

d. Write an assignment statement that changes the serving team.

10. In mathematics, the connective "or" is used inclusively, meaning "one or the other, or both." However, in everyday language "or" is often used in the exclusive sense, as in the sentence

With your order you may have french fries *or* potato salad.

Used in this way the "or" is interpreted as "one or the other, but not both." We shall use the symbol $\veebar$ to represent the connective "exclusive or."

a. Construct the truth table for $\veebar$.

b. Write a statement equivalent to $p \veebar q$ using the connectives $\sim$, $\vee$, and $\wedge$.

1.3 LOGICAL EQUIVALENCE

In mathematics, as in most subjects, there may be several different ways to say the same thing. We would like to define formally what this means for logical statements. In what follows we use the word *formula* to mean either a primitive or compound statement.

DEFINITION 1.5

Let u and v be formulas. We say that u and v are *logically equivalent,* denoted $u \equiv v$, provided u and v have the same truth-value for every possible choice of truth-values for the propositional variables involved in u and v.

The two examples of logical equivalence that follow are important in that both are used often in mathematics.

Example 1.9 Show that $\sim(p \vee q) \equiv \sim p \wedge \sim q$.

Solution We do this by constructing a truth table and comparing the columns labelled by $\sim(p \vee q)$ and $\sim p \wedge \sim q$. The truth table is shown in Figure 1.4. Since the columns headed by $\sim(p \vee q)$ and $\sim p \wedge \sim q$ agree in every row, these formulas are logically equivalent.

p	q	$\sim p$	$\sim q$	$p \vee q$	$\sim(p \vee q)$	$\sim p \wedge \sim q$
T	T	F	F	T	F	F
T	F	F	T	T	F	F
F	T	T	F	T	F	F
F	F	T	T	F	T	T

FIGURE 1.4 Truth table showing $\sim(p \vee q) \equiv \sim p \wedge \sim q$ ■

Example 1.10 Use a truth table to verify that $p \to q \equiv \sim p \vee q$.

Solution See Figure 1.5.

p	q	$\sim p$	$p \to q$	$\sim p \vee q$
T	T	F	T	T
T	F	F	F	F
F	T	T	T	T
F	F	T	T	T

FIGURE 1.5 Truth table showing $p \to q \equiv \sim p \vee q$ ■

Example 1.11 The *contrapositive* of the implication $p \to q$ is the implication $\sim q \to \sim p$. In words, the contrapositive states

If not q, then not p.

We wish to show that an implication $p \to q$ and its contrapositive $\sim q \to \sim p$ are logically equivalent. We could do it using a truth table, but we will demonstrate how it can be done without truth tables.

We shall require several facts. First, if u, v, and w are formulas with $u \equiv v$ and $v \equiv w$, then clearly $u \equiv w$. Secondly, from Example 1.10 we have that

$$s \to t \equiv \sim s \vee t \tag{1}$$

It is also not difficult to see that

$$s \vee t \equiv t \vee s \tag{2}$$

(We use the propositional variables s and t here to avoid confusion with the variables p and q, which appear in the logical equivalence we are trying to establish.) Now, by (1),

$$p \to q \equiv \sim p \vee q$$

Next, by (2),

$$\sim p \vee q \equiv q \vee \sim p$$

Finally, the hard step: by (1), with s replaced by $\sim q$ and t replaced by $\sim p$, we have that

$$q \vee \sim p \equiv \sim q \to \sim p$$

It thus follows that

$$p \to q \equiv \sim q \to \sim p$$ ■

Some formulas have the seemingly dull property of always being true. However, they are given an interesting name.

DEFINITION 1.6

A formula that is true for all possible truth-values of its constituent propositional variables is called a *tautology*. A formula that is false for all possible truth-values of its constituent propositional variables is called a *contradiction*.

Example 1.12 In Figure 1.6 is a truth table for the formula

$$(p \to q) \leftrightarrow (\sim p \vee q)$$

Since this formula is always true, it is a tautology. It is interesting to compare the result of this example with the result of Example 1.10.

p	q	$p \to q$	$\sim p \vee q$	$(p \to q) \leftrightarrow (\sim p \vee q)$
T	T	T	T	T
T	F	F	F	T
F	T	T	T	T
F	F	T	T	T

FIGURE 1.6 Truth table showing $(p \to q) \leftrightarrow (\sim p \vee q)$ is a tautology ■

Example 1.13 Verify that $[(p \wedge q) \to r] \to [\sim r \to (\sim p \vee \sim q)]$ is a tautology.

Solution Here we have our first instance of a formula that involves three propositional variables. As shown in the truth table of Figure 1.7, there are eight possible truth-value combinations that must be considered. For convenience, let u and v denote the formulas $(p \wedge q) \to r$ and $\sim r \to (\sim p \vee \sim q)$, respectively.

p	q	r	$\sim p$	$\sim q$	$\sim r$	$p \wedge q$	$\sim p \vee \sim q$	u	v	$u \to v$
T	T	T	F	F	F	T	F	T	T	T
T	T	F	F	F	T	T	F	F	F	T
T	F	T	F	T	F	F	T	T	T	T
T	F	F	F	T	T	F	T	T	T	T
F	T	T	T	F	F	F	T	T	T	T
F	T	F	T	F	T	F	T	T	T	T
F	F	T	T	T	F	F	T	T	T	T
F	F	F	T	T	T	F	T	T	T	T

FIGURE 1.7 Truth table showing $[(p \wedge q) \to r] \to [\sim r \to (\sim p \vee \sim q)]$ is a tautology

Alternate Solution It is also possible to verify that $u \to v$ is a tautology without using a truth table. We know that the only way $u \to v$ can be false is for u to be true and v to be false. Suppose that v is false; then it follows that $\sim r$ is true and $\sim p \vee \sim q$ is false. Thus r is false and p and q are both true; that is, r is false and $p \wedge q$ is true. Therefore, $(p \wedge q) \to r$, namely u, is false. Thus whenever v is false, u is also false. This shows that $u \to v$ is a tautology. ■

Example 1.14 Write the formula $\sim(p \to q)$ as a conjunction.

Solution By Example 1.10, $p \to q \equiv \sim p \vee q$. Thus $\sim(p \to q) \equiv \sim(\sim p \vee q)$. Then, using Example 1.9, $\sim(\sim p \vee q) \equiv \sim(\sim p) \wedge \sim q$. But

$$\sim(\sim p) \wedge \sim q \equiv p \wedge \sim q$$

Hence we have that

$$\sim(p \to q) \equiv p \wedge \sim q$$ ■

Note that the formula of Example 1.12 is of the form $u \leftrightarrow v$, and, as previously established in Example 1.10, $u \equiv v$. This is not just a coincidence, for it can be shown that formulas u and v are logically equivalent if and only if the compound statement $u \leftrightarrow v$ is a tautology. (See Problem 10.)

Suppose u and v represent mathematical statements and we wish to prove u to be true. If $u \equiv v$, then it suffices to prove v to be true. For example, suppose we want to prove a statement of the form $p \to q$. Since $p \to q \equiv \sim q \to \sim p$, we could instead prove $\sim q \to \sim p$ to be true. This technique, called "proof by contrapositive," is a standard proof technique in mathematics, and we shall explore it further in Section 1.6.

Here is a list of some of the logical equivalences that are frequently encountered and used in mathematics.

Logical equivalences

1. The commutative properties
   (a) $p \vee q \equiv q \vee p$ (b) $p \wedge q \equiv q \wedge p$
2. The associative properties
   (a) $(p \vee q) \vee r \equiv p \vee (q \vee r)$
   (b) $(p \wedge q) \wedge r \equiv p \wedge (q \wedge r)$
3. The distributive properties
   (a) $p \vee (q \wedge r) \equiv (p \vee q) \wedge (p \vee r)$
   (b) $p \wedge (q \vee r) \equiv (p \wedge q) \vee (p \wedge r)$
4. The idempotent laws
   (a) $p \vee p \equiv p$ (b) $p \wedge p \equiv p$
5. DeMorgan's laws
   (a) $\sim(p \vee q) \equiv \sim p \wedge \sim q$ (b) $\sim(p \wedge q) \equiv \sim p \vee \sim q$

6. Law of the excluded middle
   (a) $p \vee \sim p$ is a tautology
   (b) $p \wedge \sim p$ is a contradiction
7. An implication and its contrapositive are equivalent:

$$p \rightarrow q \equiv \sim q \rightarrow \sim p$$

8. The converse and inverse of the implication $p \rightarrow q$ are equivalent (see Exercise 3):

$$q \rightarrow p \equiv \sim p \rightarrow \sim q$$

9. Let T denote a tautology and F denote a contradiction. Then:
   (a) $p \vee \mathrm{T} \equiv \mathrm{T}$ (b) $p \wedge \mathrm{T} \equiv p$
   (c) $p \vee \mathrm{F} \equiv p$ (d) $p \wedge \mathrm{F} \equiv \mathrm{F}$

A brief comment is in order concerning statements such as $(p \wedge q) \wedge r$ or $[(p \wedge q) \wedge r] \wedge s$. In view of the fact that $(p \wedge q) \wedge r \equiv p \wedge (q \wedge r)$, it makes good sense to define $p \wedge q \wedge r$ to be $(p \wedge q) \wedge r$. It follows that $p \wedge q \wedge r \equiv p \wedge (q \wedge r)$, so we can insert parentheses in $p \wedge q \wedge r$ in both ways without loss of meaning. Similarly, we can insert parentheses in $p \wedge q \wedge r \wedge s$ in five meaningful ways, and since all the resulting expressions are logically equivalent, we shall agree to let $p \wedge q \wedge r \wedge s$ denote any one of them.

Example 1.15 The compound statement $(p \wedge (p \rightarrow q)) \rightarrow q$ is called *modus ponens*. In words it says

If both p and $p \rightarrow q$ hold, then q holds.

Use properties of logical equivalence to show that modus ponens is a tautology.

Solution We apply the properties as follows:

$$\begin{aligned}
(p \wedge (p \rightarrow q)) \rightarrow q &\equiv \sim(p \wedge (p \rightarrow q)) \vee q && \text{(by Example 1.10)}\\
&\equiv (\sim p \vee \sim(p \rightarrow q)) \vee q && \text{(by 5(b))}\\
&\equiv (\sim p \vee (p \wedge \sim q)) \vee q && \text{(by Example 1.14)}\\
&\equiv ((\sim p \vee p) \wedge (\sim p \vee \sim q)) \vee q && \text{(by 3(a))}\\
&\equiv (\mathrm{T} \wedge (\sim p \vee \sim q)) \vee q && \text{(by 1(a) and 6(a))}\\
&\equiv (\sim p \vee \sim q) \vee q && \text{(by 9(b))}\\
&\equiv \sim p \vee (\sim q \vee q) && \text{(by 2(a))}\\
&\equiv \sim p \vee \mathrm{T} && \text{(by 6(a))}\\
&\equiv \mathrm{T} && \text{(by 9(a))}
\end{aligned}$$

■

Example 1.16 Use the properties of logical equivalence to show that

$$\sim(p \leftrightarrow q) \equiv (\sim q) \leftrightarrow p$$

Solution We apply the properties as follows:

$$
\begin{aligned}
\sim(p \leftrightarrow q) &\equiv \sim((p \to q) \wedge (q \to p)) && \text{(by Definition 1.4)}\\
&\equiv \sim(p \to q) \vee \sim(q \to p) && \text{(by 5(b))}\\
&\equiv (p \wedge \sim q) \vee (q \wedge \sim p) && \text{(by Example 1.14)}\\
&\equiv ((p \wedge \sim q) \vee q) \wedge ((p \wedge \sim q) \vee \sim p) && \text{(by 3(a))}\\
&\equiv ((p \vee q) \wedge (\sim q \vee q)) \wedge ((p \vee \sim p) \wedge (\sim q \vee \sim p)) && \text{(by 1(a) and 3(a))}\\
&\equiv ((p \vee q) \wedge \mathrm{T})) \wedge (\mathrm{T} \wedge (\sim q \vee \sim p)) && \text{(by 6(a))}\\
&\equiv (p \vee q) \wedge (\sim q \vee \sim p) && \text{(by 9(b))}\\
&\equiv (q \vee p) \wedge (\sim p \vee \sim q) && \text{(by 1(a))}\\
&\equiv (\sim q \to p) \wedge (p \to \sim q) && \text{(by Example 1.10)}\\
&\equiv (\sim q) \leftrightarrow p && \text{(by Definition 1.4)} \quad \blacksquare
\end{aligned}
$$

Exercises 1.3

1. Show that $\sim(p \wedge q)$ and $\sim p \vee \sim q$ are logically equivalent.

2. Use a truth table to verify the associative property $p \vee (q \vee r) \equiv (p \vee q) \vee r$.

3. The *inverse* of the implication $p \to q$ is the implication $\sim p \to \sim q$. Show that the inverse and converse of $p \to q$ are logically equivalent.

4. Each of the conditional statements below concerns integers x and y. Find (i) the inverse, (ii) the converse, and (iii) the contrapositive of each statement.

   a. If $x = 2$, then $x^4 = 16$.
   b. If $y > 0$, then $y \neq -3$.
   c. If x and y are both odd, then xy is odd.
   d. If $x^2 = x$, then $x = 0$ or $x = 1$.
   e. If $x = 17$ or $x^3 = 8$, then x is prime.
   f. If $xy \neq 0$, then $x \neq 0$ and $y \neq 0$.

5. Find (i) the inverse, (ii) the converse, and (iii) the contrapositive of each of the following conditional statements.

   a. If quadrilateral *ABCD* is a rectangle, then the quadrilateral is a parallelogram.
   b. If triangle *ABC* is isosceles and contains an angle of 45 degrees, then the triangle is a right triangle.
   c. If quadrilateral *ABCD* is a square, then it is both a rectangle and a rhombus.
   d. If quadrilateral *ABCD* has two sides of equal length, then it is a rectangle or a rhombus.
   ★e. If the polygon *P* has the property that it is equiangular if and only if it is equilateral, then the polygon *P* is a triangle.

6. The formula $(\sim p \wedge (p \vee q)) \rightarrow q$ is known as the *disjunctive syllogism*.

a. Express the formula in words.

b. Use properties of logical equivalence to establish that the disjunctive syllogism is a tautology.

7. The formula $((p \rightarrow q) \wedge \sim q) \rightarrow \sim p$ is called *modus tollens*.

a. Express the formula in words.

b. Use properties of logical equivalence to establish that modus tollens is a tautology.

8. Write a truth table for each of the following compound statements. Which are tautologies? Which are contradictions?

a. $(\sim p \wedge q) \rightarrow p$ **b.** $p \leftrightarrow \sim(\sim p)$

c. $(p \rightarrow q) \wedge (p \rightarrow \sim q)$ **d.** $(p \rightarrow q) \vee (q \rightarrow p)$

e. $(p \rightarrow q) \rightarrow r$ **f.** $p \wedge q \wedge (p \rightarrow \sim q)$

9. Use truth tables to verify the distributive properties.

a. $p \vee (q \wedge r) \equiv (p \vee q) \wedge (p \vee r)$

b. $p \wedge (q \vee r) \equiv (p \wedge q) \vee (p \wedge r)$

10. Use properties of logical equivalence to establish each of the following.

a. $(p \rightarrow \sim q) \wedge (p \rightarrow \sim r) \equiv \sim(p \wedge (q \vee r))$

b. $(p \wedge q) \leftrightarrow p \equiv p \rightarrow q$

c. $(p \wedge q) \rightarrow r \equiv p \rightarrow (\sim q \vee r)$

d. $p \rightarrow (q \vee r) \equiv \sim q \rightarrow (\sim p \vee r)$

11. In each part, determine whether formulas u and v are logically equivalent. (Try to do it without using a truth table.)

a. u: $(p \rightarrow q) \wedge (p \rightarrow \sim q)$, v: $\sim p$

b. u: $p \rightarrow q$, v: $q \rightarrow p$

c. u: $p \leftrightarrow q$, v: $q \leftrightarrow p$

d. u: $(p \rightarrow q) \rightarrow r$, v: $p \rightarrow (q \rightarrow r)$

e. u: $(p \leftrightarrow q) \leftrightarrow r$, v: $p \leftrightarrow (q \leftrightarrow r)$

f. u: $p \rightarrow (q \rightarrow r)$, v: $(p \rightarrow q) \rightarrow (p \rightarrow r)$

g. u: $p \rightarrow (q \vee r)$, v: $(p \rightarrow q) \vee (p \rightarrow r)$

h. u: $p \vee (q \rightarrow r)$, v: $(p \vee q) \rightarrow (p \vee r)$

12. Given that the formula $(q \vee r) \rightarrow \sim p$ is false and q is false, determine the truth-values of each of r and p.

1.4 ALTERNATE STATEMENT FORMS

The formula $p \rightarrow q$ has two interpretations in words thus far:

If p, then q.

and

p implies q.

We know that $p \rightarrow q$ is logically equivalent to its contrapositive, $\sim q \rightarrow \sim p$, which is read,

If not q, then not p.

This last statement says that if q is false, then so is p. Thus p is true only under the condition that q is true. We shall write this last statement

p only if q.

It is another way of saying $p \rightarrow q$.

Another common phrase in mathematics is

p is sufficient for q.

It states that the condition that p holds is enough to guarantee that q will hold. Thus, it is another way of saying $p \rightarrow q$. Another phrase equivalent to $p \rightarrow q$ is

q is necessary for p.

which says that in order for p to be true, q must be true. Hence q being false implies that p is false, again giving us the contrapositive $\sim q \rightarrow \sim p$. To summarize, then, the following are equivalent formulas:

If p, then q.
p implies q.
p only if q.
p is sufficient for q.
q is necessary for p.

Example 1.17 Rewrite the statement

If $x = 1$, then $x^2 - 1 = 0$.

in four equivalent ways.

Solution The given statement is equivalent to any one of the following:

$x = 1$ implies $x^2 - 1 = 0$.
$x = 1$ only if $x^2 - 1 = 0$.
$x = 1$ is sufficient for $x^2 - 1 = 0$.
$x^2 - 1 = 0$ is necessary for $x = 1$. ■

Now consider the biconditional $p \leftrightarrow q$:

p if and only if q.

Recall that this is equivalent to $(p \rightarrow q) \wedge (q \rightarrow p)$, which can now be stated

p is sufficient for q and p is necessary for q.

This last statement is usually shortened to

p is necessary and sufficient for q.

Exercises 1.4

1. Identify the form of each of the following implications as being either (i) "If p, then q," (ii) "p implies q," (iii) "p only if q," (iv) "p is sufficient for q," or (v) "q is necessary for p." Then rewrite the statement in each of the other four forms.
   a. If $x = -2$, then $x^3 = -8$.
   b. Being intelligent is necessary for passing this course.
   c. Working hard is sufficient for passing this course.
   d. A number is prime only if it is not a multiple of 4.
   e. In order for triangle ABC to be a right triangle, it is necessary that the side lengths satisfy the Pythagorean theorem.
   f. Being a good student implies that one knows how to study.
   g. In order for a number to be rational, it is sufficient that its decimal expansion terminate.
   h. One can pass this course if one passes the final exam.
   i. One can pass this course only if one passes the final exam.
2. Rewrite each of the following statements in an equivalent way.
   a. $x^3 - x^2 + x - 1 = 0$ is necessary and sufficient for $x = 1$.
   b. I will pass this course if and only if you help me study for the exams.

1.5 QUANTIFIERS

Consider the sentence

The number x is prime and $x > 17$.

(Recall that a prime number is an integer $p > 1$ whose only positive divisors are itself and 1. We shall learn more about primes in Chapter 3.) The above sentence is not a statement since if $x = 19$, it is true, while if $x = 24$, it is false. Until the value of x is known we cannot determine the truth-value of the sentence. It is an example of a "propositional function."

DEFINITION 1.7

A *propositional function* is a sentence $p(x)$ about the symbol x; it becomes a statement only when x is given a particular value (or meaning). We refer to x as a *variable*. Propositional functions with the variable x will be denoted by $p(x)$, $q(x)$, $r(x)$, and so on.

Consider the sentence

If x is prime, then x is not a multiple of 4.

This sentence has the logical form

$$p(x) \rightarrow \sim q(x)$$

and its truth-value can definitely be determined for a specified value of x. (In fact, it is true for all x.) On the other hand, sentences such as

There exists an x such that x is prime and $x + 10$ is prime.

or

For all x, if x is prime, then $x^2 + 5$ is not prime.

cannot be symbolized using the logical connectives presented thus far. The reason for this is the presence of the phrases "there exists an x" and "for all x." They are used so frequently in mathematics that they warrant symbolic representation.

DEFINITION 1.8

The statement

There exists an x such that $p(x)$.

is symbolized by

$$\exists x\ p(x)$$

The symbol $\exists$ is called the *existential quantifier* and translates as "there exists." Other common phrases for $\exists$ are "for some" or "there is some." Also, the words "such that" are often replaced by "for which" or "satisfying." The statement "$\exists x\ p(x)$" is true if there is at least one value of x for which $p(x)$ is true.

DEFINITION 1.9

The statement

For all x, $p(x)$.

is symbolized by

$$\forall x\ p(x)$$

The symbol $\forall$ is called the *universal quantifier* and translates as "for all." Other phrases for $\forall$ are "for each" or "for every" or "given any." The statement "$\forall x\ p(x)$" is true only if $p(x)$ is true for every value of x.

The quantifiers $\exists$ and $\forall$, together with the logical connectives, are collectively referred to as the *logical symbols*.

Example 1.18 Consider the following statements about real numbers x:
(a) $\exists x(x^2 = 2)$ (b) $\exists x(x^2 < 0)$
(c) $\forall x(x + 1 > x)$ (d) $\forall x(\sqrt{x^2} = x)$
Statement (a) is true, for it asserts the existence of the number $\sqrt{2}$. Statement (b) is false, since every real number x has the property that $x^2 \geq 0$. Statement (c) is clearly true, while statement (d) is false, for see what happens if x is negative! ■

Notice that the statement

There is some x such that $x^2 = 2$.

is false if we are considering only rational numbers x. If the possible values of the variable x in such a statement are not assumed in advance, then we often include such assumptions as part of the statement; for example

There is some real number x such that $x^2 = 2$.

This last statement is written symbolically as

$$\exists x(x \text{ is real} \wedge x^2 = 2)$$

This leads us to consider *restricted forms of the quantifiers*.

As an example, consider the statement

For every real number x, $x^2 + 1 > 0$.

This statement uses the restricted form of the quantifier $\forall$. It says that, given any x, if x happens to be a real number, then $x^2 + 1 > 0$. In general, the statement

For every x with property $r(x)$, property $p(x)$ holds.

translates as

For every x, if $r(x)$ holds then $p(x)$ holds.

This last statement is expressed symbolically as

$$\forall x(r(x) \rightarrow p(x))$$

Next consider the statement

There is some real number x such that $x^3 - 3x + 1 = 0$.

It states that there is some x having the two properties that x is real and x is a solution of the equation $x^3 - 3x + 1 = 0$. It provides an example of the restricted use of the existential quantifier $\exists$ and is written symbolically as

$$\exists x(x \text{ is a real number} \wedge x^3 - 3x + 1 = 0)$$

Any statement of the general form

There is some x with property $r(x)$ such that property $p(x)$ holds.

translates as

There is some x such that both $r(x)$ and $p(x)$ hold.

This is written symbolically as

$$\exists x(r(x) \land p(x))$$

Example 1.19 Express each of the following statements symbolically, using the logical symbols.

(a) For every positive integer x, either x is prime or $x^2 + 1$ is prime.

(b) There is some positive integer x such that $x^2 + 1$ is prime and x is a perfect square.

Solution Statement (a) is written symbolically as

$$\forall x(x \text{ is a positive integer} \rightarrow (x \text{ is prime} \lor x^2 + 1 \text{ is prime}))$$

Notice how the restricted use of the quantifier $\forall$, "For every positive integer x . . . ," becomes an implication:

$$\forall x(x \text{ is a positive integer} \rightarrow \ldots)$$

Statement (b) is written

$$\exists x(x \text{ is a positive integer} \land x^2 + 1 \text{ is prime} \land x \text{ is a perfect square})$$

Here it is important to see how the restricted use of the quantifier $\exists$, "There is some positive integer x . . . ," becomes a conjunction:

$$\exists x(x \text{ is a positive integer} \land \ldots)$$ ■

It is also possible to have statements that involve more than one quantifier. Consider the following two statements:

(a) There exists an integer x such that for every integer y, $x + y = 4$.

(b) For every integer y there exists an integer x such that $x + y = 4$.

Using the logical symbols, these statements can be expressed as follows:

(a) $\exists x(x \text{ is an integer} \land \forall y(y \text{ is an integer} \rightarrow x + y = 4))$

(b) $\forall y(y \text{ is an integer} \rightarrow \exists x(x \text{ is an integer} \land x + y = 4))$

What can be said about the truth-values of these statements? Statement (a) asserts the existence of an integer x such that, no matter what integer y is chosen, $x + y = 4$. This statement is clearly false, for once y is chosen, $x = 4 - y$ is uniquely determined. Statement (b), on the other hand, states that for each integer y there is some integer x such that $x + y = 4$. This statement is certainly true.

Example 1.20 Write each of the following statements using logical symbols.

(a) For every positive integer n there exists a prime p such that $p > n$.
(b) There exist two primes p and q whose sum is also prime.
(c) For all rational numbers x and y, the sum $x + y$ is rational.
(d) Every even integer $n \geq 4$ is the sum of two primes p and q.

Solution

(a) $\forall n(n \text{ is a positive integer} \to \exists p(p \text{ is prime} \wedge p > n))$
(b) $\exists p\ \exists q(p \text{ is prime} \wedge q \text{ is prime} \wedge p + q \text{ is prime})$
(c) $\forall x\ \forall y((x \text{ is rational} \wedge y \text{ is rational}) \to x + y \text{ is rational})$
(d) $\forall n((n \text{ is even} \wedge n \geq 4) \to \exists p\ \exists q(p \text{ is prime} \wedge q \text{ is prime} \wedge n = p + q))$ ■

What is the negation of the statement $\exists x\ p(x)$? Of course, we can readily say, "It is not the case that there is some x such that $p(x)$." It seems clear, however, that this statement is equivalent to the statement "For every x, not $p(x)$." Hence we see that

$$\sim(\exists x\ p(x)) \equiv \forall x(\sim p(x))$$

What about the negation of $\forall x\ p(x)$? This is "It is not the case that, for every x, $p(x)$." Equivalently we can say, "There is some x such that not $p(x)$." Accordingly, we arrive at

$$\sim(\forall x\ p(x)) \equiv \exists x(\sim p(x))$$

Example 1.21 Find the negation of each of the following statements.

(a) $\forall x(x \text{ is prime} \to x^2 + 1 \text{ is even})$
(b) $\exists x(x \text{ is rational} \wedge x^2 = 3)$
(c) $\exists x\ \forall y(xy = y)$
(d) $\forall x\ \forall y(x < y \to \exists z(x < z \wedge z < y))$

Solution Statement (a) has the form $\forall x(p(x) \to q(x))$. We determine its negation as follows:

$$\sim\forall x(p(x) \to q(x)) \equiv \exists x(\sim(p(x) \to q(x))) \equiv \exists x(p(x) \wedge \sim q(x))$$

So the negation is

$$\exists x(x \text{ is prime} \wedge x^2 + 1 \text{ is odd})$$

Statement (b) has the form $\exists x(p(x) \wedge q(x))$. Hence the negation is

$$\sim\exists x(p(x) \wedge q(x)) \equiv \forall x(\sim(p(x) \wedge q(x))) \equiv \forall x(\sim p(x) \vee \sim q(x))$$

Thus the negation is

$$\forall x(x \text{ is not rational} \vee x^2 \neq 3)$$

Recalling that $p \rightarrow q \equiv \sim p \vee q$, it is preferable to restate the negation in the form

$$\forall x(x \text{ is rational} \rightarrow x^2 \neq 3)$$

For statement (c), the form is $\exists x\ \forall y(p(x,y))$, where $p(x,y)$ represents the equation $xy = y$. Its negation is $\forall x\ \exists y(\sim p(x,y))$, namely,

$$\forall x\ \exists y(xy \neq y)$$

The form of statement (d) is

$$\forall x\ \forall y(p(x,y) \rightarrow \exists z(q(x,z) \wedge r(y,z)))$$

Here is a test: verify that the negation has the form

$$\exists x\ \exists y(p(x,y) \wedge \forall z(\sim q(x,z) \vee \sim r(y,z)))$$

which translates to

$$\exists x\ \exists y(x < y \wedge \forall z(x \geq z \vee z \geq y))$$

■

Exercises 1.5

1. Let x represent a positive integer and let the propositional functions $p(x)$, $q(x)$, and $r(x)$ be defined by $p(x)$: x is prime; $q(x)$: x is even; $r(x)$: $x \geq 3$. Express each of the following statements in words.

a. $\exists x\ p(x)$ **b.** $\forall x\ r(x)$
c. $\exists x(p(x) \wedge q(x))$ **d.** $\forall x(r(x) \rightarrow (p(x) \vee q(x)))$
e. $\exists x(p(x) \wedge (q(x) \vee r(x)))$ **f.** $\forall x((p(x) \wedge q(x)) \rightarrow \sim r(x))$

2. Assume a Pascal array V is declared by

```
VAR V : ARRAY[1..30] OF REAL;
```

and that the variables I and J represent subscripts for V. Write each of the following statements using logical symbols.

a. All elements of V are positive.
b. Some element of V is equal to zero.
c. The elements of V are distinct.
d. There exist two elements of V that add to zero.
e. For each element of V there is another element that differs from it by exactly 3.
f. The array V has a unique largest element.
g. The elements of V are sorted with V[1] < V[2] < · · · < V[30].
h. Any element with an even subscript is bigger than every element with an odd subscript.

3. Write each of the following statements using logical symbols.

a. For every even integer n there exists an integer m such that $n = 2m$.
b. Every integer is a rational number.

c. There exists a right triangle T that is an isosceles triangle.
d. Given any quadrilateral Q, if Q is a parallelogram and has two adjacent sides that are perpendicular, then Q is a rectangle.
e. There exists an even prime integer.
f. There exist integers s and t such that $1 < s < t < 187$ and $st = 187$.
g. There is an x such that both $x/2$ is an integer and, for every y, $x/(2y)$ is not an integer.
h. Given any real numbers x and y, $2x^2 - xy + 5 > 0$.

4. Find the negation (in simplest form) of each of the following.
a. $\forall x(p(x) \wedge q(x))$
b. $\forall x\ \forall y(p(x,y) \rightarrow q(x,y))$
c. $\forall x\ \exists y(p(x,y) \rightarrow q(x,y))$
d. $\exists x(\forall y(p(x,y) \rightarrow q(x,y)) \wedge \exists z(r(x,z)))$

5. For each of the following, (i) write the statement symbolically, (ii) find the symbolic form of the negation, and then (iii) express the negation in words.
a. For every x and for every y, $x + y = y + x$.
b. For every x there exists y such that $y^2 = x$.
c. There exists y such that for every x, $(2x^2 + 1)/x^2 > y$.
d. There exist x and y such that $x < y$ and $x^3 - x > y^3 - y$.
e. For all x and y there exists z such that $2z = x + y$.
f. For every x and y, if $x^3 + x - 2 = y^3 + y - 2$, then $x = y$.

6. Let x represent an integer. Use the example $p(x)$: x is even; $q(x)$: x is odd, to show that
a. $\forall x(p(x) \vee q(x)) \not\equiv \forall x\ p(x) \vee \forall x\ q(x)$
b. $\exists x(p(x) \wedge q(x)) \not\equiv \exists x\ p(x) \wedge \exists x\ q(x)$
c. $\forall x(p(x) \rightarrow q(x)) \not\equiv \forall x\ p(x) \rightarrow \forall x\ q(x)$
Use the example $p(x)$: x is even; $q(x)$: $x^2 < 0$, to show that
d. $\exists x(p(x) \rightarrow q(x)) \not\equiv \exists x\ p(x) \rightarrow \exists x\ q(x)$
(The symbol "$\not\equiv$" stands for "is not logically equivalent to.")

7. Let the set of possible values for the variable x be fixed. Show that
a. $\exists x(p(x) \vee q(x)) \equiv \exists x\ p(x) \vee \exists x\ q(x)$
b. $\forall x(p(x) \wedge q(x)) \equiv \forall x\ p(x) \wedge \forall x\ q(x)$
c. $\exists x(p(x) \rightarrow q(x)) \equiv \forall x\ p(x) \rightarrow \exists x\ q(x)$

1.6 METHODS OF PROOF

It may be said that mathematics is uniquely characterized among human endeavor by the practice known as proof. To quote from *The Mathematical Experience* by Phillip Davis and Reuben Hersh:

> Mathematics, then, is the subject in which there are proofs. Traditionally, proof was first met in Euclid; and millions of hours have been spent in class

after class, in country after country, in generation after generation, proving and reproving the theorems in Euclid. After the introduction of the "new math" in the mid-nineteen fifties, proof spread to other high school mathematics such as algebra, and subjects such as set theory were deliberately introduced so as to be a vehicle for the axiomatic method and proof. In college, a typical lecture in advanced mathematics, especially a lecture given by an instructor with "pure" interests, consists entirely of definition, theorem, proof, definition, theorem, proof, in solemn and unrelieved concatenation. Why is this? If, as claimed, proof is validation and certification, then one might think that once a proof has been accepted by a competent group of scholars, the rest of the scholarly world would be glad to take their word for it and to go on. Why do mathematicians and their students find it worthwhile to prove again and yet again the Pythagorean theorem or the theorems of Lebesgue or Wiener or Kolmogoroff?

Proof serves many purposes simultaneously. In being exposed to the scrutiny and judgement of a new audience, the proof is subject to a constant process of criticism and revalidation. Errors, ambiguities, and misunderstandings are cleared up by constant exposure. Proof is respectability. Proof is the seal of authority.

Proof, in its best instances, increases understanding by revealing the heart of the matter. Proof suggests new mathematics. The novice who studies proofs gets closer to the creation of new mathematics. Proof is mathematical power, the electric voltage of the subject which vitalizes the static assertions of the theorems.

Finally, proof is ritual, and a celebration of the power of pure reason. Such an exercise in reassurance may be very necessary in view of all the messes that clear thinking clearly gets us into.

A *theorem* is a mathematical statement that is true, and a *proof* is a logical argument that verifies the truth of the theorem. Writing clear and correct proofs is an art, and to become good at it takes much practice. One of the purposes of this textbook is to help the student to become proficient at writing proofs. We hope to do this by providing examples of proofs, and also by providing many opportunities to test and improve the student's skill at writing proofs.

To start with, however, it is important to understand the proofs given in this and in other courses. To understand a proof, it is necessary to understand the method being used. In this section we shall discuss the two basic strategies of proof; these are the *direct method* and the *indirect method*. We also briefly introduce the problem of program verification.

Since a proof is a logical argument, the proof is valid if and only if the symbolic form of the argument is a tautology. Certain statements that appear in the proof are known to be true; these include axioms as well as theorems that have already been proven. Other statements that appear in the proof are hypotheses of the theorem; these are assumed to be true.

Still other statements are derived from previous ones using the rules of logical inference.

Example 1.22 We give an example of both a valid and an invalid argument.

(a) Suppose that

p: Fred passed the final exam in CS 260.

and

q: Fred passed the course CS 260.

are statements, and it is known that both p and $p \to q$ are true. Is the argument

> Fred passed the final exam in CS 260. If Fred passed the final exam, then Fred passed the course. Therefore, Fred passed the course CS 260.

a valid one?

Solution The logical form of the argument is

$$(p \wedge (p \to q)) \to q$$

This is the rule of modus ponens, which was shown to be a tautology in Example 1.15. Thus we may conclude that the argument is valid.

(b) Suppose that

p: Fred got a B or better in CS 260.

and

q: Fred passed CS 261.

are statements, and it is known that both q and $p \to q$ are true. Is the argument

> If Fred got a B or better in CS 260, then Fred passed CS 261. Fred passed CS 261. Therefore, Fred got a B or better in CS 260.

a valid one?

Solution Here the logical form of the argument is

$$((p \to q) \wedge q) \to p$$

But this is not a tautology, for it fails to hold if p is false and q is true. (Perhaps Fred got a C in CS 260, but by working hard he was able to pass CS 261.) Hence the argument is not valid. ■

Suppose that we are confronted with the problem of proving a theorem of the form $p \to q$. We must then prove that $p \to q$ is true. This will always be the case if p is false, so our problem reduces to that of showing that if p

is true, then q must also be true. This method of proof is called the *direct method,* and the basic outline for the method is as follows:

Problem: Prove $p \rightarrow q$ is true.
Direct method: (a) Assume p is true.
(b) Show that q is true.
Conclusion: $p \rightarrow q$ is true.

For the direct method of proof, the most common form of argument is known as the "syllogism." We illustrate this form with an example.

Example 1.23 Suppose the following statements are known to be true:

If you get to bed by 3 A.M., then you will get up for your 9 A.M. class.
If you win at poker, then you'll get to bed by 3 A.M.

We may then conclude that the next statement holds:

If you win at poker, then you will get up for your 9 A.M. class. ■

The type of reasoning displayed in Example 1.23 is called *syllogistic reasoning,* the logical form of the argument being

$$((p \rightarrow r) \wedge (r \rightarrow q)) \rightarrow (p \rightarrow q)$$

It is easy to verify that the above formula is a tautology (see Exercise 2). This formula is called a *syllogism.* In words it states that

If p implies r and r implies q, then p implies q.

Now suppose we wish to prove the formula $p \rightarrow q$. By itself this may be quite difficult. However, suppose we can come up with a statement s such that we can prove $(p \rightarrow s) \wedge (s \rightarrow q)$. Then the truth of $p \rightarrow q$ follows and our proof is complete. Often we may have to string several implications together, such as

$$(p \rightarrow s_1) \wedge (s_1 \rightarrow s_2) \wedge \cdots \wedge (s_{n-1} \rightarrow s_n) \wedge (s_n \rightarrow q)$$

Example 1.24 Prove the statement

If $x^2 - 4 = 0$, then $x = 2$ or $x = -2$.

Proof We are given that $x^2 - 4 = 0$, and since $x^2 - 4 = (x + 2)(x - 2)$, we have that $(x + 2)(x - 2) = 0$. If $(x + 2)(x - 2) = 0$, then either $x + 2 = 0$ or $x - 2 = 0$. It follows algebraically that either $x = -2$ or $x = 2$.

Discussion Consider the logical form of this proof. Let p, r, s, and q represent the statements

p: $x^2 - 4 = 0$
r: $(x + 2)(x - 2) = 0$
s: $x + 2 = 0$ or $x - 2 = 0$
q: $x = 2$ or $x = -2$

We wished to prove $p \to q$. The argument used was of the form

$$((p \to r) \land (r \to s) \land (s \to q)) \to (p \to q)$$

Of course, several other facts were used along the way. We factored $x^2 - 4 = (x + 2)(x - 2)$, and we used a very important property of the real numbers: if x and y are real numbers and $xy = 0$, then either $x = 0$ or $y = 0$. ■

Example 1.25 Prove that if, in triangle ABC, the perpendicular from vertex A to side BC bisects BC, then triangle ABC is an isosceles triangle. (See Figure 1.8.)

Proof Let AD be the perpendicular from A to side BC, as shown. Since D is the midpoint of BC, segments BD and DC have the same length. Now compare triangles ADB and ADC. These triangles have side AD in common, angles ADB and ADC are both right angles, and sides BD and DC have the same length. Hence, by side–angle–side, triangles ADB and ADC are congruent. Therefore, sides AB and AC have the same length, and thus triangle ABC is isosceles.

Discussion Let p, u, v, w, x, and q represent the following statements:

- p: In triangle ABC, the perpendicular AD to side BC bisects BC.
- u: Segments BD and DC have the same length.
- v: Comparing triangles ADB and ADC, they have side AD in common, angles ADB and ADC are both right angles, and sides BD and DC have the same length.
- w: Triangles ADB and ADC are congruent.
- x: Sides AB and AC have the same length.
- q: Triangle ABC is an isosceles triangle.

Then the logical form of the proof is

$$((p \to u) \land (u \to v) \land (v \to w) \land (w \to x) \land (x \to q)) \to (p \to q)$$ ■

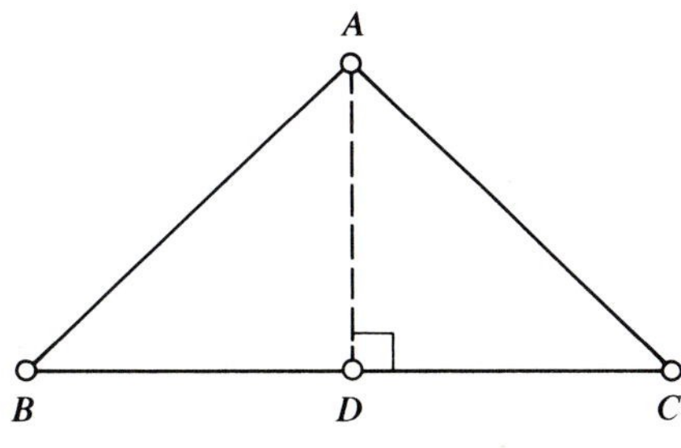

FIGURE 1.8

There are two methods of proof that we shall refer to as *indirect methods*. The first is known as "reductio ad absurdum," which means "reduction to an absurdity" and is more commonly called *proof by contradiction*. The second indirect method is called *proof by contrapositive*.

Proof by contradiction is probably the hardest method to understand. It is based on the law of the excluded middle, which, as mentioned in Section 1.3, states that if r is any statement, then $r \vee \sim r$ is a tautology, and $r \wedge \sim r$ is a contradiction. Suppose we wish to prove $p \rightarrow q$. To do this we try to find a statement r for which

$$\sim(p \rightarrow q) \rightarrow (\sim r \wedge r)$$

is true. In this case the statement $\sim(p \rightarrow q)$ implies a statement that is always false. The only way such an implication can be true is for $\sim(p \rightarrow q)$ to be false also. Hence it must be the case that $p \rightarrow q$ is true. In short, the formula

$$(\sim(p \rightarrow q) \rightarrow (\sim r \wedge r)) \rightarrow (p \rightarrow q)$$

is a tautology. Recalling that $\sim(p \rightarrow q)$ is logically equivalent to $p \wedge \sim q$, we may summarize the method of proof by contradiction as follows:

Problem: Prove $p \rightarrow q$ is true.
Method of proof by contradiction:
(a) Assume p is true and q is false.
(b) Argue to a contradiction $\sim r \wedge r$ for some statement r.
Conclusion: $p \rightarrow q$ is true.

Example 1.26 From plane geometry, recall the axiom

r: Given two distinct points, there is exactly one line containing them.

Making use of this axiom, prove that

If two distinct lines m and n intersect, then they intersect in exactly one point.

Proof Suppose that two distinct lines m and n intersect in more than one point. Let A and B denote two distinct points where m and n intersect. Then both lines m and n contain points A and B, contradicting the axiom r. Thus, if m and n intersect, then their intersection must be exactly one point.

Discussion Let p and q represent the statements

p: Lines m and n intersect.
q: The intersection of m and n is exactly one point.

We wished to prove $p \rightarrow q$. We assumed $p \wedge \sim q$, which led to $\sim r$. Thus the contradiction $r \wedge \sim r$ was obtained. Therefore, $p \rightarrow q$ is true. ■

Example 1.27 Use the method of proof by contradiction to prove

If $x = \sqrt{2}$, then x is irrational.

Proof Let $x = \sqrt{2}$ and suppose x is rational. Then there exist positive integers a and b such that $x = a/b$. Furthermore, we may assume without loss of generality that the fraction a/b is in lowest terms. Thus

$$xb = a \rightarrow x^2b^2 = a^2 \rightarrow 2b^2 = a^2 \qquad (\text{since } x^2 = 2)$$

Since $a^2 = 2b^2$, the number a^2 is even, which implies that a is even. Let $a = 2n$. Then from $a^2 = 2b^2$ we obtain

$$(2n)^2 = 2b^2 \rightarrow 4n^2 = 2b^2 \rightarrow 2n^2 = b^2$$

This implies that b^2 is even and so b is even. But now we have that both a and b are even, which contradicts our assumption that a/b was in lowest terms. Therefore, we may conclude that $\sqrt{2}$ is irrational.

Discussion Here we have the statements

p: $x = \sqrt{2}$.
q: x is irrational.
r: A rational number x can be expressed as a fraction a/b in lowest terms.

We wished to prove $p \rightarrow q$. To do so we assumed p true and q false, which led to a contradiction of the known fact r. Therefore, we may conclude that $p \rightarrow q$ is true. ■

The last method of proof we shall discuss at this point is proof by contrapositive. Recall that an implication $p \rightarrow q$ and its contrapositive $\sim q \rightarrow \sim p$ are logically equivalent. Thus, to prove $p \rightarrow q$ it suffices to prove its contrapositive. Note that once the statements $\sim q$ and $\sim p$ have been identified, the direct method is used to prove $\sim q \rightarrow \sim p$. We may summarize the method of proof by contrapositive as follows:

Problem: Prove $p \rightarrow q$ is true.
Method of proof by contrapositive:
(a) Assume q is false.
(b) Show that p is false.
Conclusion: $p \rightarrow q$ is true.

Example 1.28 Use proof by contrapositive to prove

If x is a real number and $x^3 - x^2 + x - 1 = 0$, then $x = 1$.

Then prove the same result directly and by contradiction. Compare the three methods.

Proof by contrapositive Assume $x \neq 1$. Then $x - 1 \neq 0$. Also, for any real number x, we know that $x^2 + 1 \neq 0$. Hence it follows that the product $(x - 1)(x^2 + 1) \neq 0$, which upon multiplication yields $x^3 - x^2 + x - 1 \neq 0$.

Direct proof Assume $x^3 - x^2 + x - 1 = 0$. Then $(x - 1)(x^2 + 1) = 0$, which implies that either $x - 1 = 0$ or $x^2 + 1 = 0$. But we know that $x^2 + 1 \neq 0$, and so it must be the case that $x - 1 = 0$. Hence $x = 1$.

Proof by contradiction Assume $x^3 - x^2 + x - 1 = 0$ and $x \neq 1$. Then $x - 1 \neq 0$. Since $x^3 - x^2 + x - 1 = (x - 1)(x^2 + 1)$ and $x - 1 \neq 0$, it must be that $x^2 + 1 = 0$. But this is a contradiction, given that x is a real number. ■

In this section we have discussed three basic strategies for proving a mathematical statement of the form $p \to q$: the direct method, proof by contradiction, and proof by contrapositive. As illustrated by the last example, it is often the case that more than one method can be applied. As a general rule, however, proofs by contradiction are more involved and harder to follow; they should be done with extreme care. It is perhaps a good practice to see if a direct proof or a proof by contrapositive can be found first.

Let us briefly mention the problem of showing that a statement of the form $p \to q$ is false. Since this occurs only when p is true and q is false, we must find a case where the hypothesis p holds but the conclusion q does not. Such a case is called a *counterexample* to the statement $p \to q$. For instance, consider the statement

If n is a positive integer, then $n^2 + n + 41$ is prime.

To show this statement is false, it suffices to exhibit a single positive integer n for which $n^2 + n + 41$ is not prime. Letting $n = 41$, we find that $n^2 + n + 41 = 41^2 + 41 + 41 = (41)(41 + 1 + 1) = (41)(43)$, which is clearly not prime. So the case $n = 41$ provides a counterexample to the statement.

What about the problem of proving a mathematical statement whose form is the biconditional $p \leftrightarrow q$? Recall that $p \leftrightarrow q$ is defined to be the conjunction of two implications:

$$(p \leftrightarrow q) \equiv (p \to q) \wedge (q \to p)$$

Therefore, the problem of proving $p \leftrightarrow q$ reduces to that of proving both the statements $p \to q$ and $q \to p$. Alternately, it can be shown (see Exercise 4) that

$$(p \leftrightarrow r) \wedge (r \leftrightarrow q) \to (p \leftrightarrow q)$$

Thus it is common to prove $p \leftrightarrow q$ by finding a statement r for which $p \leftrightarrow r$ and $r \leftrightarrow q$ both hold.

Before ending this section, we would like to point out a particular topic in computer science in which proofs play an important role. This pertains to the area of *algorithm* or *program verification,* in which it is proved that a given algorithm or program performs its task as specified.

Think about the process of writing, testing, and debugging a reasonably complicated program. Let's suppose, for the sake of discussion, that we have a program written in Pascal. The first thing we might try is to compile the program, but chances are the program will contain several syntactic errors and won't compile right off. *Syntactic errors* are those caused by constructs in the program that violate the rules of the Pascal language, for example, using "=" instead of ":=" in an assignment statement. With practice and the aid of good error diagnostics from the compiler, such errors are not difficult for the programmer to fix. So eventually we get the program running. Now we are interested in whether the program gives the correct results for various sets of test data. If for some set of data the program halts normally but gives incorrect results, then the program is said to contain a *logical error.* Even if the program does work properly for each set of test data, this does not necessarily mean that all logical errors have been eliminated. It is normally impossible to test all possible data sets, and although by choosing good test data we can make the probability of error quite small, we can't be entirely sure that the program is correct. (This is evidenced by the numerous examples of commercial software in which subtle, and sometimes gross, errors are found by the users.) In general, program verification requires the use of formal proof techniques not unlike those we have studied in this section.

Example 1.29 The Pascal program segment for the linear search (Example 1.5) is repeated below. We wish to prove it correct.

(*INITIAL STATE: We assume that the value of LAST is a positive integer, and that V[1], . . . , V[LAST] and KEY have values of type ELEMENTTYPE*)

(*begin search*)

```
CURRENT := 1;
FOUND := FALSE;
WHILE (NOT FOUND) AND (CURRENT <= LAST) DO
  IF V[CURRENT] = KEY THEN
    BEGIN
      FOUND := TRUE;
      POSITION := CURRENT
    END
  ELSE CURRENT := CURRENT + 1;
```

(*end search*)

(*FINAL STATE: Either FOUND = TRUE and V[POSITION] = KEY, or FOUND = FALSE and V[J] $\neq$ KEY, $1 \leq J \leq$ LAST*)

In order to show such a program segment is correct, we must have a precise statement of its intended purpose. This involves specifying the *initial state* of execution before the segment is executed and the desired *final state* of execution after the segment has been executed. Specifying the state of execution involves, among other things, specifying the values of certain relevant variables. We have included the specification of the initial and final states as comments in the program segment.

The "action" in this program segment centers about the execution of the WHILE loop. In fact, the program segment terminates when the WHILE loop is exited, which happens when the Boolean expression

```
(NOT FOUND) AND (CURRENT <= LAST)
```

evaluates to FALSE. Whenever execution of the program segment is about to evaluate this condition, we shall say that execution has reached the *top of the loop*.

For most segments that contain loops, the key to proving the segment correct is in finding a statement about the state of execution that holds whenever execution is at the top of the loop and that remains true when the loop is exited. Such a statement is known as a *loop invariant condition*. Often the most difficult part of program verification is coming up with the right loop invariant condition.

What is the right loop invariant condition in this case? A hint comes from considering the final state, since the final state should hold when the loop is exited. The final state is

```
(FOUND ∧ V[POSITION] = KEY) ∨ (NOT FOUND ∧ V[J] ≠ KEY,
1 ≤ J ≤ LAST)
```

At some intermediate stage, if we reach the top of the loop and KEY has not been found, then the most we can say is that V[J] $\neq$ KEY, $1 \leq$ J $<$ CURRENT. Thus the loop invariant condition is as follows:

Loop invariant condition (LIC)

```
(FOUND ∧ V[POSITION] = KEY) ∨ (NOT FOUND ∧ V[J] ≠ KEY,
1 ≤ J < CURRENT)
```

Proving the segment is correct now amounts to showing that LIC holds whenever execution reaches the top of the loop. This reduces to showing two things:

1. When execution first reaches the top of the loop, LIC holds.
2. If execution reaches the top of the loop and LIC holds, then either the program segment terminates or, the next time execution reaches the top of the loop, LIC still holds.

Proof of 1 When execution first reaches the top of the loop, FOUND = FALSE and CURRENT = 1, so LIC is satisfied (vacuously).

Proof of 2 Assume execution reaches the top of the loop and LIC holds. We consider three cases, depending on the values of FOUND and CURRENT.

Case 1. FOUND = TRUE. Then the program segment will terminate and, since LIC holds, V[POSITION] = KEY.

Case 2. FOUND = FALSE and CURRENT > LAST. Then the program segment will terminate and, since LIC holds, V[J] $\neq$ KEY, $1 \leq$ J $\leq$ LAST.

Case 3. FOUND = FALSE and CURRENT $\leq$ LAST. In this case execution proceeds to the body of the WHILE loop. If V[CURRENT] = KEY, then FOUND is set to TRUE and POSITION is assigned the value of CURRENT, so that when control returns to the top of the loop, LIC still holds. On the other hand, if V[CURRENT] $\neq$ KEY, then the value of CURRENT is incremented by 1, so that again when control returns to the top of the loop, LIC (in particular, (NOT FOUND $\wedge$ V[J] $\neq$ KEY, $1 \leq$ J < CURRENT)) still holds.
(Note: we have not actually shown that the program segment eventually terminates, but whenever execution reaches the top of the loop and KEY has not been found, the value of CURRENT is one greater than it was the last time execution was at the top of the loop. Thus, if KEY is never found, then eventually the value of CURRENT will exceed the value of LAST, causing the segment to terminate.) ■

As is evidenced by the preceding example, program verification is often a difficult and costly process, although perhaps not as costly as the consequences of an incorrect program. Computer scientists have been and are developing formal techniques for doing program verification. It is hoped that this work will eventually allow most program verification to be done by a computer program, thus reducing the cost and effort involved.

Exercises 1.6

1. Determine the validity of each of the following arguments by translating the argument into symbolic form and then deciding whether the symbolic form of the argument is a tautology.
   a. Ralph does not work hard. In order for Ralph to pass this course it is necessary that he work hard. Therefore, Ralph will not pass this course.
   b. If the President wins California, then he will be reelected. But the President will lose California. Therefore, he will not be reelected.

c. Either the Redskins or the Raiders will win the Super Bowl. The Raiders will not win the Super Bowl. Therefore, the Redskins will win the Super Bowl.

d. If the instructor is good, then the course is interesting. Either the assignments are challenging or the course is not interesting. The assignments are not challenging. Therefore, the instructor is not good.

e. For the President to be reelected it is sufficient that he negotiate an arms reduction treaty. He will negotiate such a treaty only if the Russians leave Afghanistan. But the Russians will not leave Afghanistan. Therefore, the President will not be reelected.

2. Verify that the syllogism $((p \rightarrow r) \wedge (r \rightarrow q)) \rightarrow (p \rightarrow q)$ is a tautology.

3. Let x and y be real numbers. Give both a direct and indirect proof of the statement

$$\text{If } x = y, \text{ then } x^2 = y^2.$$

4. Verify that the compound statement $((p \leftrightarrow r) \wedge (r \leftrightarrow q)) \rightarrow (p \leftrightarrow q)$ is a tautology.

5. Prove that if triangle ABC is isosceles, with sides AB and AC having the same length, then the angles ABC and ACB have the same measure.

6. Recall the fifth postulate (axiom) of Euclid:

> Given a line m and a point P not on m (in a plane), then there is exactly one line through P and parallel to m.

Use this axiom and the method of proof by contradiction to prove the following result: Let u, v, and w be distinct lines in a plane. If u is parallel to v and v is parallel to w, then u is parallel to w.

7. **a.** Prove directly: If n is an even integer, then n^2 is an even integer.

b. Prove indirectly: If n^2 is an even integer, then n is an even integer. (You may use the fact that an integer n is even if and only if $n = 2k$ for some integer k.)

8. Use the method of proof by contradiction to prove the following statements:

a. If $x = \sqrt{3}$, then x is irrational.

b. If $x = \sqrt[3]{2}$, then x is irrational.

9. Prove by contradiction: If x is a real number and $x^3 + 4x = 0$, then $x = 0$.

10. Prove that each of the following program segments is correct.

a. (*INITIAL STATE: A = a, B = b, where a and b are integers and $a \geq 0$*)

```
PRODUCT := 0;
WHILE A > 0 DO
  BEGIN
    PRODUCT := PRODUCT + B;
    A := A - 1
  END;
```

(*FINAL STATE: PRODUCT = ab*)

(*Hint: Use the loop invariant condition

$$A = a' \wedge \text{PRODUCT} = ab - a'b\text{*})$$

b. (*INITIAL STATE: The values of A, B, and C are of type REAL*)

```
MAX := A;
IF B > MAX THEN MAX := B;
IF C > MAX THEN MAX := C;
```

(*FINAL STATE: (MAX = A $\vee$ MAX = B $\vee$ MAX = C) $\wedge$ MAX $\geq$ A $\wedge$ MAX $\geq$ B $\wedge$ MAX $\geq$ C *)

(*Hint: Show by considering three cases that when the segment terminates, the condition

```
(MAX = A ∧ MAX ≥ B ∧ MAX ≥ C) ∨ (MAX = B ∧ MAX > A ∧ MAX ≥ C)
        ∨ (MAX = C ∧ MAX > A ∧ MAX > B)
```

holds, which implies that the final state holds.*)

11. Use a counterexample to disprove: If p is an odd prime, then $p^2 + 4$ is also prime.

CHAPTER PROBLEMS

1. The statements

p: Ralph reads The New York Times.
q: Ralph watches the MacNeil–Lehrer Report.
r: Ralph jogs 3 miles.

are those of Exercise 1 of Section 1.2. Express each of the following in words.

a. $(p \wedge q) \rightarrow r$ **b.** $\sim(p \leftrightarrow q)$
c. $\sim(p \wedge q) \rightarrow r$ **d.** $\sim p \vee \sim q \rightarrow r$
e. $p \rightarrow (q \vee r)$ **f.** $r \rightarrow (\sim(p \rightarrow q))$
g. $p \wedge q \wedge \sim r$ **h.** $(p \rightarrow q) \rightarrow r$

2. Determine the truth-value of each of the following statements.
   a. $1 + 1 = 2$ or 10 is even.
   b. $1 + 1 = 2$ and 10 is even.
   c. $1 + 1 = 2$ or 10 is odd.
   d. $1 + 1 = 2$ and 10 is odd.
   e. $1 + 1 = 3$ or 10 is odd.
   f. $1 + 1 = 3$ and 10 is odd.
   g. $1 + 1 = 2$ implies 10 is odd.
   h. $1 + 1 = 2$ implies 10 is even.
   i. $1 + 1 = 3$ implies 10 is even.
   j. $1 + 1 = 3$ implies 10 is odd.
3. Let p, q, and r denote the following statements:

   p: Fred knows BASIC.
   q: Fred knows Pascal.
   r: Fred has taken CS 260.

   Express, unambiguously, each of the following formulas in words.
   a. $r \leftrightarrow (p \vee q)$
   b. $r \rightarrow q$
   c. $r \wedge \sim p$
   d. $q \rightarrow (r \wedge \sim p)$
   e. $(p \wedge q) \vee \sim r$
   f. $p \wedge (r \rightarrow q)$
4. For each of the following statements, (i) express it symbolically, (ii) find its negation, and (iii) express the negation in words.
   a. The function f is one-to-one and onto.
   b. The number m is prime or even.
   c. Knowing that the graph G is hamiltonian is sufficient to say that G is connected.
   d. If a is related to b, and b is related to c, then a is related to c.
   e. The function f is one-to-one if and only if f is onto.
5. Find and simplify the negation of each of the following formulas.
   a. $p \wedge q \wedge r$
   b. $p \leftrightarrow q$
   c. $p \rightarrow (q \rightarrow r)$
   d. $p \wedge (q \vee r)$
   e. $p \wedge (p \rightarrow q) \wedge (q \rightarrow r)$
   f. $\sim p \wedge (q \rightarrow p)$
   g. $(p \wedge (q \rightarrow r)) \vee (\sim q \wedge p)$
   h. $p \rightarrow (q \vee r)$
6. For each of the following formulas, (i) prove the formula is a tautology and (ii) discuss how the formula provides a strategy for proving an implication involving a compound hypothesis or conclusion.
   a. $[(p \wedge q) \rightarrow r] \leftrightarrow [p \rightarrow (q \rightarrow r)]$
   b. $[(p \vee q) \rightarrow r] \leftrightarrow [(p \rightarrow r) \wedge (q \rightarrow r)]$
   c. $[p \rightarrow (q \wedge r)] \leftrightarrow [(p \rightarrow q) \wedge (p \rightarrow r)]$
   d. $[p \rightarrow (q \vee r)] \leftrightarrow [(p \wedge \sim r) \rightarrow q]$
7. Often in mathematics we wish to prove that a formula of the form

$$(p \leftrightarrow q) \wedge (q \leftrightarrow r) \wedge (r \leftrightarrow p)$$

   is true; in this case we say that the statements p, q, and r are *equivalent*. In order to prove p, q, and r equivalent, show that it suffices to prove

$$(p \rightarrow q) \wedge (q \rightarrow r) \wedge (r \rightarrow p)$$

8. For each of the following formulas, find an equivalent formula that uses only the logical operators $\wedge$ and $\sim$.

a. $p \vee q$ **b.** $p \rightarrow q$

c. $p \leftrightarrow q$ **d.** $p \vee (q \rightarrow \sim r)$

9. The *nand* operator, denoted by "$|$", is defined by $p \mid q \equiv \sim(p \wedge q)$ (hence the acronym "nand" for "not and").

a. Give the truth table for $p \mid q$.

b. Show that $p \mid p \equiv \sim p$.

Find logically equivalent formulas for

c. $p \wedge q$

d. $p \vee q$

that use only the nand operator. (This problem shows that any formula may be expressed using nand as the only logical operator.)

10. Let u and v be formulas. Prove that u and v are logically equivalent if and only if $u \leftrightarrow v$ is a tautology.

11. Find (i) the inverse, (ii) the converse, and (iii) the contrapositive of each of the following implications.

a. If x is odd, then x^2 is odd.

b. x^2 is even only if x is even.

c. If f is differentiable at x, then f is continuous at x.

d. $x^3 = y^3$ implies $x = y$.

e. $F(x) > F(y)$ only if $x > y$.

f. The conditions x is prime and $x > 2$ are sufficient for x to be odd.

g. That a function f be defined at a is necessary for f to be continuous at a.

h. Being connected and having either no cycles or one more vertex than the number of edges are together sufficient conditions for a graph to be a tree.

12. Consider the connective $\veebar$ (exclusive or), which was defined in Exercise 10 of Section 1.2. Show that $p \veebar q$ is logically equivalent to each of the following:

a. $(p \vee q) \wedge \sim(p \wedge q)$

b. $(p \wedge \sim q) \vee (q \wedge \sim p)$

c. $\sim(p \leftrightarrow q)$

13. **a.** Show that $(p \wedge q) \rightarrow r \equiv (p \wedge \sim r) \rightarrow \sim q$.

The implications $(p \wedge \sim r) \rightarrow \sim q$ and $(q \wedge \sim r) \rightarrow \sim p$ are called *partial contrapositives* of the implication $(p \wedge q) \rightarrow r$. Find both partial contrapositives of each of the following implications.

b. If n is prime and $n > 2$, then n is odd.

c. If $f'(x) = 2x + 1$ and $f(0) = 3$, then $f(x) = x^2 + x + 3$.

14. Write each of the following statements using logical symbols.

a. Given any real numbers x and y, $f(x) = f(y)$ if and only if $x = y$.

b. If d is the greatest common divisor of a and b, then there exist integers s and t such that $d = as + bt$.

c. For every positive integer n, either n is prime or there exist integers a and b, both greater than 1, such that $n = ab$.
d. There is a function f such that both f is one-to-one and f has the property that $f(x) \neq x$ for every real number x.

15. For each part of this problem, find a formula u that has the given truth table.

a.

p	q	u
T	T	F
T	F	F
F	T	F
F	F	T

b.

p	q	u
T	T	F
T	F	T
F	T	T
F	F	F

c.

p	q	u
T	T	F
T	F	T
F	T	F
F	F	T

d.

p	q	r	u
T	T	T	F
T	T	F	T
T	F	T	F
T	F	F	T
F	T	T	F
F	T	F	F
F	F	T	F
F	F	F	F

e.

p	q	r	u
T	T	T	T
T	T	F	F
T	F	T	T
T	F	F	F
F	T	T	T
F	T	F	T
F	F	T	T
F	F	F	F

f.

p	q	r	u
T	T	T	F
T	T	F	T
T	F	T	T
T	F	F	T
F	T	T	T
F	T	F	T
F	F	T	T
F	F	F	F

FIGURE 1.9

16. For each of the following pairs of statements about positive real numbers, write each statement symbolically and note the difference between them. Also, determine the truth-value of each statement.
a. For every x there exists y such that $y < x$.
There exists y such that $y < x$ for every x.
b. For every x there exists y such that $xy = 1$.
There exists y such that $xy = 1$ for every x.

17. Find and simplify the negation of each of the following.
a. $\exists y\ \forall x(xy \geq 3)$
b. $\forall x\ \exists y\ \forall z(x + y = z)$
c. $\forall x(x > 0 \to x^2 \geq x)$
d. $\exists x(x$ is rational $\wedge\ \forall y(y$ is irrational $\to x + y$ is rational$))$
e. $\forall x\ \exists y(x < y \wedge f(x) \geq f(y))$
f. $\forall x\ \forall y((x > 0 \wedge y > 0) \to \exists n(nx > y))$
g. $\forall y(y > 0 \to \exists x(\log x > y))$

18. For each of the following statements, (i) write the statement symbolically, (ii) find the symbolic form of the negation, and then (iii) write out the negation in words.

a. There exist real numbers x and y such that $x^2 + y^2 = -1$.

b. There exists a positive integer N such that for every real number $x \neq 1$, $N > 1/(x - 1)$.

c. Given any real numbers x and y, $2x^2 - xy + 5 > 0$.

d. For every real number x there is some positive integer n such that x^n is rational.

e. For every integer n, if n is a multiple of 6, then there exist integers s and t such that $n = 12s + 18t$.

f. For every positive real number ε there is some positive real number δ such that if $|x - 2| < \delta$, then $|x^2 - 4| < \varepsilon$.

19. Find the contrapositive of each of the following implications.

a. If, for all x and y, the fact that x is related to y implies that y is related to x, then the relation is symmetric.

b. If d is the greatest common divisor of a and b, then there exist integers s and t such that $d = as + bt$.

c. In order for the function f (defined on the real numbers) to be onto, it is necessary that, given any real number b, there is some real number a such that $f(a) = b$.

d. If a given relation is transitive, then, for all a, b, and c, the conditions that a is related to b and b is related to c are sufficient to imply that a is related to c.

20. a. Prove directly: If x and y are odd integers, then $x + y$ is an even integer.

b. Prove the result of part a by contradiction. (Hint: derive that 1 is even, a contradiction.)

c. Prove by contrapositive: If xy is an odd integer, then both x and y are odd integers.

21. Determine whether or not each of the following arguments is valid. (Write the argument in symbolic form and then determine if the resulting formula is a tautology.)

a. Knowing Pascal is necessary and sufficient for passing CS 260. Nancy knows Pascal; therefore she passed CS 260.

b. Fred is required to take MA 350 or MA 337. Fred will not take MA 350; therefore, he will take MA 337.

c. Al is taking both MA 350 and MA 337. If he doesn't take MA 350, then he must take MA 337. Therefore, Al isn't taking MA 337.

d. If one learns Pascal, then one passes CS 260. If one doesn't pass CS 260, then one can't take CS 261. Therefore if one is taking CS 261, then one learned Pascal.

e. Janet will pass this course if and only if she works hard. Either she'll work hard or she won't graduate. But Janet will graduate; therefore she will pass this course.

22. Prove each of the following statements about real numbers x and y by contradiction.

a. If x is rational and y is irrational, then $x + y$ is irrational.
b. If x is rational, $x \neq 0$, and y is irrational, then xy is irrational.
c. If x and y are both positive, then $\sqrt{x + y} \neq \sqrt{x} + \sqrt{y}$.

23. Prove the following statement (i) by contrapositive and (ii) by contradiction: If x is a positive real number, then $x + (1/x) \geq 2$.

24. Given lines m_1, m_2, and m_3 and angles α and β, as shown in Figure 1.10, prove that if $\alpha = \beta$, then lines m_1 and m_2 are parallel. (Hint: prove the contrapositive. If m_1 and m_2 intersect at point C, consider triangle ABC.)

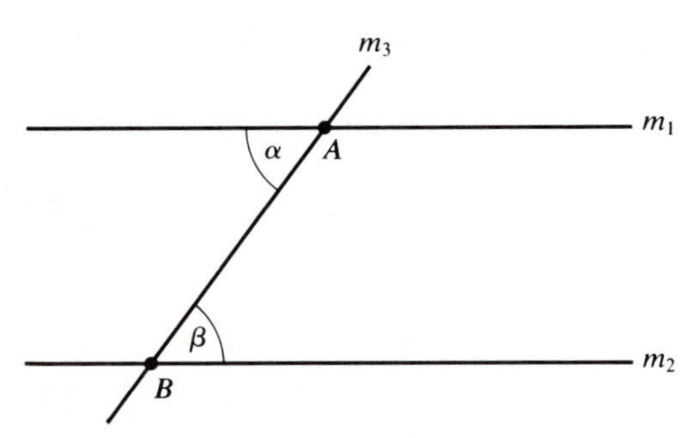

FIGURE 1.10

25. Supply a conclusion, in the form of an implication, to the following argument so as to make it valid.

If he goes to a party, he does not fail to brush his hair. To look fascinating it is necessary to be tidy. If he is an opium eater, then he has no self-command. If he brushes his hair, he looks fascinating. He wears white kid gloves only if he goes to a party. Having no self-command is sufficient to make one look untidy. Therefore, . . . [Adapted from Lewis Carroll.]

Two Set Theory

2.1 INTRODUCTION

A set can be thought of as a collection of objects. A class of students, the Greek alphabet, a baseball team, and the Euclidean plane are all examples of sets. The Greek alphabet consists of letters, a baseball team is composed of its players, and the Euclidean plane is made up of points. In general, the objects that make up a set are referred to as its *elements* or *members*. It is common practice to think of the terms "set" and "member" as undefined or primitive notions. Most of us have intuitive feelings (from examples like those just given) as to what these terms mean. We shall use the term "collection" interchangeably with "set."

Sets are the building blocks for most mathematical structures. For example, Euclidean plane geometry is based on the interpretation of the Euclidean plane as a set of points. A number of very important examples also arise in the field of abstract algebra, which studies the properties of sets on which one or more "binary operations" have been defined. For example, ordinary addition and multiplication are binary operations on the set of positive integers 1, 2, 3, ... Another example is that of a set S on which an operation $*$ is defined, satisfying

$$(a * b) * c = a * (b * c)$$

for all elements a, b, and c in S. Such a structure is called a *semigroup*. As a special example of a semigroup, consider the set of valid character strings in FORTRAN77 under the binary operation of concatenation. (When we concatenate the string 'FORT' with the string 'RAN77', we get the string 'FORTRAN77'.)

There are certain sets of numbers that will be used quite often in this text, so we shall adopt some special notations for them. The numbers 1, 2, 3, . . . are called the *positive integers* (or *natural numbers*), and we use the notation $\mathbb{N}$ to denote the set of all positive integers. The positive integers together with the numbers 0, -1, -2, -3, . . . form the set of *integers;* we denote the set of all integers by $\mathbb{Z}$. A *rational number* is any number expressible in the form m/n, where m and n are integers and $n \neq 0$. For example, $2/3$, $-5/11$, $17 = 17/1$, and $.222 \cdots = 2/9$ are rational numbers. We denote by $\mathbb{Q}$ the set of all rational numbers. Numbers like $\sqrt{2}$, $4/\sqrt{3}$, and π are not rational numbers; in general, such numbers are referred to as *irrational numbers*. The rational numbers together with the irrational numbers make up the set of *real numbers*, which is denoted by $\mathbb{R}$. To repeat, then, we have the following notational conventions:

$\mathbb{N}$—the set of positive integers
$\mathbb{Z}$—the set of integers
$\mathbb{Q}$—the set of rational numbers
$\mathbb{R}$—the set of real numbers

The basic relation between an element x and a set A is that of membership; that is, what matters is whether or not x is an element of the set A.

DEFINITION 2.1

If x is an element of the set A, then we write $x \in A$; if x is not an element of A we write $x \notin A$.

Thus, for example, $-3 \in \mathbb{Z}$, $-3 \notin \mathbb{N}$, $\sqrt{2} \in \mathbb{R}$, and $\sqrt{2} \notin \mathbb{Q}$.

Except in some very special cases, sets will be denoted by capital letters A, B, C, and so on, and elements by lower-case letters a, b, c, and so on. It is the object of this chapter to define relations and operations on sets and then to derive some basic properties. This material should provide some much-needed background for subsequent courses in mathematics and computer science.

Probably the most fundamental relation that can exist between two sets is that of "equality."

DEFINITION 2.2

Two sets A and B are called *equal*, denoted $A = B$, provided they consist of the same elements. If A and B are not equal, then we write $A \neq B$.

We remark that set equality may be restated

$$A = B \leftrightarrow \forall x(x \in A \leftrightarrow x \in B)$$

A set A is said to be *finite* if it has n elements for some nonnegative integer n. We also say that A consists of a finite number (or n) elements.

Sets that are not finite are called *infinite;* in particular, the sets $\mathbb{N}$, $\mathbb{Z}$, $\mathbb{Q}$, and $\mathbb{R}$ are infinite.

If a set A consists of a "small" number of elements, then we can exhibit A by explicitly listing its elements between braces. For example, if A is the set of prime numbers that are less than 20, then we write

$$A = \{2, 3, 5, 7, 11, 13, 17, 19\}$$

However, some sets contain too many elements to be listed this way. In many such cases we use the "three-dot notation," to mean "and so on" or "and so on up to," depending on the context. For instance, the set $\mathbb{N}$ can be exhibited as

$$\mathbb{N} = \{1, 2, 3, \ldots\}$$

Often a set A is described as consisting of all elements x satisfying a propositional function $p(x)$. Thus A consists of all elements x for which $p(x)$ is true; in this case we write

$$A = \{x \mid p(x)\}$$

Here the symbol "|" translates as "such that" or "for which."

Example 2.1 The set of positive even integers less than 100 is

$$E = \{2, 4, 6, \ldots, 98\}$$

The set of "harmonic fractions" is

$$H = \{1, \tfrac{1}{2}, \tfrac{1}{3}, \tfrac{1}{4}, \ldots\}$$

The set $\mathbb{Z}$ of integers can be written

$$\mathbb{Z} = \{\ldots, -2, -1, 0, 1, 2, \ldots\}$$

The set of primes can be exhibited as either

$$P = \{2, 3, 5, 7, 11, \ldots\}$$

or

$$P = \{p \mid p \text{ is prime}\}$$

The latter method of writing P is preferable because it is less likely to be misunderstood. ■

Suppose the set A is given by

$$A = \{x \mid p(x)\}$$

Often $p(x)$ may be of the form "$x \in S \wedge q(x)$." In this case we also allow the description

$$A = \{x \in S \mid q(x)\}$$

The above is read, "A is the set whose elements are precisely those elements x of S for which $q(x)$ holds."

Example 2.2 The set of real numbers x for which $x^2 + 2x - 3 \geq 0$ can be written $\{x \in \mathbb{R} \mid x^2 + 2x - 3 \geq 0\}$. The set of integer multiples of 3 can be written $\{n \in \mathbb{Z} \mid n \text{ is a multiple of } 3\}$. This last set can also be written $\{\ldots, -9, -6, -3, 0, 3, 6, 9, \ldots\}$. ■

There is a very special set that is much used in mathematics; this is the set that contains no elements. It is aptly called the *empty set* and is denoted by ϕ. Note that the set ϕ is uniquely determined (there is exactly one set that contains no elements), although there are numerous examples of it. For example, consider the set of former Presidents of the United States who were women, or the set of real numbers x such that x^2 is negative.

Recall that for sets A and B,

$$A = B \leftrightarrow \forall x(x \in A \leftrightarrow x \in B)$$

When is $A \neq B$? Interpreted as $\sim(A = B)$, we can obtain the logical negation as follows:

$$\begin{aligned} A \neq B &\equiv \sim\forall x(x \in A \leftrightarrow x \in B) \\ &\equiv \exists x(\sim(x \in A \leftrightarrow x \in B)) \\ &\equiv \exists x((x \in A \wedge x \notin B) \vee (x \in B \wedge x \notin A)) \end{aligned}$$

In words, $A \neq B$ if and only if there is an element x such that either $x \in A$ and $x \notin B$ or $x \in B$ and $x \notin A$. More briefly, $A \neq B$ if and only if there is an element x that belongs to one of the sets but not the other.

Example 2.3 Consider the following sets:

$$\begin{aligned} B &= \{0, 1\} \\ D &= \{0, 1, 0\} \\ A &= \{0, 1, B\} \\ G &= \{\phi, \{0\}, \{1\}, \{0, 1\}\} \\ C &= \{y \mid y = B \text{ or } y \in B\} \end{aligned}$$

Which of these sets, if any, are equal?

Solution First note that sets B and D are equal, for both contain exactly the elements 0 and 1. We would not normally write a set as we have written D; in fact, we shall adopt the convention that in listing the elements of a set, each element should be listed exactly once. The set A contains three distinct elements: 0, 1, and the set B. The set G contains four distinct elements: the empty set, and the sets $\{0\}$, $\{1\}$, and $\{0, 1\}$. Thus A and G cannot be equal, and neither is equal to B. What about C? Note that B is an element of C; also 0 and 1 belong to C since 0 and 1 belong to B. Hence $C = \{0, 1, B\} = A$. ■

Exercises 2.1

1. Write each of the following sets by listing the elements.
   a. $A = \{n \in \mathbb{Z} \mid -4 < n < 5\}$
   b. $B = \{x \in \mathbb{R} \mid x^3 - x^2 - 2x = 0\}$
2. Write each of the following sets in the form $\{x \in \mathbb{Z} \mid p(x)\}$.
   a. $\{-1, -2, -3, \ldots\}$
   b. $\{0, 1, 4, 9, 16, \ldots\}$
   c. $\{\ldots, -15, -9, -3, 3, 9, 15, \ldots\}$
   d. $\{\ldots, -4, -2, 0, 2, 4, \ldots\}$
3. Which of the following sets are equal?

$$A = \{n \in \mathbb{Z} \mid |n| < 2\}$$
$$B = \{n \in \mathbb{Z} \mid n^3 = n\}$$
$$C = \{n \in \mathbb{Z} \mid n^2 \leq 2n\}$$
$$D = \{n \in \mathbb{Z} \mid n^2 \leq 1\}$$
$$E = \{0, 1, 2\}$$

2.2 SUBSETS

As stated in the preceding section, set equality is a very basic relation in the theory of sets. Aside from this there is the possibility that every element of the set A is also an element of the set B.

DEFINITION 2.3

A set A is called a *subset* of the set B, denoted $A \subseteq B$, provided every element of A is also an element of B. If A is not a subset of B, we write $A \not\subseteq B$.

It should be noted that if A and B are sets, then

$$A \subseteq B \leftrightarrow \forall x(x \in A \to x \in B)$$

It is evident from this condition that, for any set A, both

$$\phi \subseteq A$$

and

$$A \subseteq A$$

We can also see that

$$A \not\subseteq B \leftrightarrow \exists x(x \in A \wedge x \notin B)$$

Hence to show that $A \not\subseteq B$ it suffices to show there is some $x \in A$ such that $x \notin B$.

Example 2.4 Consider the sets

$$A = \{x \in \mathbb{Z} \mid -20 \leq x \leq 20\}$$
$$B = \{n \in \mathbb{N} \mid n^2 \leq 49\}$$
$$C = \{A, B, -3, 4, 6\}$$
$$D = \{-3, 4, 6, 7\}$$

Then we have the following facts:

1. $B \subseteq A$
2. $C \not\subseteq A$, since $A \in C$ but $A \notin A$
3. $D \subseteq A$
4. $C \not\subseteq B$
5. $D \not\subseteq B$, since $-3 \in D$ but $-3 \notin B$
6. $D \not\subseteq C$, since $7 \in D$ but $7 \notin C$ ■

Frequently in mathematics we are given two sets A and B and asked to prove that $A \subseteq B$. In keeping with the above definition, it is common practice to start such an argument with the sentence

Let x be an arbitrary element of A.

Then we must proceed to show that $x \in B$. This scheme is illustrated in the proof of the following theorem.

THEOREM 2.1 Let A, B, and C be sets. If $A \subseteq B$ and $B \subseteq C$, then $A \subseteq C$.

Proof Let x be an arbitrary element of A. We must show that $x \in C$. Since $A \subseteq B$ and $x \in A$, it follows that $x \in B$. Then, since $B \subseteq C$ and $x \in B$, we may conclude that $x \in C$. Thus we have shown that for every x, if $x \in A$ then $x \in C$. Therefore $A \subseteq C$. ■

The student may observe that syllogistic reasoning is used in the above proof. In particular, the logical form of the argument is

$$((x \in A \rightarrow x \in B) \wedge (x \in B \rightarrow x \in C)) \rightarrow (x \in A \rightarrow x \in C)$$

It has been mentioned before that to prove a mathematical statement of the form $p \leftrightarrow q$, it is necessary to prove both $p \rightarrow q$ and $q \rightarrow p$. It is also possible to prove $p \leftrightarrow q$ by proving both $p \leftrightarrow r$ and r $\leftrightarrow q$ for some statement r. This technique is used to prove the following result.

THEOREM 2.2 For any sets A and B,

$$A = B \leftrightarrow (A \subseteq B \wedge B \subseteq A)$$

Proof We proceed as follows:

$$\begin{aligned} A = B &\leftrightarrow \forall x(x \in A \leftrightarrow x \in B) \\ &\leftrightarrow \forall x((x \in A \rightarrow x \in B) \wedge (x \in B \rightarrow x \in A)) \\ &\leftrightarrow \forall x(x \in A \rightarrow x \in B) \wedge \forall x(x \in B \rightarrow x \in A) \\ &\leftrightarrow (A \subseteq B) \wedge (B \subseteq A) \end{aligned}$$

■

We mentioned before that $A \subseteq A$ for any set A. This fact follows from Theorem 2.2 (with B replaced by A) and the obvious fact that $A = A$.

THEOREM 2.3 For any sets A and B,

$$A \subseteq B \leftrightarrow \forall C(C \subseteq A \rightarrow C \subseteq B)$$

Proof We first show that $A \subseteq B \rightarrow \forall C(C \subseteq A \rightarrow C \subseteq B)$. Assume $A \subseteq B$ and let C be any set such that $C \subseteq A$. Then we have $C \subseteq A$ and $A \subseteq B$, and we may conclude from Theorem 2.1 that $C \subseteq B$.

Next we must show that $\forall C(C \subseteq A \rightarrow C \subseteq B) \rightarrow A \subseteq B$. Here our hypothesis is that *for any set* C, if $C \subseteq A$, then $C \subseteq B$. In particular, suppose we let $C = A$. Then $A \subseteq A$ implies $A \subseteq B$, which is just what we wished to show. ■

THEOREM 2.4 Let A be any set. Then, for all x,

$$x \in A \leftrightarrow \{x\} \subseteq A$$

Proof See Exercise 4. ■

DEFINITION 2.4

The set A is called a *proper subset* of the set B provided $A \subseteq B$ and $A \neq B$. We write $A \subset B$.

THEOREM 2.5 For any sets A and B,

$$A \subset B \leftrightarrow (A \subseteq B \wedge \exists x(x \in B \wedge x \notin A))$$

Proof See Exercise 6. ■

Given a finite collection of sets, it is sometimes useful to have a "picture" of the subset relationships that exist among them. We construct such a picture as follows. Corresponding to each set in the collection is a point, called a *vertex,* which is labelled with the name of the set. Next suppose X and Y are two sets in the collection and $X \subset Y$. Further assume that there is no set Z in the collection such that $X \subset Z$ and $Z \subset Y$. Then in our picture, the vertex corresponding to Y is placed above the vertex corresponding to X, and a line segment, called an *edge,* is drawn joining these vertices, as shown in Figure 2.1(a). (Note: it is not necessary that Y be placed directly above X.)

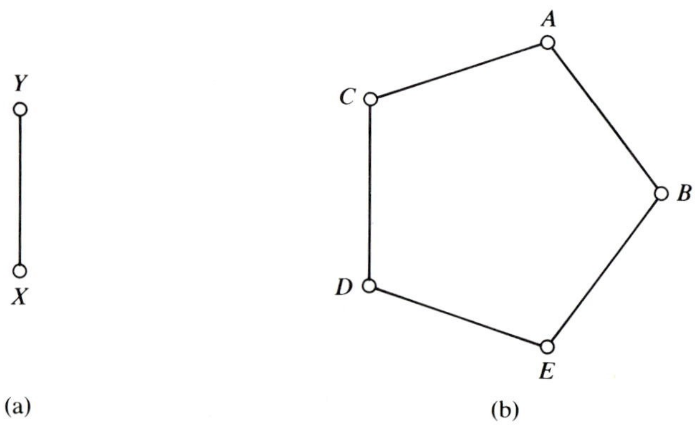

FIGURE 2.1 Subset diagrams

Example 2.5 Consider the collection of sets $\{A, B, C, D, E\}$, where $A = \{1, 2, 3, 4\}$, $B = \{1, 3\}$, $C = \{2, 3, 4\}$, $D = \{2, 3\}$, and $E = \{3\}$. For this collection we have the picture shown in Figure 2.1(b). Notice that no edge is drawn, for example, from A to D. This is due to the fact that $D \subset C$ and $C \subset A$. From the picture, and using Theorem 2.1, we can deduce that $D \subset A$. ■

A picture such as the one in Figure 2.1(b) is called a *subset diagram* for the given collection of sets.

Before proceeding further, we take time out to introduce some additional notation.

Given real numbers a and b, with $a < b$, define

$$(a, b) = \{x \in \mathbb{R} \mid a < x < b\}$$
$$[a, b] = \{x \in \mathbb{R} \mid a \leq x \leq b\}$$
$$[a, b) = \{x \in \mathbb{R} \mid a \leq x < b\}$$
$$(a, b] = \{x \in \mathbb{R} \mid a < x \leq b\}$$

These subsets of the set of real numbers are called *intervals*. Undoubtedly this notation has been encountered in previous courses. The interval (a, b) is called an *open interval* and $[a, b]$ is called a *closed interval,* while both $[a, b)$ and $(a, b]$ are called *half-open intervals*. In each case we call a and b the *endpoints* of the interval. Some other, rather special, intervals of $\mathbb{R}$ are

$$\begin{aligned}(a, \infty) &= \{x \in \mathbb{R} \mid x > a\}\\ [a, \infty) &= \{x \in \mathbb{R} \mid x \geq a\}\\ (-\infty, b) &= \{x \in \mathbb{R} \mid x < b\}\\ (-\infty, b] &= \{x \in \mathbb{R} \mid x \leq b\}\\ (-\infty, \infty) &= \mathbb{R}\end{aligned}$$

Consider next the statement

For every real number x, $\quad x^2 \geq 0$.

It has already been noted that this statement is written symbolically as

$$\forall x(x \in \mathbb{R} \rightarrow x^2 \geq 0)$$

We shall also agree to write it in each of the following ways:

$$\forall x \in \mathbb{R}(x^2 \geq 0)$$

or

$$x^2 \geq 0 \ \forall x \in \mathbb{R}$$

In general, if A is a set and $p(x)$ is a propositional function, then each of the statements

$$\forall x(x \in A \rightarrow p(x)) \qquad \textbf{(1)}$$
$$\forall x \in A(p(x))$$

and

$$p(x) \ \forall x \in A$$

all say the same thing, namely, that $p(x)$ holds for every $x \in A$. Similarly, each of the statements

$$\exists x(x \in A \wedge p(x)) \qquad \textbf{(2)}$$

and

$$\exists x \in A(p(x))$$

signify that there is some $x \in A$ for which $p(x)$ is true.

The negations of statements (1) and (2) are easily obtained if we keep in mind what each is saying. To deny that $p(x)$ holds for every $x \in A$ is to say that there is some $x \in A$ for which $p(x)$ is false. Analogously, to deny that there is some $x \in A$ for which $p(x)$ holds is to say that $p(x)$ is false for every $x \in A$. Thus we obtain

$$\begin{aligned}\sim(\forall x \in A(p(x))) &\equiv \exists x \in A(\sim p(x))\\ \sim(\exists x \in A(p(x))) &\equiv \forall x \in A(\sim p(x))\end{aligned}$$

The symbolic forms seen in the preceding paragraphs are frequently encountered in the study of mathematics and computer science. However, it should be emphasized that such statements are often encountered in verbal form. The point to be stressed is that it is important to understand what each statement means, no matter what its form.

Example 2.6 Write each of the following statements in symbolic form, negate the symbolic form of the statement, and then restate the negation in words.

(a) There exists a rational number x such that $2x^3 - 3x^2 - 4x + 6 = 0$.

(b) For every positive real number d there exists a positive real number h such that if x is a real number and $|x - 2| < h$, then $|x^2 - 4| < d$.

Solution The form of statement (a) is

$$\exists x \in \mathbb{Q}(2x^3 - 3x^2 - 4x + 6 = 0)$$

Its negation is

$$\forall x \in \mathbb{Q}(2x^3 - 3x^2 - 4x + 6 \neq 0)$$

In words this says,

For every rational number x, $2x^3 - 3x^2 - 4x + 6 \neq 0$.

Let $\mathbb{R}^+$ denote the set of positive real numbers. Statement (b) can be written symbolically as

$$\forall d \in \mathbb{R}^+ (\exists h \in \mathbb{R}^+ (\forall x \in \mathbb{R}(|x - 2| < h \rightarrow |x^2 - 4| < d)))$$

The negation is therefore

$$\exists d \in \mathbb{R}^+ (\forall h \in \mathbb{R}^+ (\exists x \in \mathbb{R}(|x - 2| < h \wedge |x^2 - 4| \geq d)))$$

In words the negation reads,

> There exists a positive real number d such that, for every positive real number h, there is some real number x such that $|x - 2| < h$ and $|x^2 - 4| \geq d$. ■

Exercises 2.2

1. Identify which of the following assertions are true and which are false.

a. $\phi \in \phi$ **b.** $1 \in \{1\}$
c. $\{1, 2\} = \{2, 1\}$ **d.** $\phi = \{\phi\}$
e. $\phi \subset \{\phi\}$ **f.** $1 \subseteq \{1\}$
g. $\{1\} \subset \{1, 2\}$ **h.** $\phi \in \{\phi\}$

2. Give examples of sets A, B, and C such that

a. $A \in B$ and $B \in C$ and $A \notin C$
b. $A \in B$ and $B \in C$ and $A \in C$
c. $A \in B$ and $A \subset B$

3. For each of the following sets, list all of its subsets.

a. ϕ **b.** $\{1\}$

c. $\{1, 2\}$ **d.** $\{1, 2, 3\}$

e. $\{\phi, 1\}$ **f.** $\{\phi, 1, \{1\}\}$

4. Prove Theorem 2.4.

5. In view of Exercise 3, make a guess as to the number of subsets of a set having n elements.

6. Prove Theorem 2.5.

7. Draw the subset diagram for the collection $\{A, B, C, D, E, F, G\}$, where $A = \{1, 2, 3, 4, 5\}$, $B = \{1, 2, 3\}$, $C = \{2, 3, 4\}$, $D = \{2, 4\}$, $E = \{3, 4\}$, $F = \{4, 5\}$, and $G = \{4\}$.

8. Specify the subset relationships that exist among the sets $\mathbb{N}$, $\mathbb{Z}$, $\mathbb{Q}$, and $\mathbb{R}$.

9. For each part of Figure 2.2, give a collection of sets having that subset diagram.

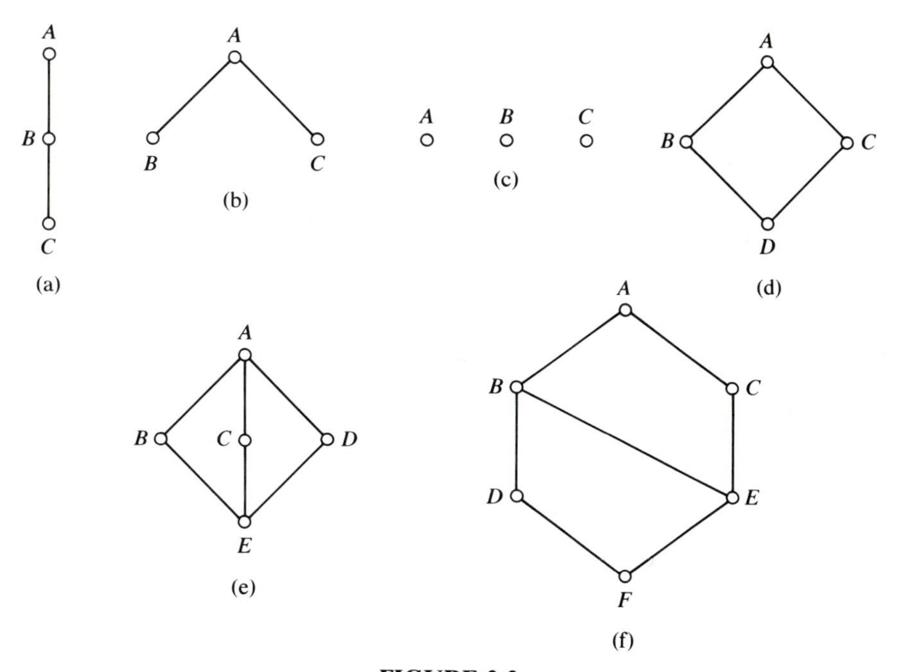

FIGURE 2.2

2.3 SET OPERATIONS

In any particular discussion of sets, we will almost always be concerned with subsets of some specified set U. The set U is called a *universal set*. Unlike ϕ, the set U is not unique; for example, in a given calculus lecture it may be that $U = \mathbb{R}$, while in a computer science class, U might be the set of all numbers representable on a particular computer.

The most frequently used operations in the theory of sets are "intersection," "union," and "difference." These are described in the following definition.

DEFINITION 2.5

Let A and B be subsets of some universal set U. We define the set operations of intersection, union, and difference as follows:

1. The *intersection of* A and B is the set $A \cap B$ defined by
$$A \cap B = \{x \mid x \in A \wedge x \in B\}$$
2. The *union* of A and B is the set $A \cup B$ defined by
$$A \cup B = \{x \mid x \in A \vee x \in B\}$$
3. The *difference set*, $A - B$, is defined by
$$A - B = \{x \mid x \in A \wedge x \notin B\}$$

We also call $A - B$ the *relative complement of* B *in* A. It is important to realize that, in general, $A - B \neq B - A$. The relative complement of A in U is called the *complement* of A and is denoted by A'. Symbolically,
$$A' = U - A$$

It is often useful to picture sets using what are called "Venn diagrams." Venn diagrams for $A \cap B$, $A \cup B$, $A - B$, and A' are shown in Figure 2.3.

Example 2.7 If $U = \{0, 1, 2, 3, 4, 5, 6\}$, $A = \{0, 2, 4, 6\}$, and $B = \{2, 3, 5\}$, determine $A \cap B$, $A \cup B$, $A - B$, $B - A$, A', and B'.

Solution

$$\begin{aligned} A \cap B &= \{2\} \\ A \cup B &= \{0, 2, 3, 4, 5, 6\} \\ A - B &= \{0, 4, 6\} \\ B - A &= \{3, 5\} \\ A' &= \{1, 3, 5\} \\ B' &= \{0, 1, 4, 6\} \end{aligned}$$

■

Example 2.8 If $U = \mathbb{N}$, $A = \{1, 3, 5, 7, \ldots\}$, $P = \{p \in \mathbb{N} \mid p \text{ is prime}\}$, and $T = \{3, 6, 9, 12, \ldots\}$, find A', P', $A \cap P$, $A - P$, $P - A$, $A' \cap T$, and $A \cup T'$.

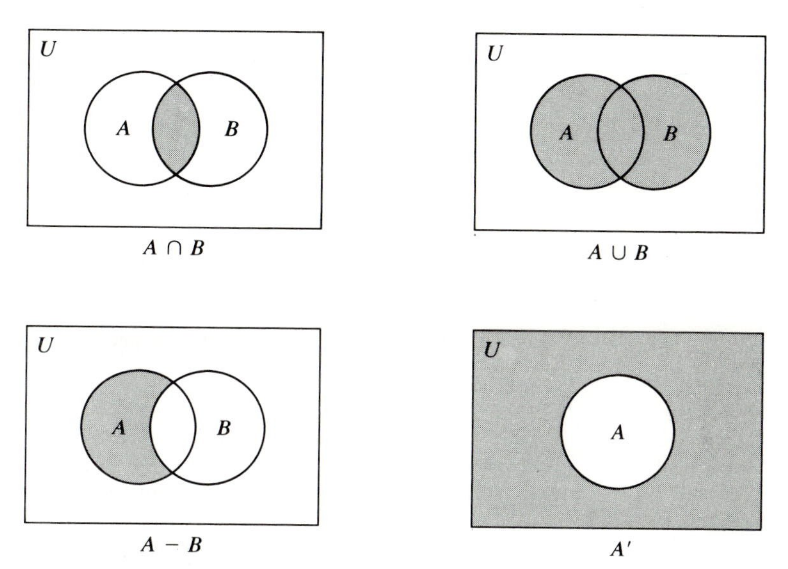

FIGURE 2.3 Venn diagrams

Solution The set A is the set of odd positive integers, so A' is the set of even positive integers: $A' = \{2, 4, 6, 8, \ldots\}$. The set P is the set of primes, so $P' = \{n \in \mathbb{N} \mid n \text{ is not prime}\} = \{1, 4, 6, 8, 9, 10, \ldots\}$. The set $A \cap P$ is the set of odd primes: $A \cap P = \{3, 5, 7, 11, 13, \ldots\} = P - \{2\}$. The set of odd positive integers that are not prime is $A - P = \{1, 9, 15, 21, 25, 27, \ldots\}$, while $P - A$ is the set of primes that are not odd, namely, $P - A = \{2\}$. The set T is the set of positive multiples of 3, so $A' \cap T$ consists of those elements of T that are even: $A' \cap T = \{6, 12, 18, 24, \ldots\}$. Finally, $A \cup T'$ includes any positive integer that is either odd or is not a multiple of 3, so $A \cup T' = \{1, 2, 3, 4, 5, 7, 8, 9, 10, 11, 13, \ldots\}$. Note that

$$A \cup T' = \{6, 12, 18, 24, \ldots\}' = (A' \cap T)'$$

■

Example 2.9 Let U, A, and B be the same as in Example 2.7, and let $C = \{0, 1, 5, 6\}$. Find and compare each of the following pairs of sets:

(a) $A \cap (B \cup C)$ and $(A \cap B) \cup (A \cap C)$

(b) $A \cup (B \cap C)$ and $(A \cup B) \cap (A \cup C)$

(c) $(A \cup B)'$ and $A' \cap B'$

(d) $(A \cap B)'$ and $A' \cup B'$

Solution For (a) we find that

$$A \cap (B \cup C) = A \cap \{0, 1, 2, 3, 5, 6\} = \{0, 2, 6\}$$

whereas

$$(A \cap B) \cup (A \cap C) = \{2\} \cup \{0, 6\} = \{0, 2, 6\}$$

Note that $A \cap (B \cup C) = (A \cap B) \cup (A \cap C)$. In (b) the reader should verify that

$$A \cup (B \cap C) = (A \cup B) \cap (A \cup C) = \{0, 2, 4, 5, 6\}$$

For (c) we have

$$(A \cup B)' = \{0, 2, 3, 4, 5, 6\}' = \{1\}$$

whereas

$$A' \cap B' = \{1, 3, 5\} \cap \{0, 1, 4, 6\} = \{1\}$$

so again the two sets are equal. And in (d),

$$(A \cap B)' = A' \cup B' = \{0, 1, 3, 4, 5, 6\}$$ ■

Example 2.9 illustrates some of the general properties of the set operations, which are among those presented in the following two theorems.

THEOREM 2.6 For subsets A, B, and C of a universal set U, the following properties hold:

1. *Commutative laws*
   (a) $A \cap B = B \cap A$
   (b) $A \cup B = B \cup A$
2. *Associative laws*
   (a) $(A \cap B) \cap C = A \cap (B \cap C)$
   (b) $(A \cup B) \cup C = A \cup (B \cup C)$
3. *Distributive laws*
   (a) $A \cap (B \cup C) = (A \cap B) \cup (A \cap C)$
   (b) $A \cup (B \cap C) = (A \cup B) \cap (A \cup C)$
4. *DeMorgan's laws*
   (a) $(A \cap B)' = A' \cup B'$
   (b) $(A \cup B)' = A' \cap B'$

Proof We shall prove 2(a), 3(b), and 4(a); the remaining parts are left to Exercise 2.

1. Proof of 2(a). Let x be an arbitrary element of $(A \cap B) \cap C$. Then

$$\begin{aligned} x \in (A \cap B) \cap C &\rightarrow (x \in A \cap B) \wedge (x \in C) && \text{(by Definition 2.5)} \\ &\rightarrow (x \in A \wedge x \in B) \wedge (x \in C) && \text{(Definition 2.5)} \\ &\rightarrow (x \in A) \wedge (x \in B \wedge x \in C) && \text{(associativity of } \wedge) \\ &\rightarrow (x \in A) \wedge (x \in B \cap C) && \text{(Definition 2.5)} \\ &\rightarrow x \in A \cap (B \cap C) && \text{(Definition 2.5)} \end{aligned}$$

Thus $x \in (A \cap B) \cap C \rightarrow x \in A \cap (B \cap C)$ and, consequently, $(A \cap B) \cap$

$C \subseteq A \cap (B \cap C)$. To show that $A \cap (B \cap C) \subseteq (A \cap B) \cap C$, simply reverse the above steps. We then have that $(A \cap B) \cap C = A \cap (B \cap C)$.

2. Proof of 3(b). In this case we prove the inclusion in both directions in one string of biconditionals. For all x,

$$\begin{aligned} x \in A \cup (B \cap C) &\leftrightarrow (x \in A) \vee (x \in B \cap C) \\ &\leftrightarrow (x \in A) \vee (x \in B \wedge x \in C) \\ &\leftrightarrow (x \in A \vee x \in B) \wedge (x \in A \vee x \in C) \quad \text{(distributivity of } \vee \text{ over } \wedge) \\ &\leftrightarrow (x \in A \cup B) \wedge (x \in A \cup C) \\ &\leftrightarrow x \in (A \cup B) \cap (A \cup C) \end{aligned}$$

Hence for all x, we have proved that $x \in A \cup (B \cap C) \leftrightarrow x \in (A \cup B) \cap (A \cup C)$, and we may conclude that $A \cup (B \cap C) = (A \cup B) \cap (A \cup C)$.

3. Proof of 4(a). For every x, we have

$$\begin{aligned} x \in (A \cap B)' &\leftrightarrow x \notin A \cap B \\ &\leftrightarrow \sim(x \in A \wedge x \in B) \\ &\leftrightarrow x \notin A \vee x \notin B \\ &\leftrightarrow x \in A' \vee x \in B' \\ &\leftrightarrow x \in A' \cup B' \end{aligned}$$

Thus $(A \cap B)' = A' \cup B'$. ■

There are three properties that are related to those in Theorem 2.6:

1. $(A')' = A$
2. $A \cup A' = U$
3. $A \cap A' = \phi$

These are left as an exercise for the reader.

DeMorgan's law $(B \cap C)' = B' \cup C'$ pertains directly to those elements of the universal set U that are not in $B \cap C$. Thus another way to express this law is as follows:

$$U - (B \cap C) = (U - B) \cup (U - C)$$

Given subsets A, B, and C of U, we can generalize the above expression by considering those elements of A that are not in $B \cap C$ (that is, $A - (B \cap C)$). A similar expression arises from considering $A - (B \cup C)$. The resulting properties are called the "generalized DeMorgan laws."

THEOREM 2.7 **Generalized DeMorgan laws** For subsets A, B, and C of a universal set U, the following hold:

1. $A - (B \cap C) = (A - B) \cup (A - C)$
2. $A - (B \cup C) = (A - B) \cap (A - C)$

Proof We present the proof of part 1; part 2 is left to Exercise 4. For all x,

$$\begin{aligned} x \in A - (B \cap C) &\leftrightarrow x \in A \wedge x \notin B \cap C \\ &\leftrightarrow x \in A \wedge (x \notin B \vee x \notin C) \\ &\leftrightarrow (x \in A \wedge x \notin B) \vee (x \in A \wedge x \notin C) \\ &\leftrightarrow (x \in A - B) \vee (x \in A - C) \\ &\leftrightarrow x \in (A - B) \cup (A - C) \end{aligned}$$

So we have that $A - (B \cap C) = (A - B) \cup (A - C)$. ■

Many times it helps to complete a Venn diagram to feel convinced that a given result is in fact true. In the case of part 1 of Theorem 2.7, the associated Venn diagrams are shown in Figure 2.4.

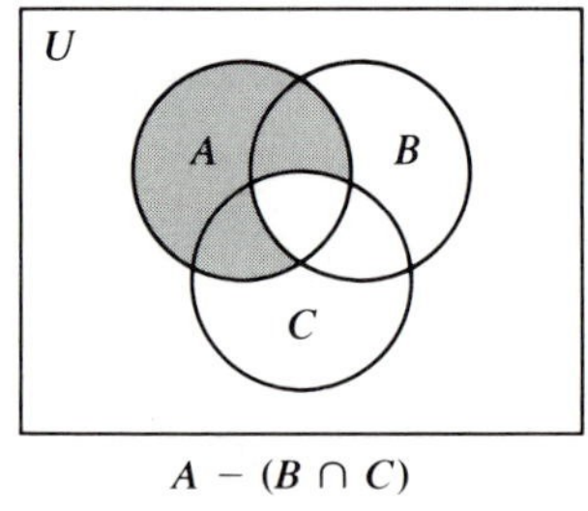

$A - (B \cap C)$

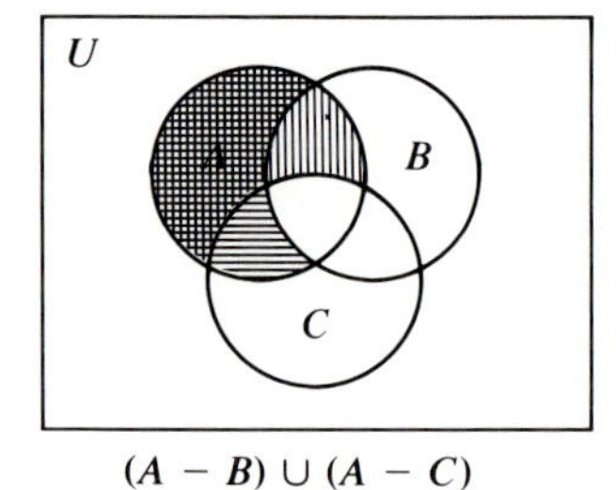

$(A - B) \cup (A - C)$

FIGURE 2.4

In the diagram for $(A - B) \cup (A - C)$, the set $A - B$ is shaded with horizontal line segments and $A - C$ is shaded with vertical segments. It should be emphasized that the use of this representation does not constitute a proof; it is merely an indication that the result holds.

In many discussions we shall be concerned with subsets of a specified set A. Of special interest is the collection of all subsets of A.

DEFINITION 2.6

The set consisting of all subsets of a set A is called the *power set* of A and is denoted by $\mathcal{P}(A)$.

Example 2.10 Describe $\mathcal{P}(A)$ if the set A has

(a) 0 elements; (b) 1 element; (c) 2 elements.

Solution (a) If the set A has no elements, then $A = \phi$ and the only subset of A is itself, so $\mathcal{P}(\phi) = \{\phi\}$. (Note that $\{\phi\}$ is not the empty set; it is a set with exactly one element, that element being ϕ.)
(b) If A has exactly one element, then the subsets of A are ϕ and A, so $\mathcal{P}(A) = \{\phi, A\}$.
(c) Suppose $A = \{x, y\}$. Then, besides ϕ and A, $\mathcal{P}(A)$ includes two nonempty proper subsets of A: $\{x\}$ and $\{y\}$. So $\mathcal{P}(A) = \{\phi, \{x\}, \{y\}, A\}$.

■

Two basic properties of $\mathcal{P}(A)$ worth remembering are that $\phi \in \mathcal{P}(A)$ and $A \in \mathcal{P}(A)$.

The order in which the elements of a set are listed is of no importance in the definition of the set; for example, it is clear that $\{a, b\} = \{b, a\}$. Thus it makes no sense to speak of a "first element" of the set $\{a, b\}$. In many cases, however, it turns out to be important to distinguish the order of appearance of two elements. This leads to the notion of an *ordered pair* of two elements a and b, denoted (a, b), where a is the *first element* and b is the *second element*. In formal set theory it is defined

$$(a, b) = \{\{a\}, \{a, b\}\}$$

so that (a, b) is in fact a set in which a and b play different roles. It then turns out that $(a, b) = (c, d)$ if and only if $a = c$ and $b = d$. We shall not dwell on this definition; instead we rely on our intuition and past experience with ordered pairs.

DEFINITION 2.7

Let A and B be subsets of some universal set U. The *Cartesian product* of A and B is the set $A \times B$ defined by

$$A \times B = \{(a, b) \mid a \in A \wedge b \in B\}$$

Example 2.11 If $A = \{0, 1, 2\}$ and $B = \{1, 3\}$, find $A \times B$ and $B \times A$.

Solution $A \times B = \{(0, 1), (0, 3), (1, 1), (1, 3), (2, 1), (2, 3)\}$

$B \times A = \{(1, 0), (1, 1), (1, 2), (3, 0), (3, 1), (3, 2)\}$

Note that $A \times B \neq B \times A$. ■

The idea of an ordered pair can be extended to more than two elements. In other words, given k elements $a_1, a_2, \ldots, a_k$, where $k \geq 3$, we can define the *ordered k-tuple* $(a_1, a_2, \ldots, a_k)$, in which a_1 is the first element (or first *coordinate*), a_2 the second, $\ldots$, a_k the kth. We can now generalize the definition of Cartesian product.

DEFINITION 2.8

If $A_1, A_2, \ldots, A_k$ are subsets of some universal set U, then the (k-fold) *product* $A_1 \times A_2 \times \cdots \times A_k$ is the set defined by

$$A_1 \times A_2 \times \cdots \times A_k = \{(a_1, a_2, \ldots, a_k) \mid a_i \in A_i \text{ for each } i, 1 \leq i \leq k\}$$

We remark that $(a_1, a_2, \ldots, a_k) = (b_1, b_2, \ldots, b_k)$ provided $a_i = b_i$ for each i, $1 \leq i \leq k$. Also notice that $A_1 \times A_2 \times \cdots \times A_k = \phi$ if any $A_i = \phi$, $1 \leq i \leq k$. In working with products of sets, we shall normally assume that the sets are nonempty.

Example 2.12 Let $A_1 = A_2 = A_3 = \{0, 1\}$. Find $A_1 \times A_2 \times A_3$.

Solution $A_1 \times A_2 \times A_3 =$

$$\{(0, 0, 0), (0, 0, 1), (0, 1, 0), (0, 1, 1), (1, 0, 0), (1, 0, 1), (1, 1, 0), (1, 1, 1)\}$$

■

The language Pascal is rather unique among high-level languages in that it provides a built-in SET data type. The elements of a Pascal set must all have the same type; this type is referred to as the *base type* and is restricted to being an enumeration or integer subrange. Pascal sets are also limited in that they may contain only up to a specified number of elements. The maximum number of elements a set is allowed to contain is implementation-dependent; it ordinarily corresponds to the number of bits in one or a few words of memory.

The Pascal declaration

```
TYPE BASETYPE = 1..30;
     SETTYPE = SET OF BASETYPE;
VAR A,B,C : SETTYPE;
```

declares sets A, B, and C whose elements are integers in the range 1 to 30. Similarly,

```
TYPE BASETYPE = CHAR;
     SETTYPE = SET OF BASETYPE;
VAR X, Y, Z : SETTYPE;
```

declares sets X, Y, and Z whose elements are characters. (Caution: such a base type is not possible in some implementations, for it would exceed the maximum number of elements allowed.)

Given A, B, and C declared as above, the assignment statements

```
A := [2,3,5,7,11];
C := [10..20];
```

assign to A the set {2, 3, 5, 7, 11} and to C the set {10, 11, 12, . . . , 20}. Notice that Pascal uses square brackets "[" and "]" rather than set braces to delimit the elements of a set. Also notice the two methods of specifying the elements to be included in a set; these two methods may be combined, as in the assignment statement

```
B := [1,3,7..11,19];
```

which assigns to B the set {1, 3, 7, 8, 9, 10, 11, 19}.

The operations that Pascal provides for sets include union (syntax "+"), intersection ("*"), and difference ("−"). For instance, with A and B as above, the statement

```
C := A * B;
```

would result in $C = A \cap B = \{3, 7, 11\}$, whereas

```
C := A + B;
```

would result in $C = A \cup B = \{1, 2, 3, 5, 7, 8, 9, 10, 11, 19\}$. The relational operators "=", "<>", "<=", and "<" are also provided. They are interpreted for sets A and B as follows:

"A = B" means "A equals B"
"A <> B" means "A does not equal B"
"A <= B" means "A is a subset of B"
"A < B" means "A is a proper subset of B"

Thus a program might contain the statement

```
IF ['A','E','I','O','U'] <= X THEN
  WRITELN('SET X CONTAINS ALL OF THE VOWELS.')
```

Perhaps the most useful set operation is the test for membership of a given element in a given set. Here the operator "IN" is used. Suppose in a particular program it is important to know whether a given input character NEXTCHR is equal to one of the operator symbols "+", "−", "*", or "/". Rather than using the cumbersome

```
IF (NEXTCHR = '+') OR (NEXTCHR = '-') OR (NEXTCHR = '*')
   OR (NEXTCHR = '/') THEN ...
```

it is much simpler and nicer to write

```
IF NEXTCHR IN ['+','-','*','/'] THEN ...
```

We have already mentioned a limitation of the Pascal SET data type, namely, that the base type must be an enumeration or integer subrange containing no more than some implementation-specified number of values. This excludes, for example, sets whose elements are of type REAL or sets whose elements are themselves sets. In addition, it should be pointed out that Pascal provides no direct facilities for input/output of sets. Thus, although it does simplify some aspects of programming, the uses of the Pascal SET data type are limited. Nevertheless, the set as a general data structure is important in the study of computer science, and we shall explore various methods for implementing sets further in the exercises and in later portions of this text.

Exercises 2.3

1. Construct a subset diagram showing the subset relationships that exist among the sets A, B, $A \cup B$, $A \cap B$, $A - B$, $B - A$, and ϕ for $A = \{1, 2, 3\}$ and $B = \{2, 4, 5\}$.

2. Prove Theorem 2.6, parts 2(b), 3(a), and 4(b).

3. Given that $U = \mathbb{Z}$, $A = \{. . . , -4, -2, 0, 2, 4, . . .\}$, $B = \{. . . , -6, -3, 0, 3, 6, . . .\}$, and $C = \{. . . , -8, -4, 0, 4, 8, . . .\}$, find the following sets.

a. $A \cap B$ **b.** $B - A$
c. $A - C$ **d.** $A \cap C'$
e. $C - A$ **f.** $B \cup C$
g. $(A \cup B) \cap C$ **h.** $(A \cup B) - C$

4. Prove Theorem 2.7, part 2.

5. Let A and B be subsets of a universal set U.

a. Prove that $A \cap B \subseteq A \cup B$.
b. Under what condition does $A \cap B = A \cup B$?

6. Let A and B be subsets of the universal set U. Prove that $A - B = A \cap B'$.

7. Each of the following statements concerns arbitrary sets A and B. Complete the statement so as to make it true.

a. $A \subseteq B \leftrightarrow A \cap B =$ _____
b. $A \subseteq B \leftrightarrow A \cup B =$ _____
c. $A \subseteq B \leftrightarrow A - B =$ _____
d. $A \subset B \leftrightarrow (A - B =$ _____ $\wedge\ B - A \neq$ _____$)$
e. $A \subset B \leftrightarrow (A \cap B =$ _____ $\wedge\ A \cap B \neq$ _____$)$
f. $A - B = B - A \leftrightarrow$ _____

8. Find each of the following power sets.
a. $\mathcal{P}(\{0, 1, 2\})$ **b.** $\mathcal{P}(\{0, \{1\}\})$ **c.** $\mathcal{P}(\mathcal{P}(\mathcal{P}(\phi)))$

9. Let $A = \{x, y\}$, $B = \{0, 1\}$, and $C = \{-1, 0, 1\}$. Find the following sets.
a. $A \times B$ **b.** $B \times C$
c. $A \times B \times C$ **d.** $(A \times B) \times C$
e. $B \times B \times B \times B$ **f.** $\mathcal{P}(B \times B)$

10. Let A, B, and C be nonempty subsets of a universal set U. Prove that $(A \cup B) \times C = (A \times C) \cup (B \times C)$.

11. Let A and B be nonempty subsets of a universal set U.
a. Under what condition does $A \times B = B \times A$?
b. Under what condition is $(A \times B) \cap (B \times A)$ empty?

12. A telephone number like 673-3459 can be thought of as a 7-tuple of digits: (6,7,3,3,4,5,9). Thus, with $D = \{0, 1, 2, \ldots, 9\}$, we can think of a telephone number as an element of the product $D \times D \times D \times D \times D \times D \times D$. Describe how each of the following can be thought of as an element of some appropriate product of sets:
a. a social security number;
b. a typical automobile license plate identification;
c. a valid Pascal identifier having exactly three characters; for example, SUM.

13. Each part of this question refers to the following declaration:

```
CONST MIN = 1; MAX = 25;
TYPE  BASETYPE = MIN..MAX;
      SETTYPE = SET OF BASETYPE;
VAR   A,B,C : SETTYPE;
```

Write Pascal assignment statements that assign to C the following:
a. the set of numbers (in the base type) that are prime;
b. the set of integers between 3 and 17, inclusive;
c. the set of numbers (in the base type) that are not in {7, 14, 21};
d. the set of those numbers that are in exactly one of the sets A or B (assume that A and B have been defined);
e. the set of numbers that are in C but not in either A or B.

14. Assume that MIN and MAX have been declared as (integer or character) constants, and we have the declaration

```
TYPE BASETYPE = MIN..MAX;
     SETTYPE = SET OF BASETYPE;
```

a. Write a procedure SETIN that inputs the elements of a set A. The procedure should first prompt the user to enter the number of elements in the set A, and then it should prompt the user to enter the elements. (Hint: if X is a variable of type BASETYPE, then

A := A + [X] assigns to A the union of A and the set containing the value of X.)

b. Write a procedure SETOUT that outputs the elements of a set A.

15. A type STRING, for variable-length character strings, may be declared in Pascal as follows:

```
CONST MAXLEN = 30;
TYPE  STRING = RECORD
         VALUE : PACKED ARRAY[1..MAXLEN] OF CHAR;
         LENGTH : 0..MAXLEN
      END;
```

Thus if WORD is a variable of type STRING, then WORD.LENGTH is the number of characters in WORD, and WORD.VALUE[J] is the Jth character in WORD, for $1 \leq J \leq$ WORD.LENGTH. Assume that SETTYPE is declared as in Exercise 14, with MIN = 'A' and MAX = 'Z' (assuming the ASCII character set).

a. Write a procedure FIND that takes a character string WORD and returns VOWS and CONS, the sets of vowels and consonants occurring in WORD, respectively. For example, if WORD = 'KALAMAZOO', then VOWS = {'A','O'} and CONS = {'K','L','M','Z'}.

b. Assuming that V1, V2, C1, C2, L1, L2, and L3 are variables of type SETTYPE, and WORD1 and WORD2 are character strings, complete the following Pascal segment so it does as the comments indicate.

```
FIND(WORD1,V1,C1);  (*V1 and C1 are the sets of vowels and
                      consonants in WORD1, respectively*)
FIND(WORD2,V2,C2);  (*V2 and C2 are the sets of vowels and
                      consonants in WORD2, respectively*)
L1 := ____________________;  (*assign to L1 the set of letters in WORD1*)
L2 := ____________________;  (*assign to L2 the set of consonants that
                               are in both WORD1 and WORD2*)
L3 := ____________________;  (*assign to L3 the set of vowels that are
                               in WORD1 but not in WORD2*)
```

16. If X is a variable of type BASETYPE and A is a set of type SETTYPE, what is the negation of the condition "X IN A"?

17. Suppose we wish to work with sets whose base type is like that of Exercise 14. Such sets may be implemented as Boolean arrays; in Pascal we could declare sets A, B, and C as follows:

```
CONST MIN = -99; MAX = 99;
TYPE  BASETYPE = MIN..MAX;
      SETTYPE = ARRAY[BASETYPE] OF BOOLEAN;
VAR   A,B,C : SETTYPE;
```

The idea here is that A[J] is TRUE if and only if J is an element of A, for MIN ≤ J ≤ MAX. Assuming that such declarations have been made, and that the sets A and B have been defined, write FOR statements of the form

```
FOR J := MIN TO MAX DO C[J] :=__________;
```

that assign to C the set

a. $A \cap B$ **b.** $A \cup B$ **c.** A'

2.4 CARDINALITY AND COUNTING

Recall that a set S is finite provided S consists of m elements, for some nonnegative integer m.

DEFINITION 2.9

Let S be a finite set. The *cardinality* of S, denoted $n(S)$, is defined to be the number of elements in S. We also say that S has *cardinal number* $n(S)$. An alternate notation for $n(S)$ is $|S|$.

Example 2.13 Let $A = \{a, b, c\}$. Then $n(A) = 3$. Determine the cardinality of $\mathcal{P}(A)$, the power set of A.

Solution $\mathcal{P}(A) = \{\phi, \{a\}, \{b\}, \{c\}, \{a, b\}, \{a, c\}, \{b, c\}, A\}$ so $n(\mathcal{P}(A)) = 8$. ■

Note that the empty set, ϕ, is the only set having cardinal number 0.

DEFINITION 2.10

Two sets A and B are said to be *disjoint* if $A \cap B = \phi$. In general, if $A_1, A_2, \ldots, A_n$ are sets and $A_i \cap A_j = \phi$ for all i and j, $1 \le i < j \le n$, then we say these sets are *pairwise disjoint*. We also call $\{A_1, A_2, \ldots, A_n\}$ a *pairwise disjoint collection* of sets.

Example 2.14 Let $A = \{-1, 2, 3, 5, 8\}$, $B = \{1, 3, 5, 7\}$, $C = \{-1, 2, 4, 9\}$, $D = \{-2, 0, 6\}$, and $E = \{-3, 8\}$. Then A and D are disjoint and $\{B, C, D, E\}$ is a pairwise disjoint collection. ■

Note that the sets A and B in Example 2.14 are not disjoint; however, it is the case that $A - B = \{-1, 2, 8\}$ and $A \cap B = \{3, 5\}$ are disjoint and that $A = (A - B) \cup (A \cap B)$. This result holds in general and is very useful; the reader will be asked to prove it in Exercise 4.

If A and B are disjoint, then

$$n(A \cup B) = n(A) + n(B)$$

In general, if $A_1, A_2, \ldots, A_k$ are pairwise disjoint, then

$$n(A_1 \cup A_2 \cup \cdots \cup A_k) = n(A_1) + n(A_2) + \cdots + n(A_k)$$

The above relation is generally referred to as the *addition principle*. A general formula for $n(A_1 \cup A_2 \cup \cdots \cup A_k)$, where $A_1, A_2, \ldots, A_k$ are any k sets, requires a detailed combinatorial argument; however, some small cases can be evaluated. To facilitate matters, recall that $A - B = A \cap B'$ (Exercise 6 of Section 2.3). We shall also need the relations

$$A \cap \phi = \phi$$

and

$$A \cap U = A$$

which hold for any subset A of the universal set U.

THEOREM 2.8 Given any subsets A and B of the universal set U, the sets $A - B$, $B - A$, and $A \cap B$ are pairwise disjoint and

$$A \cup B = (A - B) \cup (B - A) \cup (A \cap B)$$

Proof We first prove that $A - B$ and $B - A$ are disjoint; the other two cases for disjointness follow along similar lines. Observe that

$$\begin{aligned}(A - B) \cap (B - A) &= (A \cap B') \cap (B \cap A') \\ &= (A \cap (B' \cap B)) \cap A' \\ &= (A \cap \phi) \cap A' \\ &= \phi \cap A' \\ &= \phi\end{aligned}$$

which shows that $A - B$ and $B - A$ are disjoint. The proof that $A \cup B = (A - B) \cup (B - A) \cup (A \cap B)$ is as follows:

$$\begin{aligned}(A - B) \cup (B - A) \cup (A \cap B) &= (A \cap B') \cup (B \cap A') \cup (A \cap B) \\ &= (A \cap B') \cup ((B \cap A') \cup (B \cap A)) \\ &= (A \cap B') \cup (B \cap (A' \cup A)) \\ &= (A \cap B') \cup (B \cap U) \\ &= (A \cap B') \cup B \\ &= (A \cup B) \cap (B' \cup B) \\ &= (A \cup B) \cap U \\ &= A \cup B\end{aligned}$$

■

THEOREM 2.9 Given finite subsets A and B of a universal set U,

$$n(A \cup B) = n(A) + n(B) - n(A \cap B)$$

Proof From the formula $n(A) = n(A - B) + n(A \cap B)$ we obtain

$$n(A - B) = n(A) - n(A \cap B)$$

Similarly,

$$n(B - A) = n(B) - n(A \cap B)$$

Using these two facts, together with the result of Theorem 2.8, we have that

$$\begin{aligned} n(A \cup B) &= n(A - B) + n(B - A) + n(A \cap B) \\ &= (n(A) - n(A \cap B)) + (n(B) - n(A \cap B)) + n(A \cap B) \\ &= n(A) + n(B) - n(A \cap B) \end{aligned}$$

■

Example 2.15 A certain manufactured item may contain two types of defects, defect 1 or defect 2. Of 100 of these items that were inspected, fifteen had defect 1, ten had defect 2, and seven had both types of defects. How many of the items were defective in some way?

Solution Let A be the set of items having defect 1 and let B be the set of items having defect 2. We are given that $n(A) = 15$, $n(B) = 10$, and $n(A \cap B) = 7$. We are asked to determine $n(A \cup B)$, the number of items having either type of defect. By Theorem 2.9,

$$\begin{aligned} n(A \cup B) &= n(A) + n(B) - n(A \cap B) \\ &= 15 + 10 - 7 \\ &= 18 \end{aligned}$$

Thus 18 of the items were defective in some way. ■

THEOREM 2.10 If A, B, and C are finite subsets of a universal set U, then

$$\begin{aligned} n(A \cup B \cup C) = n(A) + n(B) + n(C) - n(A \cap B) - n(A \cap C) \\ - n(B \cap C) + n(A \cap B \cap C) \end{aligned}$$

Proof Making use of Theorem 2.9, we obtain

$$\begin{aligned} n(A \cup B \cup C) &= n(A \cup (B \cup C)) \\ &= n(A) + n(B \cup C) - n(A \cap (B \cup C)) \\ &= n(A) + (n(B) + n(C) - n(B \cap C)) \\ &\quad - n((A \cap B) \cup (A \cap C)) \\ &= n(A) + n(B) + n(C) - n(B \cap C) \\ &\quad - (n(A \cap B) + n(A \cap C) - n((A \cap B) \cap (A \cap C))) \\ &= n(A) + n(B) + n(C) - n(B \cap C) \\ &\quad - n(A \cap B) - n(A \cap C) + n(A \cap B \cap C) \end{aligned}$$

■

Example 2.16 An insurance company classifies its policyholders according to age, sex, and marital status. Of 500 policyholders, it was found that

350 were married,
240 were under 25,
230 were married men,
110 were married and under 25,
100 were men under 25,
40 were married men under 25, and
10 were single women 25 or older.

How many policyholders were men?

Solution Let A, B, and C denote the sets of male policyholders, married policyholders, and policyholders under 25, respectively. First of all, we are given that $n(A' \cap B' \cap C') = 10$, and therefore

$$\begin{aligned} n(A \cup B \cup C) = n[(A' \cap B' \cap C')'] &= n(U) - n(A' \cap B' \cap C') \\ &= 500 - 10 \\ &= 490 \end{aligned}$$

(Here U is the set of 500 policyholders who were surveyed.) Now apply Theorem 2.10:

$$\begin{aligned} n(A \cup B \cup C) = n(A) &+ n(B) + n(C) - n(A \cap B) - n(A \cap C) \\ &- n(B \cap C) + n(A \cap B \cap C) \end{aligned}$$

so

$$490 = n(A) + 350 + 240 - 230 - 100 - 110 + 40$$

from which it follows that $n(A) = 300$. Thus, of the 500 policyholders surveyed, 300 were men. The reader should carefully check the Venn diagram shown in Figure 2.5, which shows the cardinalities of the various disjoint subsets of U.

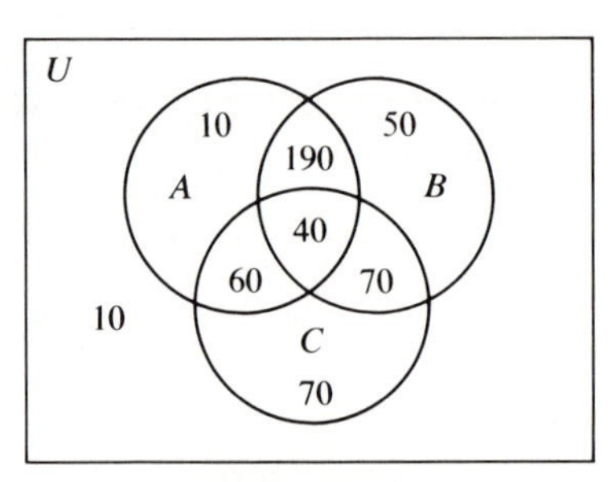

FIGURE 2.5 ■

In addition to the addition principle, there is another basic counting principle, commonly referred to as the "multiplication principle."

THEOREM 2.11 (Multiplication principle) If A and B are nonempty subsets of a universal set U such that $n(A) = r$ and $n(B) = s$, then $n(A \times B) = rs$.

Proof Assume $A = \{a_1, a_2, \ldots, a_r\}$ and $B = \{b_1, b_2, \ldots, b_s\}$. Then

$$\begin{aligned} A \times B &= \{(a_i,b_j) \mid 1 \le i \le r, 1 \le j \le s\} \\ &= \{(a_1,b_1), \ldots, (a_1,b_s), \ldots, (a_r,b_1), \ldots, (a_r,b_s)\} \\ &= \{(a_1,b_1), \ldots, (a_1,b_s)\} \cup \{(a_2,b_1), \ldots, (a_2,b_s)\} \cup \\ &\quad \cdots \cup \{(a_r,b_1), \ldots, (a_r,b_s)\} \\ &= (\{a_1\} \times B) \cup (\{a_2\} \times B) \cup \cdots \cup (\{a_r\} \times B) \end{aligned}$$

It is evident that each of the r sets $\{a_j\} \times B$, $1 \le j \le r$, has s elements. Furthermore, the sets $\{a_j\} \times B$, $1 \le j \le r$, form a pairwise disjoint collection. Hence, by the addition principle,

$$\begin{aligned} n(A \times B) &= n(\{a_1\} \times B) + n(\{a_2\} \times B) + \cdots + n(\{a_r\} \times B) \\ &= s + s + \cdots + s \\ &= rs \end{aligned}$$

■

An alternate formulation of the multiplication principle is the following.

Multiplication principle. If one object can be selected in r ways and, once this first choice has been made, a second object can be selected in s ways, then the two selections can be made together in rs ways.

For example, suppose there are 13 highways that connect Kalamazoo, Michigan, to Ottumwa, Iowa, and 17 highways connecting Ottumwa and San Francisco. Using only these highways, in how many ways can one travel (by car) from Kalamazoo to San Francisco? Using the multiplication principle, we see that the answer is $(13)(17) = 221$ ways.

A very convenient way to illustrate the multiplication principle is with the aid of a *tree diagram*. Consider the case where $A = \{a_1, a_2, a_3\}$ and $B = \{b_1, b_2, b_3, b_4\}$. The associated tree diagram for the set $A \times B$ is shown in Figure 2.6. Notice that there are three initial branches corresponding to the elements a_1, a_2, and a_3 of A. Attached to each of these initial branches are four branches corresponding to the elements b_1, b_2, b_3, and b_4 of B. To obtain an element of $A \times B$, we start at the "root" of the tree and choose an initial branch to follow. The choice of an initial branch determines the first coordinate of the ordered pair. Then, to determine the second element, we choose a secondary branch to follow. For example, the ordered pair (a_2,b_3) is obtained by choosing the initial branch labelled a_2 followed by the secondary branch labelled b_3. Note that each path from the root, along an initial branch, and then along a secondary branch to a

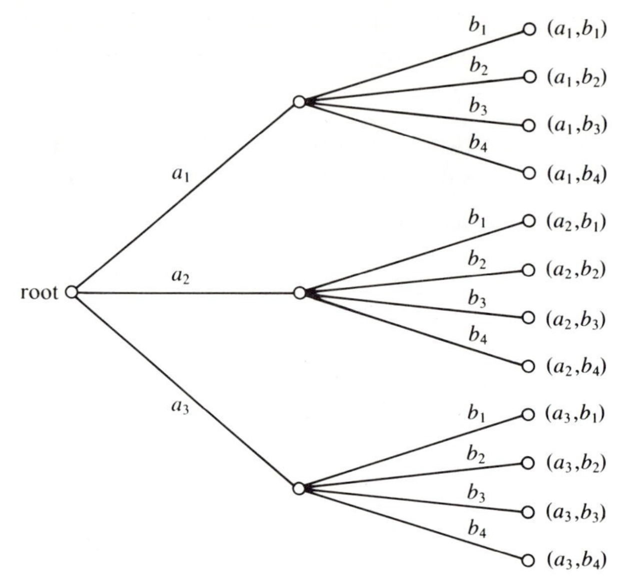

FIGURE 2.6 A tree diagram for $A \times B$

"leaf" (or "*terminus*") of the tree determines a different ordered pair; thus the leaves are labelled with the elements of $A \times B$, as shown.

The multiplication principle can be extended to an arbitrary finite number of sets. The proof, which depends on mathematical induction, will be considered at a later point.

COROLLARY 2.11.1 If $A_1, A_2, \ldots, A_m$ are finite subsets of a universal set U, then $n(A_1 \times A_2 \times \cdots \times A_m) = n(A_1)n(A_2) \cdots n(A_m)$. ■

COROLLARY 2.11.2 If A is a finite set with $n(A) = r$, and $B = A \times A \times \cdots \times A$ is the k-fold product of A with itself, then

$$n(B) = r^k$$ ■

The term "corollary" as seen above is used in the mathematical literature to indicate that the associated assertion is a direct consequence of the preceding theorem.

Example 2.17 An experiment in probability consists of tossing a fair die three times in succession. Letting $A = \{1, 2, 3, 4, 5, 6\}$, note that an outcome for this experiment can be considered as an element of the set $A \times A \times A$. For instance, the ordered triple (3,1,4) indicates that 3 was obtained on the first toss of the die, 1 on the second toss, and 4 on the third toss. Hence there are $n(A)^3 = 6^3 = 216$ possible outcomes for the experiment. How many outcomes have no number repeated?

Solution There are six possible outcomes for the first toss; call it x. Then the second toss must result in an element of $A - \{x\}$, so there are five possible results for the second toss. If the second toss is y, then the third toss must result in an element of $A - \{x, y\}$, so there are four possible results for the third toss. Therefore, by the multiplication principle, there are $(6)(5)(4) = 120$ outcomes for the experiment that have no number repeated. ■

Example 2.18 A variable name in FORTRAN consists of from one to six alphanumeric characters, where the first character must be a letter. How many such names are there? How many names have no character repeated?

Solution Let $L = \{A, B, \ldots, Z\}$ be the set of all letters and let $D = \{0, 1, \ldots, 9\}$ be the set of digits. Then $A = L \cup D$ is the set of alphanumeric characters. There are $n(L) = 26$ single-character FORTRAN variable names. A variable name having m characters, $2 \leq m \leq 6$, can be thought of as an element of the set $L \times A \times \cdots \times A$, where A is used as a factor $m - 1$ times. Hence there are $(26)(36^{m-1})$ variable names having m characters. Finally, by the addition principle, there are $26 + (26)(36) + (26)(36^2) + \cdots + (26)(36^5) = (26)(1 + 36 + \cdots + 36^5) = (26)(62,193,781) = 1,617,038,306$ possible FORTRAN variable names. For the second question, we leave it to the reader to verify that there are $(26)(35) \cdots (36 - m + 1)$ variable names having m characters with no repeated characters, $2 \leq m \leq 6$, and thus a total of 1,045,491,642 variable names that have no repeated characters. ■

Exercises 2.4

1. Of 100 manufactured items that were inspected, ten were found to have defect A, eight had defect B, and 85 were nondefective. How many of the 100 items had both defects A and B?
2. There are 30 students in a discrete mathematics class. Ten students failed the midterm exam, five students failed the final exam, and three students failed both the midterm and the final.
   a. How many students passed both the midterm and the final?
   b. How many students passed either the midterm or the final?
3. A social security number is a sequence of nine digits, for example, 080-55-1618.
   a. How many possible social security numbers are there?
   b. How many possible social security numbers do not contain the digit zero?
   c. How many possible social security numbers neither begin nor end with the digit zero?
4. Prove that if A and B are subsets of a universal set U, then the sets $A - B$ and $A \cap B$ are disjoint and $A = (A - B) \cup (A \cap B)$.
5. There are 225 computer science majors at a certain university. This

semester, 100 of these students are taking CS 260, 75 are taking CS 360, 50 are taking CS 450, seven are taking both CS 260 and CS 360, four are taking both CS 260 and CS 450, 31 are taking both CS 360 and CS 450, and one student is taking all three courses.

a. How many computer science majors are not taking any of these three courses?

b. Draw a Venn diagram like that of Figure 2.5, where U is the set of 225 computer science majors.

6. In BASIC, a legal name for a numeric variable can be either a letter or a letter followed by a digit; for example, A, Z, C2, and P7 are legal names. How many legal names are there?

7. Consider the problem of choosing, at random, a sequence of four cards from a standard deck of 52 playing cards. The cards are chosen one at a time with replacement, meaning that after a card has been chosen it is replaced and the deck is reshuffled. We are interested in the sequence of cards that is obtained (for example, 2 of clubs, ace of hearts, king of spades, 2 of clubs). Determine

a. the number of possible sequences;

b. the number of sequences that contain no spades;

c. the number of sequences in which all four cards are spades;

d. the number of sequences in which all four cards have the same suit;

e. the number of sequences in which the first card is a spade and the third card is not an ace;

f. the number of sequences in which the first card is a spade or the third card is not an ace.

8. Draw a tree diagram with branches labelled 0 and 1, such that the leaves are the eight binary integers of length three: $000 = 0$, $001 = 1$, $010 = 2, \ldots, 111 = 7$.

9. A box contains three balls colored blue, red, and green. An experiment consists of choosing a sequence of three balls, one ball at a time, with replacement. An outcome for the experiment can be written as an ordered triple; for example, (B,R,G) would denote that the first ball chosen was blue, the second ball was red, and the third ball was green. (The balls are chosen at random.)

a. How many possible outcomes for the experiment are there?

b. Draw a tree diagram that helps answer the question, "How many outcomes for the experiment are there in which no two balls chosen consecutively have the same color?"

c. Draw a tree diagram that helps answer the question, "How many outcomes for the experiment are there in which at least two of the balls chosen are blue?"

10. In 1983 the Baltimore Orioles and the Chicago White Sox played in the playoff series to determine the American League (baseball) champion. The winner in this series is the first team to win three games (best 3-out-of-5). The outcome of such a series can be represented as an ordered n-tuple, $3 \le n \le 5$; for example, (B,B,B) indicates that Baltimore won the series in three games, and (C,B,C,C) indicates that Chicago won the series in four games, with Baltimore winning the second game.

a. Draw a tree diagram showing all possible outcomes for such a series. How many are there?

Use the tree diagram to answer the following questions.

b. How many ways are there for Baltimore to win the series in four games?

c. How many ways are there for Chicago to win the series in five games?

2.5 PERMUTATIONS AND COMBINATIONS

How many different ways are there to toss an ordinary die three times in succession and have no number repeated? How many poker hands have exactly two kings?

These are questions whose solutions require careful and intricate thinking. They are classified as "counting problems," since the solutions involve the actual counting of the number of objects satisfying prescribed properties. In this section we shall develop techniques that will allow us to address such problems.

In the last section we presented both the addition principle and the multiplication principle, which enabled us to solve some general types of counting problems. In order to proceed further, several more specialized techniques will be developed.

The ideas presented below use a notation that may be unfamiliar to the reader. This is the "factorial" notation for products. If n is a positive integer, we define *n factorial* by $n! = (1)(2) \cdot \cdot \cdot (n)$. For example, $4! = (1)(2)(3)(4) = 24$. It is also convenient to define zero factorial by $0! = 1$.

DEFINITION 2.11

Let A be a finite set with $n(A) = n \ge 1$. An *r-permutation* of A, $1 \le r \le n$, is an r-tuple

$$(a_1, a_2, \ldots, a_r)$$

where each a_i is an element of A and the a_i's are distinct, that is, $a_i \ne a_j$ for all i and j, $1 \le i < j \le n$. A *permutation* of A is simply an n-permutation of A. We denote the number of r-permutations of an n-element set by $P(n, r)$.

We can also think of an r-permutation of A as an ordered selection of r distinct objects from A. For instance, suppose $A = \{1, 2, 3, 4\}$ and $r = 3$, so we desire the 3-permutations of A. If we select, in order, $a_1 = 2$, $a_2 = 1$, and $a_3 = 4$, then we obtain the 3-permutation (2,1,4). How many 3-permutations of A are there? Well, there are four choices for the first element, a_1, then three choices for a_2, and then two choices for a_3, so by the multiplication principle there are (4)(3)(2) = 24 3-permutations of A. Here is a list:

(1,2,3) (1,3,2) (1,2,4) (1,4,2) (1,3,4) (1,4,3)
(2,1,3) (2,3,1) (2,1,4) (2,4,1) (2,3,4) (2,4,3)
(3,1,2) (3,2,1) (3,1,4) (3,4,1) (3,2,4) (3,4,2)
(4,1,2) (4,2,1) (4,1,3) (4,3,1) (4,2,3) (4,3,2)

THEOREM 2.12 For each $n \in \mathbb{N}$ and $1 \le r \le n$,

$$P(n,r) = n(n-1)(n-2)\cdots(n-r+1) = \frac{n!}{(n-r)!}$$

Proof Think of making an ordered selection of r objects from a set of n distinct objects. The first object can be chosen in n ways; this done, there remain $n - 1$ objects from which the second object can be chosen. So the second object can be chosen in $n - 1$ ways; this done, there are $n - 2$ objects remaining from which the third object can be chosen, and so on, until the rth object is chosen, which can be done in $n - r + 1$ ways. Thus, by the multiplication principle, the r objects can be selected in $n(n-1)\cdots(n-r+1)$ ways. Thus

$$P(n,r) = n(n-1)\cdots(n-r+1)$$

Now note that

$$\frac{n!}{(n-r)!} = \frac{n(n-1)\cdots(n-r+1)(n-r)\cdots(1)}{(n-r)(n-r-1)\cdots(1)} = n(n-1)\cdots(n-r+1)$$ ■

It follows from Theorem 2.12 (with $r = n$) that there are $n!$ permutations of an n-element set. It is customary to denote $P(n,n)$ simply by $P(n)$; thus

$$P(n) = n!$$

Example 2.19 The number of ways to toss a die three times in succession with no number repeated is $P(6,3)$ (see Example 2.17). The number of legal FORTRAN variable names with m distinct characters, $2 \le m \le 6$, is $26 \cdot P(35, m-1)$ (see Example 2.18). Similarly, the number of seven-digit telephone numbers having no repeated digits is $P(10,7) =$

$(10)(9)(8)(7)(6)(5)(4) = 604{,}800$. (Here we are allowing the first digit to be zero; if the first digit must be nonzero, then there are $9 \cdot P(9,6) = 544{,}320$ such numbers.) ■

In choosing an r-permutation of a set A, the order in which the elements are chosen is important. For example, the telephone numbers 673-1245 and 673-1542 are different. However, there are times when, in making a series of choices, the order in which the choices are made is unimportant. For instance, consider the process of being dealt a five-card poker hand. What matters is the *set* of five cards in the final hand; the order in which the five cards are dealt does not matter.

Suppose we wish to determine the number of possible poker hands, or the number of poker hands that contain exactly three aces. Since a poker hand can be thought of as a 5-element subset of the set of 52 cards, we need to determine the number of 5-element subsets of a 52-element set. Similarly, counting the number of poker hands containing exactly three aces involves finding the number of 3-element subsets of a 4-element set (the set of aces), as well as the number of 2-element subsets of a 48-element set (the set of cards that are not aces).

In general, we denote by $C(n, r)$ the number of r-element subsets of an n-element set, where $0 \leq r \leq n$. The number $C(n, r)$ is also called the *number of combinations of n things taken r at a time.*

Another common notation for $C(n, r)$ is $\binom{n}{r}$.

To help us determine a formula for $C(n, r)$, consider an example. Let us list the 3-element subsets of $A = \{1, 2, 3, 4\}$, and next to each subset list the 3-permutations of A that use those elements:

$\{1, 2, 3\}$ (1,2,3), (1,3,2), (2,1,3), (2,3,1), (3,1,2), (3,2,1)
$\{1, 2, 4\}$ (1,2,4), (1,4,2), (2,1,4), (2,4,1), (4,1,2), (4,2,1)
$\{1, 3, 4\}$ (1,3,4), (1,4,3), (3,1,4), (3,4,1), (4,1,3), (4,3,1)
$\{2, 3, 4\}$ (2,3,4), (2,4,3), (3,2,4), (3,4,2), (4,2,3), (4,3,2)

Note that each of the four 3-element subsets of A yields six 3-permutations of A. Thus we have that

$$P(4,3) = 24 = (6)(4) = P(3)C(4,3)$$

This example suggests a general relationship between $P(n,r)$ and $C(n,r)$.

THEOREM 2.13 For $n \in \mathbb{N}$ and $0 \leq r \leq n$, we have

$$C(n,r) = \frac{n!}{r!(n-r)!}$$

Proof Consider any r-element subset of an n-element set A. Such a subset gives rise to $r!$ r-permutations of A, since the r elements in the subset can be ordered in $r!$ ways. Moreover, every r-permutation of A is so determined. Thus we obtain the relation

$$P(n,r) = r!C(n,r)$$

which holds for $1 \leq r \leq n$. Applying Theorem 2.12 then yields

$$C(n,r) = \frac{n!}{r!(n-r)!}$$

for $1 \leq r \leq n$. For $r = 0$, it follows easily that $C(n,0) = \dfrac{n!}{0!n!} = 1$. ■

Example 2.20 The number of distinct poker hands is

$$C(52,5) = \frac{52!}{(5!)(47!)} = 2{,}598{,}960$$

How many of these hands contain exactly three aces? To answer such a question we employ the multiplication principle. There are $C(4,3)$ ways to select a subset of three of the four aces. Then, there are $C(48,2)$ ways to select the other two cards from the 48 cards that are not aces. Thus, by the multiplication principle, $C(4,3) \cdot C(48,2)$ is the number of ways to obtain a poker hand containing exactly three aces. ■

Example 2.21 A lot of 100 items from a manufacturing process is known to contain 10 defective items (and 90 items that are nondefective). A sample of seven items is to be selected at random and checked. How many samples contain (a) exactly three defective items? (b) at least one defective item?

Solution (a) If the sample contains exactly three defective items, then it contains exactly four nondefective items; so by the multiplication principle, $C(10,3) \cdot C(90,4)$ samples contain exactly three defective items.
(b) The best way to answer this question is to consider instead the complement of the set of samples containing at least one defective item. This is equivalent to asking the question, how many samples contain no defective items? To answer this, first notice that there are altogether $C(100,7)$ possible samples, since a sample is just a 7-element subset of the set of 100 items. If a sample contains no defectives, then it constitutes a subset of the 90 nondefective items. Thus there are $C(90,7)$ samples containing no defective items. It follows that $C(100,7) - C(90,7)$ samples contain at least one defective item. ■

We have previously defined the power set of a set A to be the set of all subsets of A. Suppose $n(A) = m$. Since $C(m,r)$ is the number of r-element subsets of A, we have that

$$n(\mathcal{P}(A)) = C(m,0) + C(m,1) + \cdots + C(m,m) = \sum_{r=0}^{m} C(m,r)$$

However, in Example 2.10 we found that if $n(A) = 0, 1$, or 2, then $n(\mathcal{P}(A)) = 1, 2$, or 4, respectively. And in Example 2.13 we found that if $n(A) = 3$, then $n(\mathcal{P}(A)) = 8$. It seems reasonable to conjecture that if $n(A) = m$, then $n(\mathcal{P}(A)) = 2^m$. We next prove this result.

THEOREM 2.14 Let A be a finite set having m elements. Then the power set of A has 2^m elements; symbolically,

$$n(A) = m \to n(\mathcal{P}(A)) = 2^m$$

Proof Let $A = \{a_1, a_2, \ldots, a_m\}$. Consider the problem of placing each element of A into one of two boxes. For each such placement we may define a subset B of A by letting B consist of those elements placed in the first box. Note that there are exactly as many subsets B as there are different ways to place the elements of A into the two boxes. Now, in order to place the elements of A into the two boxes, we must decide for each element a_i, $1 \leq i \leq m$, whether to put a_i into the first box or the second box. Thus there are, by the multiplication principle, 2^m ways to place the elements of A into the boxes, since there are m decisions to be made, and for each decision there are two choices. This shows that A has 2^m subsets. ■

COROLLARY 2.14 For each positive integer n,

$$C(n,0) + C(n,1) + \cdots + C(n,n) = 2^n$$

■

Example 2.22 The purpose of this example is to point out a very common error that is made in counting problems. Consider a box containing seven distinct colored balls: three blue, two red, and two green. A subset of three of the balls is to be selected at random. How many such subsets contain at least two blue balls?

Incorrect solution We want a subset that contains at least two blue balls. Start by choosing two blue balls. Since there are three blue balls to choose from, this may be done in $C(3,2)$ ways. Now any one of the remaining five balls may be chosen as the third ball in the subset. Thus, by the multiplication principle, the number of subsets containing at least two blue balls is

$$C(3,2) \cdot 5 = (3)(5) = 15$$

Correct solution We first solve the problem by "brute force"; the number of balls was purposely kept small in this problem to enable listing all the 3-element subsets with at least two blue balls. Let the set of balls be

$$\{b_1, b_2, b_3, r_1, r_2, g_1, g_2\}$$

where b_1, b_2, and b_3 are blue. Also let x denote an arbitrary one of the four nonblue balls. Then the 3-element subsets with at least two blue balls are $\{b_1, b_2, b_3\}$, the four subsets of the form $\{b_1, b_2, x\}$, the four subsets of the form $\{b_1, b_3, x\}$, and the four subsets of the form $\{b_2, b_3, x\}$. Hence there are 13 such subsets, not 15. We can also obtain the correct answer by properly applying our counting techniques. If a 3-element subset is to contain at least two blue balls, then either it contains exactly two blue balls or it contains exactly three blue balls. There are $C(3,2) \cdot 4 = 12$ 3-element subsets with exactly two blue balls. Also, there is one 3-element subset with exactly three blue balls. Therefore, by the addition principle, $12 + 1 = 13$ 3-element subsets contain at least two blue balls. (The reader is encouraged to find the flaw in the incorrect solution.) ■

Exercises 2.5

1. Seven runners are entered in the mile race at a track meet. Different trophies will be awarded to the first, second, and third place finishers. In how many ways can the trophies be awarded?

2. How many possible social security numbers have no repeated digits?

3. Three (distinct) mathematics and two (distinct) computer science final examinations must be scheduled during a five-day period. How many ways are there to schedule the exams if

a. no two of the exams may be scheduled on the same day?

b. no two mathematics exams may be scheduled the same day, nor can the two computer science exams?

c. only the two computer science exams may be given on the same day?

4. A box contains 12 distinct colored balls, which are numbered 1 through 12. Balls 1 through 3 are red, balls 4 through 7 are blue, and balls 8 through 12 are green. An experiment consists of choosing three balls at random from the box, one at a time, *without replacement;* the outcome is the sequence of balls chosen; for example, (7,4,9).

a. How many outcomes are there?

In how many ways can it happen that

b. the first ball is red, the second ball is blue, and the third is green?

c. exactly two balls are green?

d. one ball of each color is chosen?

e. all three balls chosen are the same color?

f. at least one of the balls chosen has an odd number?

5. A hand in a game of bridge consists of 13 of the 52 cards in a regular deck.

a. How many different bridge hands are there?

How many bridge hands contain

b. exactly two aces?

c. exactly three aces, two kings, one queen, and no jacks?

d. at least one heart?

e. four spades, three hearts, three diamonds, and three clubs?

f. four cards in one suit and three cards in each of the other three suits?

g. at least one heart and at least one spade?

h. exactly six hearts or exactly seven diamonds?

6. How many eight-letter "words," constructed using the 26 letters of the alphabet,

a. contain exactly three a's?

b. contain either three or four vowels (a,e,i,o,u)?

c. have no letter repeated?

d. contain an even number of e's?

7. In the game of Mastermind, one player, the "codemaker," selects as the "code" a sequence of four colors, each chosen from a set of six possible colors. For example, (green,red,blue,red) and (white,black, yellow,red) are possible codes.

a. How many such codes are there?

b. For each x, $1 \le x \le 4$, how many codes contain exactly x colors?

8. In how many ways can 12 different homework problems be distributed among 20 students if no student gets more than one problem and no problem is assigned to more than one student?

9. A man has n friends who enjoy playing bridge, and he is able to invite a different subset of three of them over to his home every Wednesday night for two years. How large (at least) must n be?

10. Show that $C(n,k) = C(n,n-k)$ for $0 \le k \le n$.

11. The Department of Mathematics and Computer Science has four full professors, eight associate professors, and three assistant professors. A four-person search committee is to be chosen. How many ways are there to select the committee if

a. at least one full professor must be chosen?

b. the committee must contain at least one professor of each rank?

c. the committee must contain exactly one assistant professor or it must contain exactly two associate professors?

12. Prove that $C(m + n,k) - C(n,k) =$

$$C(m,1)C(n,k - 1) + C(m,2)C(n,k - 2) + \cdots + C(m,k)C(n,0)$$

for $k \le m \le n$. (Hint: Consider a sample of k items taken from a lot of $m + n$ items, m of which are defective. How many such samples contain at least one defective item? See Example 2.21.)

13. a. Show that

$$C(n,k) = \frac{(n - k + 1) \cdot C(n,k - 1)}{k}$$

for $1 \le k \le n$.

b. Use the result of part a and the fact that $C(n,0) = 1$ to write a Pascal function COMBO that returns the value of $C(N,K)$, where N and K are integers and $0 \le K \le N$.

14. Here is an algorithm to generate the $n!$ permutations of the set $A = \{1, 2, \ldots, n\}$:

Step 1. (initialize) Set a_i equal to i for $1 \leq i \leq n$, and output the permutation $(a_1, a_2, \ldots, a_n)$. Set k equal to 1 (k will count the number of permutations generated so far).

Step 2. Set m equal to $n - 1$. While $a_m > a_{m+1}$, decrement m by 1.

Step 3. Set p equal to n. While $a_m > a_p$, decrement p by 1.

Step 4. Interchange the values of a_m and a_p.

Step 5. For $i = m + 1$ to $(m + n)$ DIV 2, interchange the values of a_i and $a_{m+n+1-i}$.

Step 6. Output the permutation $(a_1, a_2, \ldots, a_n)$. Replace k by $k + 1$. If $k = n!$, stop; otherwise repeat steps 2 through 6.

a. Trace the algorithm for $n = 3$.
b. Suppose that $n = 7$ and the algorithm has just output the permutation (1,6,2,7,5,4,3). Trace steps 2 through 6 and determine which permutation the algorithm will output next.
c. The key to this algorithm is that it outputs the permutations in a definite order. Given two permutations $(x_1, x_2, \ldots, x_n)$ and $(y_1, y_2, \ldots, y_n)$ of A, describe how to determine which one will be output first by the algorithm.
d. Implement the algorithm as a Pascal program that inputs n and outputs the permutations.

2.6 ARBITRARY UNIONS AND INTERSECTIONS

In Section 2.3, the union $A \cup B$ and the intersection $A \cap B$ were defined for two sets A and B. In this section these notions are extended to cover unions and intersections of any number of sets. In the following discussion, all sets discussed are understood to be subsets of some universal set U.

If A_1 and A_2 are sets, then

$$A_1 \cup A_2 = \{x \mid x \in A_1 \lor x \in A_2\}$$
$$A_1 \cap A_2 = \{x \mid x \in A_1 \land x \in A_2\}$$

If A_1, A_2, and A_3 are any three sets, then by the associative laws we have

$$(A_1 \cup A_2) \cup A_3 = A_1 \cup (A_2 \cup A_3)$$
$$(A_1 \cap A_2) \cap A_3 = A_1 \cap (A_2 \cap A_3)$$

Thus there is no confusion if we simply write $A_1 \cup A_2 \cup A_3$ or $A_1 \cap A_2 \cap A_3$. We have then

$$\begin{aligned} A_1 \cup A_2 \cup A_3 &= \{x \mid x \in A_1 \vee x \in A_2 \vee x \in A_3\} \\ &= \{x \mid x \in A_i \text{ for some } i,\ 1 \le i \le 3\} \end{aligned}$$

and

$$\begin{aligned} A_1 \cap A_2 \cap A_3 &= \{x \mid x \in A_1 \wedge x \in A_2 \wedge x \in A_3\} \\ &= \{x \mid x \in A_i \text{ for each } i,\ 1 \le i \le 3\} \end{aligned}$$

DEFINITION 2.12

Given any n sets $A_1, A_2, \ldots, A_n$, we define

$$A_1 \cup A_2 \cup \ldots \cup A_n = \{x \mid x \in A_i \text{ for some } i,\ 1 \le i \le n\}$$

and

$$A_1 \cap A_2 \cap \ldots \cap A_n = \{x \mid x \in A_i \text{ for each } i,\ 1 \le i \le n\}$$

We shall express these sets as

$$\bigcup_{i=1}^{n} A_i \quad \text{and} \quad \bigcap_{i=1}^{n} A_i$$

respectively.

Example 2.23 For $i \in \{1, 2, \ldots, 10\}$, define $A_i = [-i, 10 - i]$. Then $A_1 = [-1, 9]$, $A_2 = [-2, 8], \ldots, A_{10} = [-10, 0]$. Hence

$$\bigcup_{i=1}^{10} A_i = [-10, 9]$$

and

$$\bigcap_{i=1}^{10} A_i = [-1, 0]$$

■

Example 2.24 For $k \in \{1, 2, \ldots, 100\}$, define $B_k = \left\{r \in \mathbb{Q} \mid \frac{-1}{k} \le r \le \frac{1}{k}\right\}$. Then $B_1 = \{r \in \mathbb{Q} \mid -1 \le r \le 1\}$, $B_2 = \{r \in \mathbb{Q} \mid -\frac{1}{2} \le r \le \frac{1}{2}, \ldots,$

$B_{100} = \{r \in \mathbb{Q} \mid -\frac{1}{100} \leq r \leq \frac{1}{100}\}$. Notice that $B_{100} \subset B_{99} \subset \cdots \subset B_2 \subset B_1$. It follows that

$$\bigcup_{k=1}^{100} B_k = B_1$$

and

$$\bigcap_{k=1}^{100} B_k = B_{100}$$

■

DEFINITION 2.13

Given sets $A_1, A_2, \ldots, A_n, \ldots$, we define

1. their *union* to be the set

$$\bigcup_{n=1}^{\infty} A_n = \{x \mid x \in A_r \text{ for some } r \in \mathbb{N}\}$$

2. their *intersection* to be the set

$$\bigcap_{n=1}^{\infty} A_n = \{x \mid x \in A_r \text{ for all } r \in \mathbb{N}\}$$

Example 2.25 Given that $A_n = \left[0, \frac{1}{n}\right]$ for $n \in \mathbb{N}$, find

$$S = \bigcup_{n=1}^{\infty} A_n \qquad \text{and} \qquad T = \bigcap_{n=1}^{\infty} A_n$$

Solution We first note that $A_2 \subset A_1$, $A_3 \subset A_2$, and so on. Thus $S = A_1 = [0, 1]$. To find the intersection, T, first notice that $0 \in A_n$ for every n, so $0 \in T$. We claim that $T = \{0\}$. To verify this, we must show that if x is any positive real number, then $x \notin T$. Let $x > 0$. Consider the harmonic fractions 1, 1/2, 1/3, . . . ; eventually there is a fraction $1/m$ such that $1/m < x$. Since $A_m = [0, 1/m]$, it follows that $x \notin A_m$. Therefore, $x \notin T$. ■

It should be pointed out that a very important principle of the real numbers was used in the above example. It is the *Archimedean principle,* and states that if x and y are any two positive real numbers, then there exists a positive integer m such that $mx > y$. It was applied above in concluding that there is a positive integer m such that $1/m < x$ (here $y = 1$).

Example 2.26 For $n \in \mathbb{N}$, define

$$A_n = \left(\frac{-1}{n}, \frac{2n-1}{n}\right)$$

Then $A_1 = (-1, 1)$, $A_2 = (-1/2, 3/2)$, A_3 $(-1/3, 5/3)$, and so on. Notice that the numbers $-1/n$ are increasing and approaching 0 (from the negative side), while the numbers $(2n-1)/n$ are increasing and approaching 2 (through numbers that are less than 2). Using these facts it can be argued that

$$\bigcup_{n=1}^{\infty} A_n = (-1, 2) \qquad \text{and} \qquad \bigcap_{n=1}^{\infty} A_n = [0, 1]$$ ■

If A and B are sets and $A \subseteq B$, then we also write $B \supseteq A$. Also, if $A_2 \subseteq A_1$, $A_3 \subseteq A_2$, etc., as in Example 2.25, then we write

$$A_1 \supseteq A_2 \supseteq A_3 \supseteq \cdots$$

THEOREM 2.15 Given sets $A_1, A_2, \ldots, A_n, \ldots$, the following hold:

1. If $A_1 \subseteq A_2 \subseteq A_3 \subseteq \cdots$, then

$$\bigcap_{n=1}^{\infty} A_n = A_1$$

2. If $A_1 \supseteq A_2 \supseteq A_3 \supseteq \cdots$, then

$$\bigcup_{n=1}^{\infty} A_n = A_1$$

Proof We supply the proof of part 1 only; the proof of part 2 is similar. Let

$$T = \bigcap_{n=1}^{\infty} A_n$$

and let x be an arbitrary element of T. Then x belongs to A_k for every $k \in \mathbb{N}$, which certainly implies that $x \in A_1$. So $T \subseteq A_1$. Next let x be any element of A_1. Since $A_1 \subseteq A_2 \subseteq A_3 \subseteq \ldots$, it is clear that $A_1 \subseteq A_k$ for every $k \in \mathbb{N}$. Thus $x \in A_k$ for every $k \in \mathbb{N}$, which means that $x \in T$. Thus $T \supseteq A_1$. Therefore, $T = A_1$. ■

Example 2.27 For each $n \in \mathbb{N}$, let $A_n = \{m \in \mathbb{Z} \mid -n \leq m \wedge 2^m \leq n\}$. Then $A_1 = \{-1, 0\}$, $A_2 = \{-2, -1, 0, 1\}$, $A_3 = \{-3, -2, -1, 0, 1\}$, $A_4 = \{-4, -3, -2, -1, 0, 1, 2\}$, and so on. Note that $A_1 \subseteq A_2 \subseteq A_3 \subseteq \ldots$ Hence, by Theorem 2.15,

$$\bigcap_{n=1}^{\infty} A_n = A_1$$

We leave it to the reader to verify that

$$\bigcup_{n=1}^{\infty} A_n = \mathbb{Z}$$ ■

We now define the operations of union and intersection in the most general setting, for any collection of sets.

DEFINITION 2.14

Let I be a nonempty set and assume that there is associated with each $i \in I$ a set A_i. We then call I an *index set* for the collection

$$\mathcal{A} = \{A_i \mid i \in I\}$$

1. The *union* of the collection of sets $\mathcal{A}$ is defined to be the set
$$\bigcup_{i \in I} A_i = \{x \mid x \in A_i \text{ for some } i \in I\}$$
2. The *intersection* of the collection $\mathcal{A}$ is defined to be the set
$$\bigcap_{i \in I} A_i = \{x \mid x \in A_i \text{ for all } i \in I\}$$

Note that Definitions 2.12 and 2.13 are just special cases of Definition 2.14. In Definition 2.12, $I = \{1, 2, \ldots, n\}$, while in Definition 2.13, $I = \mathbb{N}$.

Example 2.28 Let $I = (0, 1)$ and for $i \in I$ define $A_i = (-i, i)$. For instance, $A_{1/2} = (-1/2, 1/2)$ and $A_{\sqrt{2}/3} = (-\sqrt{2}/3, \sqrt{2}/3)$. Notice that if $0 < i < j < 1$, then $\{0\} \subset A_i \subset A_j \subset (-1, 1)$. It follows that

$$\bigcup_{i \in I} A_i = (-1, 1) \quad \text{and} \quad \bigcap_{i \in I} A_i = \{0\}$$ ■

Example 2.29 Let $I = \mathbb{R}$. For $r \in \mathbb{R}$ define $A_r = \{(x,y) \mid y = rx\}$. Geometrically, A_r is a line in the plane that passes through the origin and has slope r. The only point that is on every such line is the origin; hence

$$\bigcap_{r \in \mathbb{R}} A_r = \{(0,0)\}$$

On the other hand, if $a \neq 0$, then the point (a,b) lies on the line $y = \dfrac{b}{a}$, and so $(a,b) \in A_{b/a}$. Thus

$$\bigcup_{r \in \mathbb{R}} A_r = \{(a,b) \mid a \neq 0 \vee a = b = 0\}$$

In other words, $\bigcup_{r \in \mathbb{R}} A_r$ includes the origin plus all other points in the plane except those on the y-axis. ■

Example 2.30 Let S be the set of valid Pascal character strings, and for $s \in S$, let C_s be the set of all valid Pascal character strings that contain the string s as a substring. For instance, if s = 'MA', then the strings 'MATRIX' and 'KALAMAZOO' both belong to C_s. The intersection

$$\bigcap_{s \in S} C_s$$

is empty, since no character string of finite length contains every possible substring. On the other hand,

$$\bigcup_{s \in S} C_s = S$$

since $s \in C_s$ (every character string contains itself as a substring). ■

THEOREM 2.16 (Extended DeMorgan laws) Let $\mathscr{A} = \{A_i \mid i \in I\}$ be a collection of sets indexed by the nonempty set I. Then

$$\left(\bigcup_{i \in I} A_i\right)' = \bigcap_{i \in I} A_i' \qquad \textbf{(1)}$$

$$\left(\bigcap_{i \in I} A_i\right)' = \bigcup_{i \in I} A_i' \qquad \textbf{(2)}$$

Proof We prove identity (1) only. For each x,

$$x \in \left(\bigcup_{i \in I} \mathrm{A_i}\right)' \leftrightarrow \sim\left(x \in \bigcup_{i \in I} A_i\right)$$

$$\leftrightarrow \sim (x \in A_i \text{ for some } i \in I)$$

$$\leftrightarrow x \notin A_i \text{ for all } i \in I$$

$$\leftrightarrow x \in A_i' \text{ for all } i \in I$$

$$\leftrightarrow x \in \bigcap_{i \in I} A_i'.$$

Thus

$$\left(\bigcup_{i \in I} A_i\right)' = \bigcap_{i \in I} A_i'.$$ ■

Exercises 2.6

1. For $k \in \{1, 2, \ldots, 50\}$, define $C_k = (-1/k, 2k)$. Find

$$\bigcap_{k=1}^{50} C_k \quad \text{and} \quad \bigcup_{k=1}^{50} C_k$$

2. Prove: If $A_1 \subseteq A_2 \subseteq \cdots \subseteq A_m$, then

$$\bigcap_{n=1}^{m} A_n = A_1 \qquad \text{and} \qquad \bigcup_{n=1}^{m} A_n = A_m$$

3. For each of the following collections $\{A_n \mid n \in \mathbb{N}\}$, find

$$\bigcap_{n=1}^{\infty} A_n \qquad \text{and} \qquad \bigcup_{n=1}^{\infty} A_n$$

   a. $A_n = \{k \in \mathbb{Z} \mid -n \leq k \leq 2n\}$
   b. $A_n = (-1/n, (n + 1)/n)$ if n is odd, $A_n = ((1 - n)/n, 1/n)$ if n is even
   c. $A_1 = \mathbb{N}$ and, for $n \geq 2$, $A_n = \{m \in \mathbb{N} \mid m \geq n \wedge m/n \notin \mathbb{N}\}$
   d. $A_1 = \{-2, 2\}$ and, for $n \geq 2$, $A_n = \{2nm \mid m \in \mathbb{Z}\}$

4. Prove Theorem 2.15, part 2.
5. For each of the following collections $\{A_i \mid i \in I\}$, find

$$\bigcap_{i \in I} A_i \qquad \text{and} \qquad \bigcup_{i \in I} A_i$$

   a. $U = \mathbb{R}$, $I = (0, 1)$, $A_i = [1 - i, 1/i]$
   b. $U = \mathbb{R} \times \mathbb{R}$, $I = [0, \infty)$, $A_i = \{(x,y) \mid x^2 + y^2 = i^2\}$
   c. $U = \{(a_0, a_1, \ldots, a_n) \mid \text{each } a_i = 0 \text{ or } 1\} = \{\text{binary words of length } n + 1\}$,
   $I = \{0, 1, \ldots, n\}$, $A_i = \{(a_0, a_1, \ldots, a_n) \mid a_i = 1\}$
   d. U = {identifiers in a given Pascal program P},
   I = {procedures in P (including the main procedure)},
   A_i = {identifiers that may be referenced in procedure i}
   e. For a given nonempty set B, $U = \mathcal{P}(B)$, $I = B$, $A_i = \{C \in \mathcal{P}(B) \mid i \in C\}$

6. Prove Theorem 2.16, identity (2).

CHAPTER PROBLEMS

1. In separate Venn diagrams like the one shown in Figure 2.7, shade the region corresponding to each of the following sets.

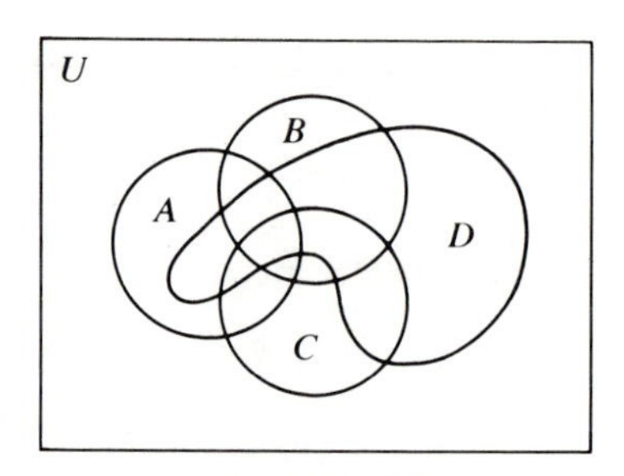

FIGURE 2.7

a. $A \cap B$
b. $B \cap C' \cap D$
c. $(A \cup B) \cap D'$
d. $(A \cap B') \cup (C \cap D)$
e. $(A \cup B \cup C \cup D)'$
f. $((A' \cup B) \cap C)'$

2. Let U be the set of those people who voted in the 1984 U.S. presidential election. Define the following subsets of U:

$$D = \{x \in U \mid x \text{ registered as a Democrat}\}$$
$$R = \{x \in U \mid x \text{ voted for Ronald Reagan}\}$$
$$W = \{x \in U \mid x \text{ belonged to a union}\}$$

Describe each of the following sets in terms of D, R, and W and draw an appropriate Venn diagram for each.

a. The set of people who did not vote for Reagan
b. The set of union members who voted for Reagan
c. The set of registered Democrats who voted for Reagan but did not belong to a union
d. The set of union members who either were not registered as Democrats or voted for Reagan
e. The set of people who voted for Reagan but were not registered as Democrats and were not union members
f. The set of people who either registered as Democrats, were union members, or did not vote for Reagan

3. An automobile insurance company classifies its policyholders by using the following sets:

A = {policyholders who drive subcompact cars}
B = {policyholders who drive cars more than 5 years old}
C = {policyholders who are married}
D = {policyholders over 20 years of age}
E = {male policyholders}

Express each of the following sets in terms of A, B, C, D, and E.

a. The set of female policyholders who are over 20 years of age
b. The set of policyholders who are either male or drive cars that are more than 5 years old
c. The set of female policyholders over 20 years of age who drive subcompact cars
d. The set of male policyholders who are either married or over 20 years old, but who do not drive a subcompact car

4. Let A, B, and C be sets, and consider the conditions (i) $A \cap B = A \cap C$,

(ii) $A \cup B = A \cup C$, and (iii) $B - A = C - A$. For each possible pair of these conditions,

a. (i) and (ii)

b. (i) and (iii)

c. (ii) and (iii)

either prove that the two conditions together imply that $B = C$, or give an example where the two conditions hold but $B \neq C$.

5. Let A, B, C, and D be nonempty subsets of a universal set U. Prove that

$$(A \times B \subseteq C \times D) \leftrightarrow (A \subseteq C \wedge B \subseteq D)$$

6. Let $A \subseteq C$ and $B \subseteq D$. Prove that

a. $A \cap B \subseteq C \cap D$

b. $A \cup B \subseteq C \cup D$

c. $A - D \subseteq C - B$

7. Let A, B, and C be subsets of a universal set U. Prove that

a. $A \times (B \cap C) = (A \times B) \cap (A \times C)$

b. $A \times (B - C) = (A \times B) - (A \times C)$

8. Prove each of the following.

a. $A \cap B = A - (A - B)$

b. $(A - B) - C = (A - C) - B$

c. $A - (B - C) = A \cap (B' \cup C)$

d. $(A - B) - C = A - (B \cup C)$

e. $(A - B) - C = (A - C) - (B - C)$

9. Prove or give a counterexample to each of the following statements.

a. $\mathcal{P}(A \cap B) = \mathcal{P}(A) \cap \mathcal{P}(B)$

b. $\mathcal{P}(A \cup B) = \mathcal{P}(A) \cup \mathcal{P}(B)$

c. $\mathcal{P}(A - B) = \mathcal{P}(A) - \mathcal{P}(B)$

10. The *symmetric difference* of A and B is the set $A * B$ defined by

$$A * B = (A - B) \cup (B - A)$$

Prove that

a. $A * B = B * A$

★**b.** $(A * B) * C = A * (B * C)$

c. $A * B = (A \cup B) - (A \cap B)$

d. $A \cap (B * C) = (A \cap B) * (A \cap C)$

e. $A \cap B = \phi \leftrightarrow A * B = A \cup B$

11. Let A, B, and C be subsets of a universal set U. Prove:

a. If $C \subseteq A$ and $C \subseteq B$, then $C \subseteq A \cap B$.

b. If $A \subseteq C$ and $B \subseteq C$, then $A \cup B \subseteq C$.

12. Let A, B, and C be finite subsets of a universal set U, as shown in the Venn diagram of Figure 2.8.

a. Express each of the regions 1 through 7 in terms of A, B, and C.

b. Give a formula for $n(A \cup B \cup C)$ in terms of the cardinalities of regions 1 through 7.

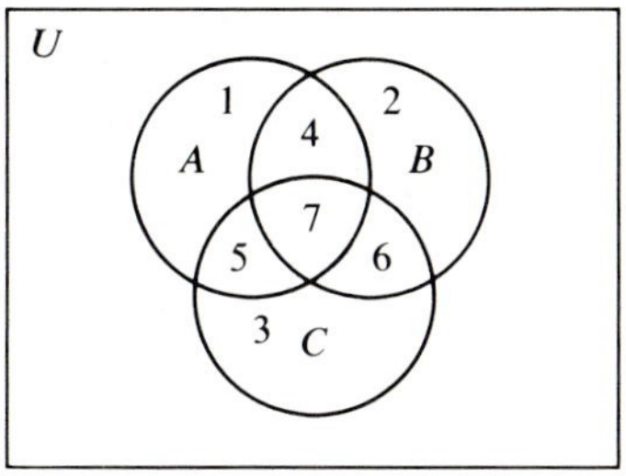

U
A B C
1 2 3 4 5 6 7

FIGURE 2.8

13. In a survey of 100 students, the following information was obtained:

12 are taking courses in English, mathematics, and biology.
22 are taking courses in English and mathematics, but not biology.
3 are taking courses in English and biology, but not mathematics.
7 are taking courses in mathematics and biology, but not English.
20 are taking courses in English, but not biology or mathematics.
17 are taking courses in biology, but not English or mathematics.

Also, all 100 students were taking a course in at least one of these three areas. How many were taking courses only in mathematics? (Hint: use the result of Problem 12.)

14. The registrar at a certain college has observed that 67% of the students have a 2 P.M. class, 22% have an 8 A.M. class, 52% have a 9 A.M. class, 35% have a 9 A.M. and a 2 P.M. class, 12% have an 8 A.M. and a 9 A.M. class, 9% have an 8 A.M. and a 2 P.M. class, and 5% of the students have a class during all three of these hours.

a. What percentage of students has a least one class during these three hours?

b. What percentage of students has no classes during any of these three hours?

15. For each of the following words, determine the number of distinct ways in which the letters may be rearranged.

a. PIE **b.** APPLE **c.** QUEUE **d.** KALAMAZOO

★**e.** In general, if $L_1, L_2, \ldots, L_m$ are distinct letters, how many different "words" can be formed using n_1 L_1's, n_2 L_2's, . . . , and n_m L_m's, where each n_i is a positive integer and $n_1 + n_2 + \cdots + n_m = n$?

★**16.** There are six ways to express the number 5 as a sum of three positive integers: $1 + 1 + 3$, $1 + 2 + 2$, $1 + 3 + 1$, $2 + 1 + 2$, $2 + 2 + 1$, and $3 + 1 + 1$. Show that there are $C(n - 1, m - 1)$ ways to express the positive integer n as a sum of m positive integers.

17. Show that there are $C(m + n - 1, m - 1)$ ways to express the positive integer n as the sum of m nonnegative integers. (Hint: If $x_1 + \cdots +$

$x_m = n$, with each x_i nonnegative, let $y_i = x_i + 1$, $1 \le i \le m$. Then the y_i's are positive and $y_1 + y_2 + \cdots + y_m = n + m$. Now apply the result of Problem 16.)

18. A bakery sells seven kinds of donuts. How many ways are there to put together a box of one dozen donuts if
a. there are no restrictions?
b. there must be at least one donut of each kind?
(Hint: use the results of Problems 16 and 17.)

19. How many poker hands contain
a. exactly one pair?
b. exactly three of a kind?
c. exactly two pairs?
d. a full house (three of one kind, a pair of another)?
e. a straight (for example, 7, 8, 9, 10, jack)?
f. a flush (all the same suit)?

★20. Let P_n be the set of polynomials of degree n with coefficients $a_0, a_1, \ldots, a_n$ (a_i is the coefficient of x^i) chosen from the set $\{-5, -4, \ldots, 4, 5\}$. What is the cardinality of the set P_n? (Note that $a_n \neq 0$.)

21. Five married couples are to be seated at a round table for dinner. Two seating arrangements will be considered the same if every person is seated next to the same person on his/her left and the same person on his/her right. How many ways are there to seat these ten people if
a. there are no restrictions?
b. every man sits next to two women?
c. each husband must be seated next to his wife?
d. the five men sit consecutively?

22. a. In how many ways can eight identical pawns be placed on a chessboard so that no two are in the same row or column?
b. In how many ways can the four rooks (two white, two black) be placed so that no rook may attack one of the opposite color? (Consider rooks of the same color to be identical.)

23. A professor has six calculus books, five linear algebra books, and four discrete mathematics books to arrange together on a shelf. In how many ways can this be done if
a. there are no restrictions?
b. books on the same subject must be placed next to one another?

24. The mathematics club at a small college has six senior, five junior, four sophomore, and three freshmen members. Five members are to be chosen to represent the club on the mathematics department curriculum committee. In how many ways can these five students be chosen if
a. each class is represented?
b. at least two seniors are chosen?
c. at most one freshman is chosen?

d. exactly two sophomores are chosen?
e. at least two seniors or exactly two sophomores are chosen?

25. An urn contains three red balls, four orange balls, five blue balls, and six green balls, numbered 1 through 18, respectively. A subset of four balls is to be chosen from the urn at random. How many such subsets contain

a. exactly two orange balls and one green ball?
b. one ball of each color?
c. balls of at least two different colors?
d. at least two balls that are not blue?

26. a. How many integers between 1 and 600 are not divisible by 2 or 3?
b. How many ways are there to (i) shuffle a regular deck of 52 cards and (ii) deal five-card poker hands to four players?
c. How many ways are there to choose four pairs of people from a group of 20 people?
d. Among ten students who like beer, five drink Kochs and one each drinks Budweiser, Miller, Stroh's, Schaefer, and Coors. How many orders are possible if a subset of five of these students goes into a local pub and each orders a beer?
e. How many permutations of the letters in "computer" have the vowels in alphabetical order?
f. Given n points in a plane, no three on a line, how many triangles do they determine?
g. A regular six-sided die is tossed six times and the sequence of the outcomes is recorded. How many such sequences contain one 2, two 4's, and three 6's?
h. How many ways can six coins be selected from six quarters, six dimes, six nickels, and six pennies?

27. Given the collection $\{A_i \mid i \in I\}$, prove:

a. $(\forall i \in I(B \subseteq A_i)) \leftrightarrow \left(B \subseteq \bigcap_{i \in I} A_i\right)$

b. $(\forall i \in I(A_i \subseteq B)) \leftrightarrow \left(\bigcup_{i \in I} A_i \subseteq B\right)$

28. Given the collection $\{A_i \mid i \in I\}$, let $B = \bigcup_{i \in I} A_i$ and $C = \bigcap_{i \in I} A_i$.

Prove that

$$C = B - \bigcup_{i \in I} (B - A_i)$$

29. Assuming the declaration

```
TYPE BASETYPE = MIN..MAX;
     SETTYPE = SET OF BASETYPE;
```

(where MIN and MAX are constants), write a Pascal function CARD, that returns the cardinality of a set A of type SETTYPE.

30. Prove that $C(n,k) = C(n-1, k) + C(n-1, k-1)$ for $1 \leq k < n$.

31. Let $A = \{1, 2, \ldots, n\}$. In this problem we describe a procedure for generating the power set of A. In particular, we wish to generate the 2^n subsets of A; call them $B_1, B_2, \ldots, B_{2^n}$, where $B_1 = \phi$ and $B_{2^n} = A$. The algorithm works as follows. Suppose m is an integer between 1 and n; at a certain point in the algorithm, suppose we have generated the $t = 2^{m-1}$ subsets of the set $\{1, \ldots, m-1\}$: $B_1 = \phi, B_2, \ldots, B_t$. We will then form t new subsets $B_{t+1}, B_{t+2}, \ldots, B_{t+t}$, where $B_{t+k} = B_k \cup \{m\}$, $1 \leq k \leq t$. Here is a Pascal-like description of the algorithm:

```
B1 := φ; (*initially set B1 equal to the empty set*)
t := 1; (*t is the number of subsets generated so far*)
FOR m := 1 to n DO BEGIN  (*consider the next element m of A*)
  FOR k := 1 to t DO Bt+k := Bk ∪ {m}; (*add the element m to each of the
                                          subsets generated so far*)
   t := 2 * t  (*this doubles the number of subsets so far generated*)
   END;
```

a. Trace the algorithm for $n = 3$.

b. Implement the algorithm as a Pascal procedure that finds the power set of a Pascal set A of type SETTYPE. (You may store the power set as an array of sets.)

32. A very general method of implementing sets in Pascal is as follows:

```
CONST MAX = 50;   (*MAX is the maximum cardinality*)
TYPE BASETYPE = REAL;   (*or whatever type the elements are*)
     SETTYPE = RECORD
       ELEMENT : ARRAY[1..MAX] OF BASETYPE;
       CARD : 0..MAX
     END;
```

Thus if A is a set of type SETTYPE, then A.CARD is the cardinality of A, and if A is nonempty, then A.ELEMENT[1], A.ELEMENT[2], . . . , A.ELEMENT[A.CARD] are the elements of A.

a. Write a Boolean function MEMBOF that determines whether an element X belongs to a set A.

Use MEMBOF to write procedures

b. INTERSECTION **c.** UNION **d.** DIFFERENCE

to find $A \cap B$, $A \cup B$, and $A - B$, respectively, for two sets A and B.

Three Number Theory and Mathematical Induction

3.1 INTRODUCTION

Number theory is that area of mathematics that deals with the properties of the integers under the ordinary arithmetic operations of addition, subtraction, multiplication, and division. It is one of the oldest and, without dispute, one of the most beautiful branches of mathematics. Its problems have been studied by mathematicians, scientists, and others for over two thousand years. In a large measure, the subject is characterized by the simplicity with which many difficult problems can be stated and the ease with which they can be understood by someone without much mathematical background. Thus it should come as no surprise that such problems have attracted the attention of professional mathematicians and amateurs alike.

Several very interesting and basic problems in number theory involve prime numbers. Here is an example of one such problem:

> Are there infinitely many primes of the form $n^2 + 1$, where n is an integer?

This question is certainly easy to understand and yet, to this day, no one has found the correct answer to it. Another unsolved problem, due to the very famous French mathematician Pierre de Fermat (1601–1665), goes like this:

> If n is an integer and $n \geq 3$, then the equation $x^n + y^n = z^n$ has no positive integral solutions.

Fermat had a practice of making notes in his copy of the works of the Greek mathematician Diophantus (circa A.D. 300) and would quite often state, without proof, a result he had discovered. The above problem is one such discovery; in fact, it is the only one of his discoveries that the mathematical community has been unable to prove. Fermat himself wrote, "For this I have discovered a truly wonderful proof, but the margin is too small.'' Because of this claim the problem has come to be called ''Fermat's last theorem.'' It should be mentioned that attempts to solve the problem have given rise to a good portion of modern algebra.

This short chapter is intended to acquaint the student with some topics that will be encountered in subsequent course work in the areas of modern algebra, combinatorial mathematics, and number theory. In addition, several of the notions introduced will provide some material that is pertinent to Chapter 4.

One of the most basic principles used in discrete mathematics, especially in number theory, is the ''principle of well-ordering.'' We state it here as an axiom.

Principle of well-ordering. Every nonempty set of positive integers has a smallest element.

It is impossible to prove this statement using the ordinary properties satisfied by the integers under addition and multiplication. However, after a little thought, the principle should seem truly self-evident. At a later point in this chapter, we shall introduce the student to the ''principle of mathematical induction,'' which is a very useful, logically equivalent form of the principle of well-ordering.

In general, a subset T of $\mathbb{R}$ is said to be *well-ordered* if any nonempty subset S of T has a smallest element. Thus the principle of well-ordering states that $\mathbb{N}$ is well-ordered.

Given a nonempty finite set S of positive integers, it is instructive to see that the following algorithm will (eventually) determine the smallest element of S.

```
BEGIN
 N := 1;  (* initialize N to 1 *)
 WHILE NOT (N IN S) DO
  N := N + 1;  (* as long as N is not an element of S,
                  increment N *)
 WRITELN('THE SMALLEST ELEMENT OF THE SET IS', N)
END.
```

Example 3.1 Find the smallest element in each of the following sets.

(a) $S = \{n \in \mathbb{N} \mid n \text{ is prime}\}$
(b) $S = \{n \in \mathbb{N} \mid n \text{ is a multiple of } 7\}$
(c) $S = \{n \in \mathbb{N} \mid n = 110 - 17m \wedge m \in \mathbb{Z}\}$
(d) $S = \{n \in \mathbb{N} \mid n = 12x + 18y \wedge x, y \in \mathbb{Z}\}$

Solution
(a) The smallest prime number is 2.
(b) The smallest positive multiple of 7 is 7.
(c) Here we must find the smallest positive number n of the form 110 −

$17m$, where m is an integer. The number $110 = 110 - (17)(0)$ is of this form, and as m increases, n decreases. As m takes on the values 1, 2, 3, . . . , the values of n form the sequence

$$93, 76, 59, \ldots, 8, -9, \ldots$$

Hence the smallest element of S is 8. The number 8 just happens to be the remainder when 110 is divided by 17. This is more than just a coincidence, as we shall see in the next section when we discuss the "division algorithm."

(d) We are looking for the smallest positive number n of the form $12x + 18y$, where x and y are integers. Note that if $n = 12x + 18y$, then $n = 6(2x + 3y)$, and so n must be a multiple of 6. Furthermore, $6 = (12)(-1) + (18)(1)$, and so 6 is an element of S. This shows that 6 is the smallest element of S. The number 6 happens to be the greatest common divisor of 12 and 18, a notion we shall explore further in Section 3.3. ■

It should be pointed out that the principle of well-ordering is not a property that holds in general for subsets of $\mathbb{R}$. In particular, $\mathbb{R}$ itself has no smallest element and thus is not well-ordered. Also, any interval of the form $[a,b)$, where $a < b$, fails to be well-ordered. This is because $(a,b) \subset [a,b)$, and the interval (a,b) does not contain a smallest element. As other examples of sets that fail to contain a smallest element, consider the set of positive rational numbers, or the set of negative integers.

We will frequently make use of the following slight extension of the principle of well-ordering.

THEOREM 3.1 If S is a nonempty set of nonnegative integers, then S contains a smallest element.

Proof If $0 \in S$, then clearly 0 is the smallest element of S. If $0 \notin S$, then S is a nonempty set of positive integers and thus, by the principle of well-ordering, S has a smallest element. Hence in either case, S has a smallest element. ■

Exercises 3.1

1. Decide whether each of the following sets T is well-ordered; in other words, determine whether every nonempty subset S of T has a smallest element.

a. $T = \mathbb{Z}$

b. $T = \{t \in \mathbb{Q} \mid t \geq 0\}$

c. T is the set of primes

d. $T \subseteq \mathbb{Z}$ and T itself has a smallest element t_0

e. T is a nonempty, finite subset of $\mathbb{R}$

2. Let S and T be nonempty subsets of $\mathbb{R}$ such that $S \subseteq T$. Prove: If T is well-ordered, then so is S.

3. Find the smallest element of each of the following subsets of $\mathbb{N}$.

a. $A = \{n \in \mathbb{N} \mid n = m^2 - 10m + 28 \wedge m \in \mathbb{Z}\}$
b. $B = \{n \in \mathbb{N} \mid 2$ is the remainder when n is divided by $5\}$
c. $C = \{n \in \mathbb{N} \mid n = -150 - 19m \wedge m \in \mathbb{Z}\}$
d. $D = \{n \in \mathbb{N} \mid n = 5x + 8y \wedge x \in \mathbb{Z} \wedge y \in \mathbb{Z}\}$

3.2 DIVISIBILITY AND THE DIVISION ALGORITHM

One of the fundamental concepts included in any introductory-level course in number theory is that of factorization of integers. In particular, we are interested in expressing an arbitrary integer $n > 1$ as a product of primes (for example, $6 = 2 \cdot 3$). Is this always possible? Can it be done in possibly several ways? Before these questions can be answered it is necessary to define formally the terms "divisor" (or factor) and "prime number."

DEFINITION 3.1

Let a and b be integers with $a \neq 0$.

1. We say that a *divides* b, denoted $a \mid b$, if there is an integer c such that $ac = b$. We say that a is a *divisor* (or *factor*) of b and we call b a *multiple* of a. If a does not divide b we write $a \nmid b$.
2. An integer d is called a *proper divisor* of b if $d \mid b$ and $1 < d < |b|$.
3. An integer $p > 1$ is called a *prime number* (or simply a *prime*) if p has no proper divisors. An integer $c > 1$ that is not prime is called *composite*.

In view of the definition of a prime, it is clear that an integer $p > 1$ is a prime if and only if it is impossible to express p as $p = ab$, where a and b are integers and both $1 < a < p$ and $1 < b < p$. An integer $c > 1$ is composite if and only if there exist integers a and b such that $c = ab$ and both $1 < a < c$ and $1 < b < c$. It's also worth mentioning that if a is a nonzero integer, then a divides zero, but zero is not admissible as a divisor.

Example 3.2 (a) $2 \mid 6$, since $(2)(3) = 6$.
(b) $-3 \mid 27$, since $(-3)(-9) = 27$.

(c) $12 \mid -72$, since $(12)(-6) = -72$.
(d) $4 \nmid 7$, since $4c \neq 7$ for any $c \in \mathbb{Z}$.
(e) $8 \nmid 28$, since $8c \neq 28$ for any $c \in \mathbb{Z}$. ■

Example 3.3 Find all divisors of 126 and factor 126 as a product of primes.

Solution The positive divisors of 126 are 1, 2, 3, 6, 7, 9, 14, 18, 21, 42, 63, and 126. If m is a divisor of n, so is $-m$, and hence $-1, -2, -3, -6, \ldots, -63$, and -126 are also divisors of 126. The proper divisors of 126 are 2, 3, 6, 7, 9, 14, 18, 21, 42, and 63; 2, 3, and 7 are the prime factors of 126. To factor 126 as a product of primes, we write $126 = 2 \cdot 3 \cdot 3 \cdot 7$. ■

There are several basic properties of the relation divides that we shall have occasion to use in this chapter. The next theorem lists a few of these properties.

THEOREM 3.2 For integers a, b, and c, with $a \neq 0$, the following hold:

1. If $a \mid b$, then $a \mid bd$ for any integer d.
2. If $a \mid b$ and $b \mid c$ (where $b \neq 0$), then $a \mid c$.
3. If $a \mid b$ and $a \mid c$, then $a \mid (bx + cy)$ for any integers x and y.
4. If $b \neq 0$ and both $a \mid b$ and $b \mid a$, then $a = b$ or $a = -b$.
5. If $a \mid b$ and both a and b are positive, then $a \leq b$.

Proof The proofs of parts 2 and 3 provide a good illustration of how the remaining parts should be proved.

For the proof of 2, if $a \mid b$ and $b \mid c$, then by definition there exist integers d and e such that $ad = b$ and $be = c$. Thus by substitution we obtain $(ad)e = c$, or $a(de) = c$. It follows that $a \mid c$.

For the proof of 3, from $a \mid b$ and $a \mid c$ we have that $as = b$ and $at = c$ for some $s, t \in \mathbb{Z}$. For any $x, y \in \mathbb{Z}$, we see that $a(sx + ty) = (as)x + (at)y = bx + cy$. Therefore, $a \mid (bx + cy)$. ■

Once again the meaning of the statement "a divides b" should be emphasized, namely, that there is an integer c such that $ac = b$. No doubt the reader recalls the process of "long division." Here one integer b is divided by another integer a, obtaining a "quotient" and a "remainder." For example, if $a = 7$ and $b = 23$, then the relation $23 = 7 \cdot 3 + 2$ is obtained. In this case 3 is the quotient and 2 is the remainder. In general, if $a > 0$, then it will happen that there exist integers q and r, called the *quotient* and *remainder*, respectively, such that $b = aq + r$ and $0 \leq r < a$. We refer to this property as the *division algorithm*.

THEOREM 3.3 **Division algorithm** Given integers a and b with $a > 0$, there exist integers q and r such that $b = aq + r$, where $0 \leq r < a$. Moreover, q and r are uniquely determined by a and b.

Proof We first show that there exist integers q and r such that $b = aq + r$ and $0 \leq r < a$. In order to do this, we apply the principle of well-ordering (the extended version) to the set

$$S = \{b - ax \mid x \in \mathbb{Z} \wedge b - ax \geq 0\}$$

So S is a set of nonnegative integers. In order to apply the well-ordering principle, we must first show that $S \neq \phi$. If $b \geq 0$ then, choosing $x = 0$, we obtain $b \in S$. On the other hand, if $b < 0$ then, since $a \geq 1$, choosing $x = b$ yields $b - ab = b(1 - a) \geq 0$. Hence in this case $b - ab \in S$. So in either case, $S \neq \phi$. It follows by the principle of well-ordering that S has a smallest element, say r. Thus there is some $x \in \mathbb{Z}$, say $x = q$, such that $r = b - aq$, or $b = aq + r$.

Since $r \in S$, it follows from the definition of S that $r \geq 0$, so it remains to show that $r < a$. We proceed by contradiction and assume that $r \geq a$. Then $r = a + t$ for some $t \in \mathbb{Z}$ with $t \geq 0$. Therefore, $t = r - a < r$ and

$$t = r - a = (b - aq) - a = b - a(q + 1)$$

Since $t \geq 0$ and $t = b - a(q + 1)$, we see that $t \in S$. But then we have $t \in S$ and $t < r$, contradicting the fact that r is the smallest element of S. Thus $r < a$ and we have shown that $b = aq + r$, where $0 \leq r < a$.

We next show that the integers q and r in the previous paragraph are uniquely determined. For suppose that $b = aq + r$ and $b = ap + t$, where q, r, p, and t are integers and both $0 \leq r < a$ and $0 \leq t < a$. We must show that $p = q$ and $t = r$. Assume, without loss of generality, that $t \leq r$; hence $r - t \geq 0$. Since $ap + t = b = aq + r$, we obtain $a(p - q) = r - t$. Thus $a \mid (r - t)$. But $r - t < a$, and so it must be the case that $r - t = 0$; thus $r = t$. Then to see that $p = q$, observe that $a(p - q) = r - t = 0$ and $a \neq 0$, so that $p - q = 0$. ■

COROLLARY 3.3 Given integers a and b with $a \neq 0$, there exist integers q and r such that $b = aq + r$, where $0 \leq r < |a|$. The integers q and r are uniquely determined.

Proof See Exercise 4. ■

The division algorithm will turn out to be very useful in determining further divisibility properties of the integers. Its corollary is used, somewhat implicitly, in many programming languages. For example, the operators "DIV" and "MOD" in Pascal have the effect of yielding q and r, respectively, when applied to positive integers b and a.

DEFINITION 3.2

Let a and b be integers with $a \neq 0$, and suppose $b = aq + r$, where q and r are integers and $0 \leq r < |a|$. Define the operators DIV and MOD by b DIV $a = q$ and b MOD $a = r$. The application of DIV to two integers is referred to as *integer division*.

Example 3.4 Find b DIV a and b MOD a for the given pairs of numbers.

(a) $a = 17, b = 110$
(b) $a = 7, b = -59$
(c) $a = -11, b = 41$
(d) $a = -5, b = -27$

Solution (a) As we saw in Example 3.1(c),

$$110 = (17)(6) + 8$$

Hence 110 DIV 17 = 6 and 110 MOD 17 = 8.

(b) Here many people make the mistake of saying -59 DIV $7 = -8$ and -59 MOD $7 = -3$. But remember, b MOD a must always satisfy

$$0 \leq b \text{ MOD } a < |a|$$

The correct answers are -59 DIV $7 = -9$ and -59 MOD $7 = 4$.

(c) Since $41 = (-11)(-3) + 8$, it follows that 41 DIV $-11 = -3$ and 41 MOD $-11 = 8$.

(d) Similarly, $-27 = (-5)(6) + 3$, so -27 DIV $-5 = 6$ and -27 MOD $-5 = 3$. ■

Given any integer x, the number x MOD 5 belongs to the set $\{0, 1, 2, 3, 4\}$. Define the sets A_r, $0 \leq r \leq 4$, by

$$A_r = \{x \in \mathbb{Z} \mid x \text{ MOD } 5 = r\}$$

Then A_r is the set of all those integers x that yield a remainder of r when divided by 5. For instance, 11 MOD 5 = 1, so $11 \in A_1$, whereas -13 MOD 5 = 2, so $-13 \in A_2$. By the division algorithm, each integer x belongs to exactly one of the sets A_r. It follows that

1. $A_0 \cup A_1 \cup A_2 \cup A_3 \cup A_4 = \mathbb{Z}$, and
2. $\{A_0, A_1, A_2, A_3, A_4\}$ is a pairwise-disjoint collection of nonempty sets.

Because of properties 1 and 2, we say that the collection $\{A_0, A_1, A_2, A_3, A_4\}$ is a "partition" of the set $\mathbb{Z}$; the important concept of partition is one we shall study further in Chapter 4.

Explicitly, what is the set A_r? First consider A_0. If $x \in A_0$, then x MOD $5 = 0$, and so there exists an integer q such that $x = 5q$. This is the same as saying that x is a multiple of 5. Conversely, if $x = 5q$, then x MOD $5 = 0$. Thus

$$A_0 = \{5q \mid q \in \mathbb{Z}\}$$

More generally, $x \in A_r$ if and only if there exists an integer q such that $x = 5q + r$. Thus

$$A_r = \{5q + r \mid q \in \mathbb{Z}\}$$

Explicitly,

$$\begin{aligned}
A_0 &= \{. . . , -10, -5, 0, 5, 10, . . .\} \\
A_1 &= \{. . . , -9, -4, 1, 6, 11, . . .\} \\
A_2 &= \{. . . , -8, -3, 2, 7, 12, . . .\} \\
A_3 &= \{. . . , -7, -2, 3, 8, 13, . . .\} \\
A_4 &= \{. . . , -6, -1, 4, 9, 14, . . .\}
\end{aligned}$$

Example 3.5 Let the sets A_r, $0 \le r \le 4$, be as defined in the preceding discussion, and let x and y be integers such that $x \in A_2$ and $y \in A_3$. To which set A_r do the following quantities belong?

(a) $x + y$ (b) xy

Solution We are given that $x \in A_2$ and $y \in A_3$, so x MOD $5 = 2$ and y MOD $5 = 3$. Hence there exist integers s and t such that

$$x = 5s + 2 \qquad \text{and} \qquad y = 5t + 3$$

So for (a) we have that

$$\begin{aligned}
x + y &= (5s + 2) + (5t + 3) \\
&= 5s + 5t + 5 \\
&= 5(s + t + 1)
\end{aligned}$$

Thus $x + y$ is a multiple of 5, that is, $(x + y)$ MOD $5 = 0$, and so

$$x + y \in A_0$$

For (b) we have that

$$\begin{aligned}
xy &= (5s + 2)(5t + 3) \\
&= 25st + 15s + 10t + 6 \\
&= 25st + 15s + 10t + 5 + 1 \\
&= 5(5st + 3s + 2t + 1) + 1
\end{aligned}$$

Thus xy has the form $5q + 1$, and hence $xy \in A_1$. ■

Exercises 3.2

1. Find b DIV a and b MOD a for the following pairs of numbers:
   a. $a = 11, b = 297$ **b.** $a = 9, b = -63$
   c. $a = 8, b = 77$ **d.** $a = 6, b = -71$
   e. $a = -5, b = 35$ **f.** $a = 6, b = 39$
2. Find the proper divisors of 297 and factor 297 as a product of primes.
3. Given that x MOD 7 = 2 and y MOD 7 = 6, find
   a. $(x + 5)$ MOD 7 **b.** $2x$ MOD 7
   c. $-y$ MOD 7 **d.** $(x + y)$ MOD 7
   e. $(2x + 3y)$ MOD 7 **f.** xy MOD 7
4. Prove Corollary 3.3.
5. Prove that, given any three consecutive integers, one of them is a multiple of 3.
6. Prove Theorem 3.2, parts 1, 4, and 5.
7. The division algorithm is an "algorithm" because its proof suggests a method of performing integer division using only the operations of addition and subtraction. Develop such an algorithm that inputs integers a and b, with $a \neq 0$, and computes the quotient q and the remainder r when b is divided by a. (Hint: First consider the case where a and b are both positive. Initialize q to 0 and r to b. Now let q count the number of times a can be subtracted from r until the value of r is less than a.)
8. For some (if not all) Pascal compilers, the operators DIV and MOD give results that sometimes do not agree with Definition 3.2. Write a Pascal program to test b DIV a and b MOD a for various values of a and b. (The problem seems to occur when the dividend is negative. For $b > 0$, Pascal defines b DIV a to be the integer part of b/a and $-b$ DIV a to be $-(b$ DIV $a)$.

3.3 THE EUCLIDEAN ALGORITHM

In this section we define the greatest common divisor of two integers (not both zero) and exhibit a method for finding it, given the integers.

DEFINITION 3.3

An integer $c \neq 0$ is called a *common divisor* of the integers a and b if $c \mid a$ and $c \mid b$. If a and b are not both zero, then we define the *greatest common divisor* of a and b to be the largest common divisor of a and b. If d is the greatest common divisor of a and b, we write

$$d = \gcd(a, b)$$

Example 3.6 (a) $\gcd(12, 18) = 6$
(b) $\gcd(-5, 10) = 5$
(c) $\gcd(-60, -24) = 12$ ■

Given integers a and b, not both zero, we can make some observations about $\gcd(a, b)$ that are easily verified. First, since $1 \mid a$ and $1 \mid b$, we see that $1 \leq \gcd(a, b)$. Also, it is readily seen that $\gcd(a, b) = \gcd(b, a)$ and that $\gcd(-a, b) = \gcd(a, b)$. If $b \neq 0$, then it is clear that $\gcd(0, b) = |b|$. Thus, in order to determine $\gcd(a, b)$ it suffices, without loss of generality, to consider only the case $0 < a \leq b$. In this case notice that $1 \leq \gcd(a, b) \leq a$.

If $0 < a \leq b$, then a simple method for finding $\gcd(a, b)$ is to scan the integers $1, 2, \ldots, a$, seeking the largest of these that is a common divisor of a and b. This method is presented in the following Pascal segment.

ALGORITHM 3.1

```
(*begin: assume A and B are integers and 0 < A ≤ B*)
D := 1;  (*initialize D*)
FOR T := 2 TO A DO
  IF (A MOD T = 0) AND (B MOD T = 0) THEN D := T;  (*if a larger common
                                                      divisor is found,
                                                      update D*)
(*end: at this stage D = GCD (A,B)*)
```
■

Algorithm 3.1 takes essentially a steps to compute $d = \gcd(a, b)$, since d is initialized to 1 and then the body of the FOR loop is repeated $a - 1$ times. If a is quite large, then this algorithm is inefficient.

A faster method, developed by Euclid, is based on repeated application of the division algorithm. Given a and b, there exist integers q and r such that $b = aq + r$, where $0 \leq r < a$. Suppose $d = \gcd(a, b)$ and $e = \gcd(r, a)$. Since $b - aq = r$, and d is a common divisor of a and b, we see that $d \mid r$. Thus d is a common divisor of a and r and we conclude that $d \leq e$. Similarly, since e is a common divisor of r and a and $b = aq + r$, we obtain $e \mid b$. So e is a common divisor of a and b and thus $e \leq d$. Hence we conclude that $d = e$. If $r = 0$, then $\gcd(a, b) = \gcd(0, a) = a$; otherwise, we can apply the division algorithm to a and r to get $a = q_1 r + r_1$, where $0 \leq r_1 < r$. Thus $\gcd(a, b) = \gcd(r, a) = \gcd(r_1, r)$. If $r_1 = 0$, then $\gcd(a, b) = \gcd(0, r) = r$ and the process stops. If $r_1 \neq 0$, then we continue to apply the division algorithm to r and r_1 to obtain $r = r_1 q_2 + r_2$, where $0 \leq r_2 < r_1$. Noting that $0 \leq r_2 < r_1 < r < a$, we see that this process must eventually terminate with a zero remainder. Indeed, if r_{k+1} is the first zero remainder ($r_{k+1} = 0$ and $r_k \neq 0$), then $\gcd(a, b) = r_k$. This method is known as the *Euclidean algorithm*.

The equations that we get from applying the division algorithm are

$$\begin{aligned} b &= aq + r \\ a &= rq_1 + r_1 \\ r &= r_1q_2 + r_2 \\ &\vdots \\ r_{k-2} &= r_{k-1}q_k + r_k \\ r_{k-1} &= r_kq_{k+1} \end{aligned}$$

Notice that the number of equations exhibited is $k + 2$; this represents the number of steps required to determine gcd(a, b) using the Euclidean algorithm. Much is known about this algorithm, and it turns out to be significantly faster than the method of Algorithm 3.1. Indeed, a result of G. Lani (1845) states that

$$10^{k+2} \le a^5$$

Thus, for example, if a = 1,000,000 = 10^6, then the above inequality yields

$$10^{k+2} \le (10^6)^5 = 10^{30}$$

from which it follows that $k + 2 \le 30$. Certainly 30 steps is far better than a million!

An implementation of the Euclidean algorithm in Pascal is presented next.

ALGORITHM 3.2 Euclidean algorithm

```
(*begin: assume A and B are integers and 0 < A ≤ B*)
OLDR := A;  (*OLDR is the remainder at the previous step;
               initially, OLDR = A*)
R := B MOD A;  (*R is the current remainder*)
WHILE R > 0 DO
 BEGIN  (*update R and OLDR*)
  TEMP := R;
  R := OLDR MOD R;
  OLDR := TEMP
 END;
(*end: at this stage OLDR = GCD(A,B)*)
```

■

Example 3.7 Use Algorithm 3.2 to find gcd(228, 528).

Solution

$$\begin{aligned} 528 \text{ MOD } 228 &= 72 \\ 228 \text{ MOD } 72 &= 12 \\ 72 \text{ MOD } 12 &= 0 \end{aligned}$$

Hence gcd(228, 528) = 12. ■

The next theorem provides some important characterizations of $\gcd(a, b)$. The theorem applies the division algorithm in a strong way.

THEOREM 3.4 Let a and b be integers, not both zero. Then the following statements are equivalent:

1. d is the least positive integer such that $d = as + bt$ for some $s, t \in \mathbb{Z}$.
2. d is a positive integer satisfying
   (a) $d \mid a$ and $d \mid b$,
   (b) if $e \mid a$ and $e \mid b$, then $e \mid d$.
3. $d = \gcd(a, b)$.

Outline of proof Here we are to prove a proposition of the form

$$(p \leftrightarrow q) \wedge (q \leftrightarrow r) \wedge (r \leftrightarrow p)$$

Recall that to prove such a proposition it suffices to prove that $(p \rightarrow q) \wedge (q \rightarrow r) \wedge (r \rightarrow p)$ holds. So the proof will be done in three parts:

prove $1 \rightarrow 2$

prove $2 \rightarrow 3$

prove $3 \rightarrow 1$

Proof of 1 → 2. Assume that d is the least positive integer for which there exists integers s and t such that

$$d = as + bt$$

We must show that $d \mid a$, $d \mid b$ and, if $e \mid a$ and $e \mid b$, then $e \mid d$. We show that $d \mid a$; the proof that $d \mid b$ is identical if a is replaced by b. By the division algorithm applied to a and d, there exist integers q and r such that $a = dq + r$, where $0 \leq r < d$. Thus, since $d = as + bt$,

$$r = a - dq = a - (as + bt)q = a(1 - sq) + b(-tq)$$

So r is expressible as $r = ax + by$, where x and y are integers. Since d is the least positive integer that can be so expressed (and since $r < d$), it must be that $r = 0$. Hence $d \mid a$.

Next suppose that $e \mid a$ and $e \mid b$ for some $e \in \mathbb{Z}$, $e \neq 0$. Then by Theorem 3.2, part 3, $e \mid (ax + by)$ for all $x, y \in \mathbb{Z}$; in particular, $e \mid (as + bt)$. Hence $e \mid d$.

Proof of 2 → 3. Assume d is a positive integer that satisfies

(a) $d \mid a$ and $d \mid b$,

(b) if $e \mid a$ and $e \mid b$, then $e \mid d$.

We must show that $d = \gcd(a, b)$. By (a), d is a common divisor of a and

b. Assume that e is some other common divisor of a and b. We must show that $e \le d$. This is clearly true if $e < 0$, so assume e is positive. Then by (b) we have that $e \mid d$. Thus by Theorem 3.2, part 5, $e \le d$. Therefore $d = \gcd(a, b)$.

Proof of 3 → 1. Assume $d = \gcd(a, b)$. It must then be shown that d is the least positive integer expressible in the form $d = as + bt$, where $s, t \in \mathbb{Z}$. We apply the principle of well-ordering to the set

$$S = \{ax + by \mid x, y \in \mathbb{Z} \wedge ax + by > 0\}$$

Since a and b are not both zero, we can assume, without loss of generality, that $a \ne 0$. With the choice of $x = a$ and $y = 0$ we see that $(a)(a) + (b)(0) = a^2 > 0$, so $a^2 \in S$. Thus $S \ne \phi$ and, by the principle of well-ordering, S has a least element, say c. So $c = as + bt$ for some integers s and t; it follows that c is the least positive integer expressible in the form $as + bt$ for some $s, t \in \mathbb{Z}$. It suffices to show that $c = d$. Just as in the proof of 1 → 2, we can show that $c \mid a$ and $c \mid b$. But $d = \gcd(a, b)$, so it must be that $c \le d$. Moreover, since $d \mid a$ and $d \mid b$, it follows from Theorem 3.2, part 3, that $d \mid (as + bt)$. Hence $d \mid c$ and, since $c > 0$ and $d > 0$, $d \le c$. Therefore $c = d$ and the proof is complete. ■

Example 3.8 Use the Euclidean algorithm to find gcd(119, 154) and to express it as a linear combination of 119 and 154.

Solution We wish to express $d = \gcd(119, 154)$ as a linear combination of 119 and 154. Now, d is the last nonzero remainder obtained during the process of applying the Euclidean algorithm. Thus if we take care to express, at each stage, the current remainder as a linear combination of 119 and 154, then, when the process terminates, we shall have d so expressed.

Since 154 MOD 119 = 35 and 154 DIV 119 = 1, we have that 35 = 154 − 119. Next, 119 MOD 35 = 14 and 119 DIV 35 = 3, so

$$\begin{aligned} 14 &= 119 - (3)(35) \\ &= 119 - (3)(154 - 119) \\ &= (119)(4) + (154)(-3) \end{aligned}$$

Next we find that 35 MOD 14 = 7 and 35 DIV 14 = 2, so

$$\begin{aligned} 7 &= 35 - (2)(14) \\ &= (154 - 119) - (2)[(119)(4) + (154)(-3)] \\ &= (119)(-9) + (154)(7) \end{aligned}$$

Finally, 14 MOD 7 = 0, so 7 = gcd(119, 154). Furthermore, the equation

$$7 = (119)(-9) + (154)(7)$$

expresses 7 as a linear combination of 119 and 154. ■

COROLLARY 3.4 If a and b are integers, not both zero, then $\gcd(a, b) = 1$ if and only if $as + bt = 1$ for some integers s and t.

Proof See Exercise 4. ■

The above corollary contains a rather interesting property, namely, if a and b are integers and $as + bt = 1$ for some integers s and t, then $\gcd(a, b) = 1$. It should be remarked that this statement does not generalize! In other words, if $d > 1$ and a and b are integers such that $d = as + bt$ for some integers s and t, then it does not follow that $d = \gcd(a, b)$. For example, $10 = (3)(10) + (2)(-10)$ but $\gcd(2, 3) \neq 10$.

DEFINITION 3.4

Two integers a and b are called *relatively prime* if $\gcd(a, b) = 1$.

Note that a and b are relatively prime if and only if 1 is the only common positive divisor of a and b. For instance, 10 and 21 are relatively prime, as are 63 and 88.

Example 3.9 Show that $5n + 3$ and $7n + 4$ are relatively prime for any $n \in \mathbb{Z}$.

Solution The trick here is to apply Corollary 3.4. Notice that

$$7(5n + 3) - 5(7n + 4) = 1$$

and so by Corollary 3.4, the two numbers $5n + 3$ and $7n + 4$ are relatively prime. ■

There are many interesting, intriguing, and useful results concerning two integers that are relatively prime. We present two of these, with further applications presented as exercises.

THEOREM 3.5 **Euclid's lemma** If a, b, and c are integers such that c and a are relatively prime and $c \mid ab$, then $c \mid b$.

Proof Since a and c are relatively prime, there exist integers s and t such that $as + ct = 1$. It follows that $bas + bct = b$, which can be rewritten as $(ab)s + c(bt) = b$. By hypothesis $c \mid ab$, and also $c \mid c$. Thus Theorem 3.2, part 3, applies and we conclude that $c \mid ((ab)s + c(bt))$, or $c \mid b$. ■

COROLLARY 3.5 If a and b are integers and p is a prime such that $p \mid ab$, then $p \mid a$ or $p \mid b$.

Proof Either $p \mid a$ or $p \nmid a$. If $p \mid a$, then the conclusion follows, so assume $p \nmid a$. It is then the case that $\gcd(p, a) = 1$. Now Theorem 3.5 (with $c = p$) can be applied to give that $p \mid b$. ■

The above corollary can be extended to include any finite number of factors; in particular, if $a_1, a_2, \ldots, a_n$ are integers and p is a prime such that $p \mid (a_1 a_2 \cdots a_n)$, then $p \mid a_i$ for some i, $1 \le i \le n$. We shall give the proof of this fact as an exercise at a later point.

Exercises 3.3

1. Use the Euclidean algorithm to find $d = \gcd(412, 936)$ and integers s and t such that $d = 412s + 936t$.

2. Develop an algorithm that inputs a and b, with $0 < a \le b$, and finds (i) $d = \gcd(a, b)$; (ii) the values of s and t such that $d = as + bt$. (Hint: Generalize the process employed in Example 3.8. Let $r_0, r_1, \ldots, d$ be the sequence of remainders found during application of the Euclidean algorithm, where $r_0 = a$, $r_1 = b - aq_1$, etc. We wish to express each remainder as a linear combination of a and b as the algorithm progresses, so that when d is found, s and t will have been found also. This can be done as follows. Suppose

$$r_{i-2} = as_2 + bt_2$$

and

$$r_{i-1} = as_1 + bt_1$$

Then, since $r_i = r_{i-2} - r_{i-1}q_i$, we find that

$$\begin{aligned} r_i &= (as_2 + bt_2) - (as_1 + bt_1)q_i \\ &= a(s_2 - s_1 q_i) + b(t_2 - t_1 q_i) \end{aligned}$$

In other words, the "new" values for s and t can be computed from the two previous values for each, something like

```
SNEW := OLDS - Q * CURRENTS;
TNEW := OLDT - Q * TCURRENT;
```

as expressed in Pascal.)

3. Apply the algorithm of Exercise 2 to find $d = \gcd(189, 520)$ and to write d as a linear combination of 189 and 520.

4. Prove Corollary 3.4.

5. Suppose that during application of the Euclidean algorithm to find $\gcd(a, b)$, we obtain a remainder r_i that is exactly 1 less than the previous remainder r_{i-1}. What does this imply?

6. Given integers a and b, with $0 < a < b$, prove or disprove:

$d = \gcd(a, b)$ if and only if (i) $d \mid a$, $d \mid b$ and (ii) there exist integers s and t such that $d = as + bt$.

7. Show that any two consecutive positive integers are relatively prime.

8. Let m and n be positive integers and let p be a prime such that $p \nmid n$. Prove that $\gcd(n, p^m) = 1$.
9. In applying the Euclidean algorithm to find $d = \gcd(a, b)$, it may happen that we obtain a certain remainder r_i that we recognize to be prime.
   a. Show that $d = r_i$ or $d = 1$.
   b. How can we tell which of the above is the case?
   c. Use the above result to determine gcd(40, 371).
10. Let p be a prime and let n be a positive integer such that p does not divide n. Show that $\{n \text{ MOD } p, 2n \text{ MOD } p, \ldots, (p-1)n \text{ MOD } p\} = \{1, 2, \ldots, p-1\}$.
11. **a.** Show that no integer of the form $n^2 + 1$ is a multiple of 11.
    b. Find all integers n such that $13 \mid (n^2 + 1)$.
12. Let a, b, c, and d be integers and let k and n be positive integers. Suppose that $a \text{ MOD } n = b \text{ MOD } n$ and $c \text{ MOD } n = d \text{ MOD } n$.
    a. Show that $(a + c) \text{ MOD } n = (b + d) \text{ MOD } n$.
    b. Show that $ac \text{ MOD } n = bd \text{ MOD } n$.
    As a consequence of part b, it can be shown that $a^k \text{ MOD } n = b^k \text{ MOD } n$. Use this result to find the following.
    c. 3^{40} MOD 7
    d. $(1^3 + 2^3 + 3^3 + \cdots + 100^3)$ MOD 4
13. Prove: If $a \text{ MOD } n_1 = b \text{ MOD } n_1$, $a \text{ MOD } n_2 = c \text{ MOD } n_2$, and $n = \gcd(n_1, n_2)$, then $b \text{ MOD } n = c \text{ MOD } n$. (Here n, n_1, and n_2 are positive integers.)
14. Prove: If $a \text{ MOD } n = b \text{ MOD } n$, then $\gcd(a, n) = \gcd(b, n)$.

3.4 THE FUNDAMENTAL THEOREM OF ARITHMETIC

It was stated in an earlier section that one of the basic notions in number theory is that of factorization of an integer $a > 1$ as a product of primes. We prove this result in this section, along with the fact that such a factorization is, in a certain sense, unique.

THEOREM 3.6 **Fundamental theorem of arithmetic** Given any integer $a \geq 2$, there exist primes $p_1, p_2, \ldots, p_m$, with $p_1 \leq p_2 \leq \cdots \leq p_m$, such that

$$a = p_1 p_2 \cdots p_m \qquad (*)$$

Furthermore, the above factorization is unique in the sense that, if $q_1, q_2, \ldots, q_n$ are primes with $q_1 \leq q_2 \leq \cdots \leq q_n$ and $a = q_1 q_2 \cdots q_n$, then $n = m$ and $q_i = p_i$ for $i = 1, 2, \ldots, m$.

Proof We first prove the existence of such a factorization. We proceed by contradiction and assume that the statement of the theorem is false. This means that there is some integer $b > 1$ that is not expressible in the form

(*). Let S be the set of all integers $c > 1$ that are not expressible in the form (*). Since $b \in S$, the set S is a nonempty set of positive integers. Hence, by the principle of well-ordering, S has a least element, say d. It must then be the case that d is not prime; hence d is composite and thus expressible as $d = s \cdot t$, where s and t are integers such that $1 < s \leq t < d$. But then neither s nor t belongs to S, and hence both are expressible in the form (*), say $s = p_1 p_2 \cdots p_k$ and $t = q_1 q_2 \cdots q_m$, where $p_1, p_2, \ldots, p_k$ and $q_1, q_2, \ldots, q_m$ are primes, with $p_1 \leq p_2 \leq \cdots \leq p_k$ and $q_1 \leq q_2 \leq \cdots \leq q_m$. This yields $d = (p_1 p_2 \cdots p_k)(q_1 q_2 \cdots q_m)$. Hence, after a possible rearrangement of factors, d is expressible in the form (*). However, this is a contradiction since $d \in S$. Therefore, we are forced to conclude that S is empty and hence that every integer $a > 1$ is expressible in the form (*).

Next we prove uniqueness. Again we proceed by contradiction and assume that there is some integer $b > 1$ for which

$$b = p_1 p_2 \cdots p_m = q_1 q_2 \cdots q_n$$

where $p_1, p_2, \ldots, p_m, q_1, q_2, \ldots, q_n$ are primes such that $p_1 \leq p_2 \leq \cdots \leq p_m$ and $q_1 \leq q_2 \leq \cdots \leq q_n$, and it is not true that both $m = n$ and $p_i = q_i$ for each i, $1 \leq i \leq m$. Without loss of generality, assume $m \leq n$. Now compare p_i with q_i for $i = 1, 2, \ldots, m$, looking for the smallest value of i for which $p_i \neq q_i$. First of all, it might happen that we find no such i; that is, that $p_i = q_i$ for each i, $1 \leq i \leq m$. Then, since $p_1 p_2 \cdots p_m = q_1 q_2 \cdots q_n$, it must be that $m = n$. Thus we may suppose that there is a smallest value i, $1 \leq i \leq m$, such that $p_i \neq q_i$. Now observe that since $p_i \cdots p_m = q_i \cdots q_n$, it follows that $p_i \mid (q_i \cdots q_n)$ and $q_i \mid (p_i \cdots p_m)$. Assume, without loss of generality, that $p_i < q_i$. Then, by the extended version of Euclid's lemma (Corollary 3.5), $p_i \mid q_j$ for some j, $i \leq j \leq n$. But then, since q_j is prime, $p_i = q_j$. However, $q_j \geq q_i$, which contradicts our assumption that $p_i < q_i$. It follows that, indeed, $m = n$ and $p_i = q_i$ for each i, $1 \leq i \leq m$. ∎

Suppose now that the integer $a > 1$ is expressed as a product of primes, $a = q_1 q_2 \cdots q_m$. The primes $q_1, q_2, \ldots, q_m$ need not be distinct; however, we can collect all equal prime factors and express a in the form

$$a = p_1^{\alpha_1} p_2^{\alpha_2} \cdots p_r^{\alpha_r}$$

where $p_1, p_2, \ldots, p_r$ are primes, $\alpha_i > 0$ for each i, and $p_1 < p_2 < \cdots < p_r$. We call this expression the *canonical factorization* of a. For example, $72 = 2^3 \cdot 3^2$ is the canonical factorization of 72. As an interesting sidelight, consider the problem of finding gcd(504, 300). The canonical factorizations of these numbers are $504 = 2^3 \cdot 3^2 \cdot 7$ and $300 = 2^3 \cdot 3 \cdot 5^2$; however, it becomes convenient to write them as $504 = 2^3 \cdot 3^2 \cdot 5^0 \cdot 7^1$ and $300 =$

$2^2 \cdot 3^1 \cdot 5^2 \cdot 7^0$, so that each factorization includes the same primes. Then note that

$$\begin{aligned}\gcd(504, 300) &= \gcd(2^3 \cdot 3^2 \cdot 5^0 \cdot 7^1, 2^2 \cdot 3^1 \cdot 5^2 \cdot 7^0)\\ &= 2^2 \cdot 3^1 \cdot 5^0 \cdot 7^0\\ &= 12\end{aligned}$$

Thus for each of the primes involved, choose the smaller of the two exponents to determine its contribution to gcd(504, 300). This procedure can be formulated in general terms without much difficulty (see Exercise 3). It should be mentioned that there exist additional applications of the above notation.

Before ending this section we should, for the sake of completeness, prove that the number of primes is infinite. The reader is no doubt aware of this fact but perhaps has never seen a proof.

THEOREM 3.7 The number of primes is infinite.

Proof We proceed by contradiction. To deny that there is an infinite number of primes is to assert that the number of primes is finite. Suppose that $p_1, p_2, \ldots, p_n$ is a list of all the primes. Consider the integer $m = (p_1 p_2 \cdots p_n) + 1$. Clearly $m \geq 2$. We leave it to the reader to verify that m MOD $p_i = 1$ for each i, $1 \leq i \leq n$, and thus none of the p_i's is a factor of m. However, by the fundamental theorem of arithmetic, m can be factored as a product of primes; let p be a prime factor of m. Then $p \neq p_i$ for any i, $1 \leq i \leq n$, contradicting the assumption that $p_1, p_2, \ldots, p_n$ were all the primes. This proves the result. ■

Exercises 3.4

1. Find the canonical factorizations of each of the following integers.
   a. 4725 **b.** 9702 **c.** 25625
2. Develop an algorithm that inputs an integer $n \geq 2$ and outputs its prime factorization.
3. Given integers a and b, with $1 < a < b$, let $P_{a,b} = \{p \mid p \text{ is prime} \wedge p \mid ab\}$. Thus $P_{a,b}$ is the set of primes that divide a or b. Suppose that $P_{a,b} = \{p_1, p_2, \ldots, p_n\}$, where $p_1 < p_2 < \cdots < p_n$; further, suppose that

   $$a = p_1^{\alpha_1} p_2^{\alpha_2} \cdots p_n^{\alpha_n}$$

   and

   $$b = p_1^{\beta_1} p_2^{\beta_2} \cdots p_n^{\beta_n}$$

   where $\alpha_i \geq 0$ and $\beta_i \geq 0$ for each i. (For example, $P_{300,504} = \{2, 3, 5, 7\}$, since $300 = 2^2 \cdot 3^1 \cdot 5^2 \cdot 7^0$ and $504 = 2^3 \cdot 3^2 \cdot 5^0 \cdot 7^1$.)
   a. Give the formula for $\gcd(a, b)$ in terms of $p_1, p_2, \ldots, p_n$.

b. Use the result of part a to find gcd(4725, 9702).

c. The *least common multiple* of a and b is denoted lcm(a, b) and is defined as the smallest positive integer c such that $a \mid c$ and $b \mid c$. Give a formula for lcm(a, b) in terms of $p_1, p_2, \ldots, p_n$. (Hint: $\text{lcm}(300, 504) = 12600 = 2^3 \cdot 3^2 \cdot 5^2 \cdot 7^1$.)

d. Use the result of part c to find lcm(4725, 9702).

e. Show that $\gcd(a, b) \cdot \text{lcm}(a, b) = ab$.

4. Develop an algorithm that inputs an integer $n \geq 1$ and finds the first n primes.

5. a. Prove: If a MOD $n_1 = b$ MOD n_1 and a MOD $n_2 = b$ MOD n_2, then a MOD $n = b$ MOD n, where $n = \text{lcm}(n_1, n_2)$.

b. What does the result in part a say if n_1 and n_2 are relatively prime?

6. Develop an algorithm that inputs an integer $n > 2$ and outputs the canonical factorization of n factorial.

3.5 THE PRINCIPLE OF MATHEMATICAL INDUCTION

In mathematics there are many problems having the following general form:

1. Let $P(n)$ be a statement about the positive integer n.
2. Prove that $P(n)$ is true for every $n \in \mathbb{N}$.

A number of examples, taken from various areas of mathematics, will give an idea of the frequency with which such problems occur.

Example 3.10 The following are statements about an arbitrary positive integer n.

(a) $1 + 2 + 3 + \cdots + n = \dfrac{n(n+1)}{2}$

(b) For any real numbers x and y,

$$(x + y)^n = \sum_{k=0}^{n} C(n, k)x^{n-k}y^k$$

(This statement is known as the binomial theorem.)

(c) If the length of a longest chain in a finite poset $(A, \preceq)$ is n, then A can be partitioned into n antichains.

(d) $1 + \dfrac{1}{4} + \dfrac{1}{9} + \cdots + \dfrac{1}{n^2} \leq 2 - \dfrac{1}{n}$

(e) If A is a finite set with n elements, then $\mathcal{P}(A)$ has 2^n elements.

(f) $6 \mid (n^3 + 5n)$ ■

How do we solve problems of this general form? Suppose that we could prove that the following two conditions hold:

1. $P(1)$ is true.
2. If $P(k)$ is true for some arbitrary $k \in \mathbb{N}$, then $P(k + 1)$ is true.

Condition 2, stated another way, says that the truth of $P(k)$ for an arbitrary $k \in \mathbb{N}$ implies the truth of $P(k + 1)$. How can these conditions aid in proving that $P(n)$ is true for every $n \in \mathbb{N}$? Condition 1 tells us that $P(1)$ is true, hence condition 2, applied with $k = 1$, states that $P(2)$ is true. Well then, since $P(2)$ is true, condition 2, applied with $k = 2$, gives us that $P(3)$ is true. Since $P(3)$ is true, $P(4)$ is true, and so on. It definitely seems inviting now to state that $P(n)$ is true for every $n \in \mathbb{N}$. This is exactly what the "principle of mathematical induction" will allow us to do. It can be shown that this principle is logically equivalent to the principle of well-ordering, which we restate here.

Principle of well-ordering. Every nonempty set of positive integers has a smallest element.

THEOREM 3.8 **Principle of mathematical induction** Let S be a set of positive integers such that

1. $1 \in S$;
2. if $k \in S$, then $k + 1 \in S$.

Then $S = \mathbb{N}$.

Proof We proceed by contradiction; assume that S is a set of positive integers satisfying conditions (1) and (2) of the theorem, but $S \neq \mathbb{N}$. Let $A = \mathbb{N} - S$; then $A \neq \phi$ and $\mathbb{N} = A \cup S$. Since $A \neq \phi$, the principle of well-ordering implies that A has a smallest element, say a. Since it is given that $1 \in S$, we see that $a \geq 2$ and hence that $a - 1 \geq 1$. Since a is the smallest element of A, it follows that $a - 1 \notin A$. Also, since $a - 1 \in \mathbb{N}$ and $\mathbb{N} = A \cup S$, it must be that $a - 1 \in S$. Now condition (2) of the theorem, applied with $k = a - 1$, implies that $k + 1 = a \in S$. But this is a contradiction. It follows that $S = \mathbb{N}$. ■

With the principle of mathematical induction (PMI) established, we now describe how it can be used to prove statements like those presented in Example 3.10. It must first be determined whether a given problem has the general form exhibited in the opening paragraph of this section. If this is the case, and a decision has been made to use the PMI to prove the statement, then the following outline is strongly recommended.

Outline of a proof by mathematical induction

Proceed by induction on n, and let S be the set of positive integers for which the statement $P(n)$ is true.

Step 1. Show $1 \in S$.

Step 2. Assume $k \in S$ for some arbitrary $k \geq 1$.

Step 3. Show $k + 1 \in S$.

It follows that $S = \mathbb{N}$ and hence that $P(n)$ is true for all $n \in \mathbb{N}$.

Notice the use of the phrase, "Proceed by induction on n." It is standard practice to use this wording when using the PMI to prove the statement $P(n)$. Step 1 is commonly referred to as the *anchor step;* once it has been shown that $1 \in S$, we will quite often say that the induction is "anchored." Steps 2 and 3, taken together, make up the implication

$$\forall k \in \mathbb{N}(k \in S \rightarrow k + 1 \in S)$$

For emphasis and clarity we state the premise and conclusion separately. We call Step 2 the *induction hypothesis,* IHOP for short, and we call Step 3 the *inductive step*. Several examples will serve to illustrate the convenience of this outline when employing the PMI.

Example 3.11 Use the PMI to prove that the statement $P(n)$,

$$1 + 2 + \cdots + n = \frac{n(n + 1)}{2}$$

holds for each $n \in \mathbb{N}$.

Proof Proceed by induction on n and let S be the set of positive integers for which $P(n)$ is true.

1. Show $1 \in S$.
   Since $1 = 1(1 + 1)/2$, it follows that $1 \in S$.
2. Assume $k \in S$ for some arbitrary $k \geq 1$.

$$\text{IHOP:} \quad 1 + 2 + \cdots + k = \frac{k(k + 1)}{2}$$

3. Show that $k + 1 \in S$; in other words, show that

$$1 + 2 + \cdots + (k + 1) = \frac{(k + 1)(k + 2)}{2}$$

Adding $k + 1$ to both sides of the IHOP equation yields

$$\begin{aligned} 1 + 2 + \cdots + k + (k + 1) &= \frac{k(k + 1)}{2} + (k + 1) \\ &= \frac{k(k + 1) + 2(k + 1)}{2} \\ &= \frac{(k + 1)(k + 2)}{2} \end{aligned}$$

This shows that $k + 1 \in S$.

It follows that $S = \mathbb{N}$ and hence that $P(n)$ is true for each $n \in \mathbb{N}$. ■

Example 3.12 Use the PMI to prove that the statement $P(n)$,

$$1 + \frac{1}{4} + \frac{1}{9} + \cdots + \frac{1}{n^2} \leq 2 - \frac{1}{n}$$

holds for each $n \in \mathbb{N}$.

Proof Proceed by induction on n and let S be the set of positive integers for which $P(n)$ is true.

1. Show $1 \in S$.
   Since $1 \leq 2 - \frac{1}{1}$, it follows that $1 \in S$.
2. Assume $k \in S$ for some arbitrary $k \geq 1$.

$$\text{IHOP:} \quad 1 + \frac{1}{4} + \cdots + \frac{1}{k^2} \leq 2 - \frac{1}{k}$$

3. Show that $k + 1 \in S$; that is, show that

$$1 + \frac{1}{4} + \cdots + \frac{1}{(k + 1)^2} \leq 2 - \frac{1}{k + 1}$$

Adding the quantity $1/(k + 1)^2$ to both sides of the IHOP inequality yields

$$1 + \frac{1}{4} + \cdots + \frac{1}{k^2} + \frac{1}{(k + 1)^2} \leq 2 - \frac{1}{k} + \frac{1}{(k + 1)^2}$$

Now,

$$\begin{aligned} 2 - \frac{1}{k} + \frac{1}{(k + 1)^2} &= 2 - \frac{k^2 + k + 1}{k(k + 1)^2} \\ &\leq 2 - \frac{k^2 + k}{k(k + 1)^2} \qquad (\text{since } k^2 + k < k^2 + k + 1) \\ &= 2 - \frac{k(k + 1)}{k(k + 1)^2} \\ &= 2 - \frac{1}{k + 1} \end{aligned}$$

Hence we have that

$$1 + \frac{1}{4} + \cdots + \frac{1}{(k+1)^2} \leq 2 - \frac{1}{k+1}$$

which shows that $k + 1 \in S$.

It follows that $S = \mathbb{N}$ and hence that $P(n)$ is true for each $n \in \mathbb{N}$. ■

In many instances we will encounter problems that ask us to prove that a statement $P(n)$ is true for all integers $n \geq n_0$, where n_0 is some given fixed integer. For example, we might be asked to show that $n! > 2^n$ for all integers $n \geq 4$. (Note that this inequality fails for $0 \leq n \leq 3$.) In cases such as this, the following corollary to Theorem 3.8 will provide the necessary proof technique.

COROLLARY 3.8 Let $n_0 \in \mathbb{Z}$ and let $M = \{n \in \mathbb{Z} \mid n \geq n_0\}$. Let S be a subset of M such that

1. $n_0 \in S$;
2. if $k \in S$, then $k + 1 \in S$.

Then $S = M$. ■

The proof of Corollary 3.8 is left as an exercise. In applying the corollary to solve a problem, it is necessary only to replace 1 by n_0 and $\mathbb{N}$ by M in Steps 1 through 3 of the suggested outline.

Example 3.13 Prove that $n! > 2^n$ for all $n \geq 4$.

Proof Proceed by induction on n; let $M = \{n \in \mathbb{N} \mid n \geq 4\}$ and let $S = \{n \in M \mid n! > 2^n\}$.

1. Show $4 \in S$.
   Since $4! = 24 > 16 = 2^4$, it follows that $4 \in S$.
2. Assume $k \in S$ for some arbitrary $k \geq 4$.

   $$\text{IHOP:} \quad k! > 2^k$$

3. Show that $k + 1 \in S$; that is, show that $(k+1)! > 2^{k+1}$.
   We proceed as follows:

   $$\begin{aligned} (k+1)! &= (k+1) \cdot k! \\ &> (k+1) \cdot 2^k \qquad \text{(by IHOP)} \\ &> 2 \cdot 2^k \qquad \text{(since } k + 1 > 2\text{)} \\ &= 2^{k+1} \end{aligned}$$

   Thus $k + 1 \in S$.

It follows that $S = M$. ■

Example 3.14 Use the PMI to prove that there are 2^m subsets of a finite set A with m elements, $m \geq 0$.

Proof We proceed by induction on m. We wish to prove that $n(\mathcal{P}(A)) = 2^m$.

1. If $m = 0$, then $A = \phi$ and $\mathcal{P}(A) = \{\phi\}$, so $n(\mathcal{P}(A)) = 1$. Thus the result holds when $m = 0$.
2. Assume that the result holds when $m = k$, for some arbitrary $k \geq 0$, namely, that there are 2^k subsets of any set having k elements.
3. Let A be a set that has $k + 1$ elements, say $A = \{a_1, a_2, \ldots, a_{k+1}\}$. Consider the set $B = A - \{a_{k+1}\}$. Since $n(B) = k$, we have by the induction hypothesis that $n(\mathcal{P}(B)) = 2^k$. Furthermore, to each subset C of B there correspond two subsets of A, namely, C and $C \cup \{a_{k+1}\}$. In other words, A has twice as many subsets as B has. Therefore,

$$n(\mathcal{P}(A)) = 2 \cdot n(\mathcal{P}(B)) = 2 \cdot 2^k = 2^{k+1}$$

Hence the result holds when $m = k + 1$.

Thus, by the PMI, the result holds for every $m \geq 0$. ■

Let us now consider an application of mathematical induction to the problem of proving program correctness. (See the end of Section 1.6 for a discussion of this topic.)

Consider the following Pascal program, which inputs a real value for X and a nonnegative integer for N and computes and outputs $Y = X^N$.

```
PROGRAM EXPO(INPUT,OUTPUT);
VAR X,Y : REAL;
      J : INTEGER;
      N : 0..MAXINT;
BEGIN
  READ(X,N);
  Y := 1;  (*initialize Y to 1*)
  FOR J := N DOWNTO 1 DO
    Y := Y * X;  (*find the product of X with itself N
                   times*)
  WRITELN(Y)
END.
```

The key feature of this program is that it contains a loop, in this case a FOR loop. The loop control variable is J, whose initial value is the value of N. Since N can, in theory, take on any nonnegative-integer value, the body of the loop may be executed 0, 1, 2, or, in general, any nonnegative-integer number of times. Thus, there is a natural connection between mathematical induction and verifying the correctness of programs that contain loops.

Before analyzing a program such as EXPO, it is often helpful to represent the program as a flowchart. This is because flowcharts provide a more general setting in which to analyze programs, free from the particular syntactic features of a given language. (The program EXPO would look somewhat different were it written in FORTRAN.) Also, a high-level language such as Pascal uses statements that intentionally hide much of what is going on. For example, the Pascal FOR loop hides the incrementing or decrementing, and the exit-testing steps that are performed with the loop control variable. We have translated the program EXPO into the flowchart of Figure 3.1.

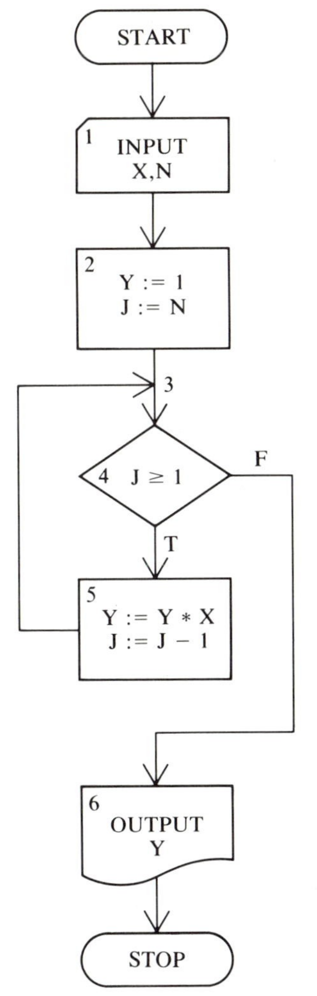

Explanation

1. Input values for X (real) and for N (nonnegative integer).
2. Initialize Y to 1 and J to the value of N.
3. This is the top of the loop.
4. Compare the value of J with 1.
5. If J ≥ 1, execute the body of the loop. Multiply the current value of Y by the value of X, and assign the result to Y. Decrement the value of J by 1.
6. If J < 1, output the value of Y and halt.

FIGURE 3.1 Flowchart for the program EXPO

We wish to prove the following assertion about the program represented by the flowchart of Figure 3.1.

Assertion A. If a real value is input for X and a nonnegative integer is input for N, then the program will compute and output the value of $Y = X^N$.

The key to proving Assertion A is to focus attention on the loop in the program. Suppose execution of the program reaches the top of the loop, and the variables X, Y, and J have the specific values x, y, and n, respectively. In this case, what value will be output for Y? What we wish to show is that the answer to this question is $y * x^n$.

Assertion B. If execution of the program reaches the top of the loop and the variables X, Y, and J have the values x, y, and n, respectively, then the program will output the value $y * x^n$.

The truth of Assertion A will follow if we can prove Assertion B. For, suppose the values x and n are input for X and N, respectively. Then Y is initialized to 1, J is initialized to n, and when we first reach the top of the loop $X = x$, $Y = y = 1$, and $J = n$. Assertion B then says that the program will output $y * x^n = 1 * x^n = x^n$.

Proof of Assertion B We proceed by induction on n, the value of J.

1. If $J = n = 0$, then the condition $J \geq 1$ is false. Thus the value of Y, namely y, will be output. This agrees with the conclusion of Assertion B since $y * x^n = y * x^0 = y * 1 = y$.
2. Assume the assertion holds when the value of J is k, where k is an arbitrary nonnegative integer. IHOP: If execution of the program reaches the top of the loop with $X = x$, $Y = y$, and $J = k$, then the program will output the value $y * x^k$. (Note: in IHOP the values x and y are fixed but arbitrary; they can be *any* real numbers.)
3. Suppose execution reaches the top of the loop with $X = x$, $Y = y$, and $J = k + 1$. Since $k + 1 \geq 1$, the body of the loop will be executed. This gives Y the value $y * x$ and decrements the value of J to k. Control is then transferred back to the top of the loop. But now the value of J is k, and so IHOP can be applied. Since the value of X is x and the value of Y is $y * x$, we may conclude by IHOP that the value $(y * x) * x^k = y * (x * x^k) = y * x^{k+1}$ will be output. This shows that the assertion holds when $J = k + 1$.

It follows from the PMI that Assertion B holds for any nonnegative integer n.

Exercises 3.5

1. Prove: $1 + 3 + \cdots + (2n - 1) = n^2$ for each $n \in \mathbb{N}$.
2. Prove: $1^2 + 2^2 + \cdots + n^2 = n(n + 1)(2n + 1)/6$ for each $n \in \mathbb{N}$.
3. Use the PMI to prove that $6 \mid (n^3 + 5n)$ for each $n \in \mathbb{N}$.
4. Let $P(n)$ be the statement that $n^2 + n + 41$ is prime.
   a. Show that $P(n)$ holds for $1 \leq n \leq 10$.
   b. Is $P(n)$ true for every $n \in \mathbb{N}$?

5. Prove that for real numbers a and r, $r \neq 1$, and $n > 0$,

$$a + ar + \cdots + ar^n = \frac{a(1 - r^{n+1})}{1 - r}$$

6. For each $n \in \mathbb{N}$, let $P(n)$ be the statement that $3 \mid (3n + 2)$. Show that if $P(k)$ holds for some arbitrary $k \in \mathbb{N}$, then $P(k + 1)$ holds. Why can't we conclude that $P(n)$ is true for every $n \in \mathbb{N}$?

7. **a.** Use Corollary 3.8 to prove that $2n + 1 < 2^n$ for all $n \geq 3$.
b. Use Corollary 3.8 and part a to prove that $n^2 < 2^n$ for all $n \geq 5$.

8. Prove Corollary 3.8.

9. Use the PMI to prove that $2 + (1 + 1/\sqrt{2} + \cdots + 1/\sqrt{n}) > 2\sqrt{n + 1}$ for all $n \in \mathbb{N}$.

10. Use the PMI to prove that $2(1 + 1/8 + \cdots + 1/n^3) < 3 - 1/n^2$ for all $n \in \mathbb{N}$.

11. The flowchart in Figure 3.2 represents a program to compute the product of an integer A and a nonnegative integer B. (It works by adding A to itself B times.) Use the PMI to prove that if the integer value a and the nonnegative-integer value b are input for A and B, respectively, then the program will output the value $a * b$.

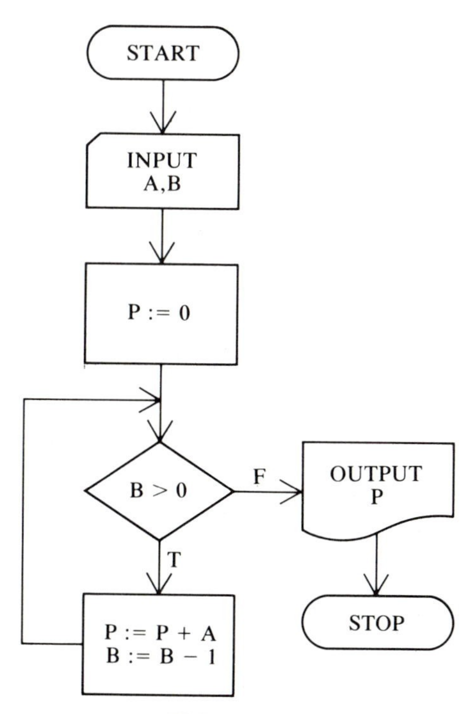

FIGURE 3.2

★**12.** Prove the principle of well-ordering is implied by the principle of mathematical induction. (This, together with the proof of Theorem 3.8, shows that these two principles are logically equivalent.)

3.6 THE STRONG FORM OF INDUCTION

In some cases the method of induction employed in the last section cannot be applied so nicely. To see this, consider the sequence of numbers 1, 1, 2, 3, 5, 8, 13, ... These numbers are known as the *Fibonacci numbers*. Letting f_n denote the nth Fibonacci number, $n \geq 1$, we note that

(a) $f_1 = f_2 = 1$
(b) for $n \geq 3, f_n = f_{n-2} + f_{n-1}$

Suppose we are asked to prove that $f_n < 2^n$ for all positive integers n. The statement of the problem suggests using induction; let's see what happens. As usual, let S be the set of all positive integers n for which the statement $f_n < 2^n$ holds. Clearly $1 \in S$. Assume $k \in S$ for some arbitrary $k \geq 1$, so that $f_k < 2^k$. It must be shown that $k + 1 \in S$, that is, that $f_{k+1} < 2^{k+1}$. What we do know is that $f_{k+1} = f_{k-1} + f_k$ if $k \geq 2$. It appears that we are in a bit of a bind, as we have no direct information concerning f_{k-1}.

Problems such as this can be handled by using an alternate form of induction, called the *strong form of induction*. Like the PMI, it is equivalent to the principle of well-ordering. We demonstrate only the implication by the principle of well-ordering.

THEOREM 3.9 **Strong form of induction** Let S be a set of positive integers such that

1. $1 \in S$;
2. for any $n > 1$, if $k \in S$ for all integers k such that $1 \leq k < n$, then $n \in S$.

Then $S = \mathbb{N}$.

Proof Suppose, on the contrary, that S satisfies conditions (1) and (2) of the theorem, but $S \neq \mathbb{N}$. Then $\mathbb{N} - S \neq \phi$ and $\mathbb{N} = (\mathbb{N} - S) \cup S$. Since $\mathbb{N} - S \neq \phi$, it has a smallest element, say m, and since $1 \in S$, $m > 1$. Since m is the smallest element of $\mathbb{N} - S$, the integers $1, 2, \ldots, m - 1$ must all belong to S. Thus condition (2) of the theorem applies to give that $m \in S$. But this is a contradiction, since $m \in \mathbb{N} - S$. Hence it follows that $S = \mathbb{N}$. ■

To illustrate the strong form of induction, we use it to prove that the nth Fibonacci number f_n satisfies $f_n < 2^n$ for each $n \in \mathbb{N}$. As usual, let S be the set of positive integers n such that $f_n < 2^n$.

1. Show $1 \in S$ and $2 \in S$.
   Since $f_1 = 1 < 2^1$ and $f_2 = 1 < 2^2$, we have $1,2 \in S$. (The reason for anchoring the induction at both $n = 1$ and $n = 2$ is so we can have $n \geq 3$ in the inductive step, allowing us to apply the formula $f_n = f_{n-2} + f_{n-1}$.)
2. For an arbitrary $n > 2$, assume that $k \in S$ for all integers k such that $1 \leq k < n$, that is, $f_k < 2^k$ for $1 \leq k < n$.
3. Show $n \in S$; that is, show $f_n < 2^n$.
   We know that $f_n = f_{n-2} + f_{n-1}$, and by the induction hypothesis, $f_{n-2} < 2^{n-2}$ and $f_{n-1} < 2^{n-1}$. Hence

$$\begin{aligned} f_n = f_{n-2} + f_{n-1} &< 2^{n-2} + 2^{n-1} \\ &= 2^{n-2}(1 + 2) \\ &= 2^{n-2} \cdot 3 \\ &< 2^{n-2} \cdot 4 \\ &= 2^n \end{aligned}$$

Thus $n \in S$ and therefore $S = \mathbb{N}$.

Example 3.15 The flowchart shown in Figure 3.3 represents a program to compute the integer quotient N DIV D for a nonnegative integer N and a positive integer D. It works by counting the number of times D must be subtracted from N in order to obtain a number R satisfying $0 \leq \text{R} < \text{D}$ (R is the remainder: R = N MOD D). Use the strong form of induction to prove that if a nonnegative integer n is input for N and the positive integer d is input for D, then the value n DIV d will be output for Q.

Solution The correctness of the program will be shown if we can prove the following assertion:

> If execution of the program reaches the top of the loop and the values of R, Q, and D are n, q, and d, respectively, then the value q + (n DIV d) will be output.

(When we first reach the top of the loop, R = n, Q = 0, and D = d, so the truth of the assertion implies that q + (n DIV d) = 0 + (n DIV d) = n DIV d will be output.)

We shall prove the assertion by induction (the strong form) on the nonnegative integer n.

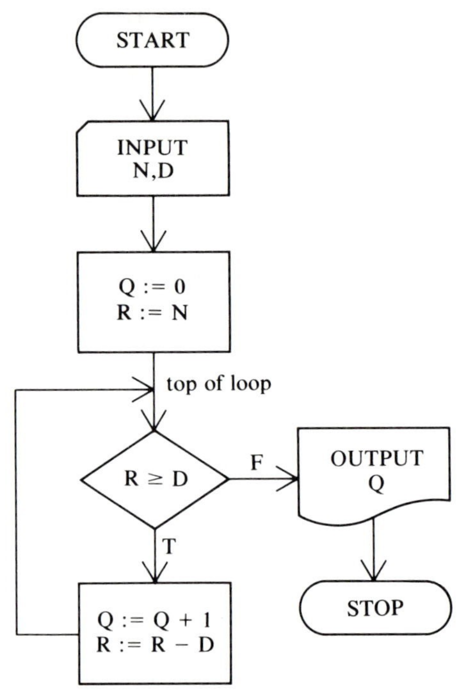

FIGURE 3.3 Flowchart for Example 3.15

1. If execution reaches the top of the loop with R $= n = 0$, Q $= q$, and D $= d$, then, since d is positive, the condition R ≥ D is false. Hence the value q will be output, which is correct since $q = q + 0 = q + (0 \text{ DIV } d) = q + (n \text{ DIV } d)$.
2. For an arbitrary $n > 0$, assume that for all integers k, $0 \leq k < n$, if execution reaches the top of the loop with R $= k$, Q $= q$, and D $= d$, then the value $q + (k \text{ DIV } d)$ will be output.
3. Show that the assertion holds when R $= n$. Suppose execution reaches the top of the loop with R $= n$, Q $= q$, and D $= d$. If $n < d$ then the condition R ≥ D is false. In this case the value q will be output, which is correct since $n < d$ implies that n DIV $d = 0$. On the other hand, if $n \geq d$ then the condition R ≥ D is true. In this case the body of the loop will be executed, giving Q the value $q + 1$ and R the value $n - d$. Execution then returns to the top of the loop. Now the value of R is $n - d$, and $n - d < n$, so the induction hypothesis applies. Since $R = n - d$, Q $= q + 1$, and D $= d$, we may conclude that the value $(q + 1) + ((n - d) \text{ DIV } d)$ will be output. To see that this value equals $q + (n \text{ DIV } d)$, use the fact (see Chapter Problem 5) that $(n - d) \text{ DIV } d = (n \text{ DIV } d) - 1$:

$$\begin{aligned}(q + 1) + ((n - d) \text{ DIV } d) &= (q + 1) + (n \text{ DIV } d) - 1 \\ &= q + (n \text{ DIV } d)\end{aligned}$$

Therefore, our assertion holds for all nonnegative integers n. ■

Exercises 3.6

1. Use the strong form of induction to prove the existence part of the fundamental theorem of arithmetic: Every positive integer $n \geq 2$ can be factored as a product of primes.

2. Define the sequence of numbers $t_1, t_2, t_3, \ldots, t_n, \ldots$ by

(a) $t_1 = 1, \quad t_2 = 2, \quad t_3 = 3;$

(b) $t_n = t_{n-3} + t_{n-2} + t_{n-1}$ for all $n \geq 4$

Use the strong form of induction to prove that $t_n < 2^n$ for every $n \in \mathbb{N}$.

3. The flowchart shown in Figure 3.4 represents a program that inputs integers A and B, with $0 < \text{A} \leq \text{B}$, and claims to output GCD = gcd(A, B). (It uses the Euclidean algorithm.) Use the strong form of induction to prove that if proper values a and b are input for A and B, respectively, then the program will output the value of gcd(a, b). (Hint: prove that if execution reaches the top of the loop and the values of GCD and R are n and r, respectively, and $0 \leq r < n$, then the value gcd(r, n) will be output.)

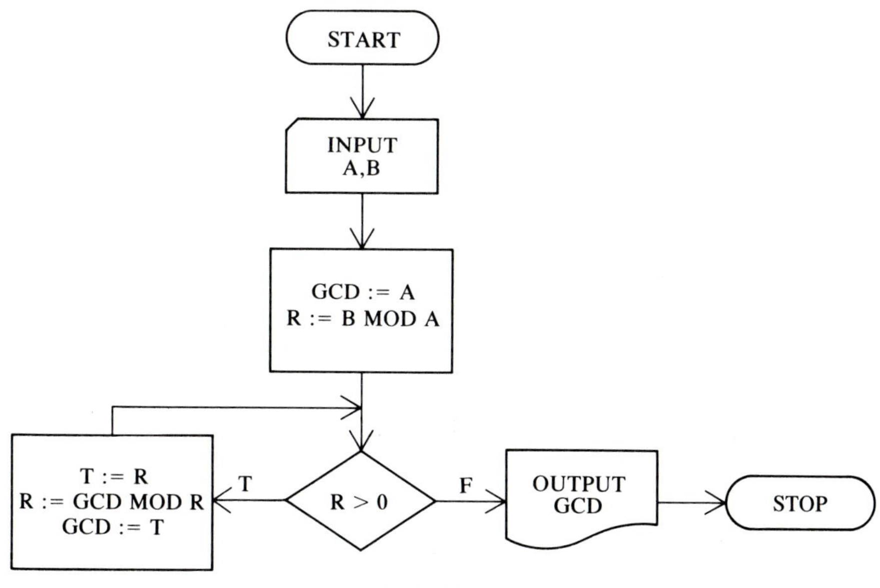

FIGURE 3.4

CHAPTER PROBLEMS

1. Let a, b, and c be integers with $a \neq 0$. Prove: If $a \mid b$ and $a + b = c$, then $a \mid c$.

2. Show that for any positive integer n,
a. $2 \mid (n^2 - n)$ **b.** $4 \nmid (n^2 + 1)$ **c.** $6 \mid [n(n + 1)(2n + 1)]$

3. Let m and n be integers. Show that if $m > 1$ and $m \mid (35n + 26)$ and $m \mid (7n + 3)$, then $m = 11$.

4. Show that if n MOD $6 = 5$, then n MOD $3 = 2$. What about the converse?

5. Assume that $n \geq d \geq 1$. Show that $(n - d)$ DIV $d = (n$ DIV $d) - 1$. Does this result hold if n and d are any integers and $d \neq 0$?

6. Let $n = d_m d_{m-1} \cdots d_1 d_0$ be the usual decimal representation of the positive integer n, where d_i is the 10^is digit. Then

$$n = d_m \cdot 10^m + d_{m-1} \cdot 10^{m-1} + \cdots + d_1 \cdot 10 + d_0$$

where each $d_i \in \{0, 1, \ldots, 9\}$ and $d_m \neq 0$. Prove that $9 \mid n$ if and only if $9 \mid (d_0 + d_1 + \cdots + d_m)$.

7. Show that, if the integer n is both a perfect square and a perfect cube (for example, $64 = 8^2 = 4^3$), then n MOD $7 \in \{0, 1\}$.

8. Let a, b, and c be integers with $b \neq 0$. Prove: If $b \mid c$, then $\gcd(a, b) = \gcd(a + c, b)$.

9. Prove: If c is a positive integer, then $\gcd(ca, cb) = c \cdot \gcd(a,b)$.

10. Prove that any positive integer n is uniquely expressible in the form

$$n = 2^m + b_{m-1} \cdot 2^{m-1} + \cdots + b_1 \cdot 2 + b_0$$

where $m \geq 0$ and each $b_i \in \{0, 1\}$. The representation $1b_{m-1} \cdots b_1 b_0$ is called the *binary representation* of n. (For example, the binary representation of 25 is 11001.)

11. Find $d = \gcd(a, b)$ and integers s and t such that $d = as + bt$.
a. $a = 357$, $b = 629$ **b.** $a = 1109$, $b = 4999$

12. How many integers between 1 and $6n$, inclusive, are relatively prime to 6? (Hint: let $U = \{1, 2, \ldots, 6n\}$, $A = \{2, 4, \ldots, 6n\}$, and $B = \{3, 6, \ldots, 6n\}$; find $n(A' \cap B')$.)

13. Prove: If $a \mid c$, $b \mid c$, and $d = \gcd(a, b)$, then $ab \mid cd$. (It follows that if $a \mid c$, $b \mid c$, and a and b are relatively prime, then $ab \mid c$.)

14. Prove: If $n \in \mathbb{N}$, then $\gcd(a, a + n) \mid n$.

15. Prove: If $\gcd(a, b) = d$, then $\gcd(a/d, b/d) = 1$.

16. If S is a subset of $\{1, 2, \ldots, 2m\}$ such that $n(S) = m + 1$, show that S contains two integers a and b that are relatively prime.

17. An equation of the form

$$ax + by = c$$

where a, b, and c are integers, is called a *Diophantine equation;* a solution is a pair of integers x and y that satisfy the equation. Show that such an equation has a solution if and only if $\gcd(a, b) \mid c$.

18. Given an integer $m \geq 2$, define the set E_m by

$$E_m = \{k \mid 1 \leq k \leq m \wedge \gcd(k, m) = 1\}$$

In other words, E_m is the set of positive integers between 1 and m that are relatively prime to m. In this problem we are interested in $n(E_m)$. For example, if $m = 15$, then $E_{15} = \{1, 2, 4, 7, 8, 11, 13, 14\}$ and $n(E_{15}) = 8$. (In number theory, the number $n(E_m)$ is denoted $\Phi(m)$; Φ is called the "Euler phi function.") Let p and q be distinct primes.

a. Find $n(E_p)$.
b. Find $n(E_{p^k})$, where $k \in \mathbb{N}$.
c. Find $n(E_{pq})$.
★d. Let a and b be integers with $a \geq 2$, $b \geq 2$. It can be proven that if a and b are relatively prime, then $n(E_{ab}) = n(E_a) \cdot n(E_b)$. Use this result to give a formula for $n(E_m)$, assuming that $m = p_1^{a_1} p_2^{a_2} \cdots p_k^{a_k}$ is the canonical factorization of m.

19. If a, b, and c are positive integers such that $a^2 + b^2 = c^2$, then a, b, and c are said to form a *Pythagorean triple*. Prove:

a. If a, b, and c form a Pythagorean triple, then a or b is even.
b. If a, b, and c form a Pythagorean triple, and d is a positive integer, then da, db, and dc form a Pythagorean triple.

A Pythagorean triple is said to be *primitive* if $\gcd(a, b) = 1$. (In this case it also happens that $\gcd(a, c) = \gcd(b, c) = 1$.)

c. Develop an algorithm that finds all primitive Pythagorean triples with $1 < a < b < c < 100$.

20. Given three integers a, b, and c, not all zero, we may define their *greatest common divisor* to be the largest common divisor of all three. Denote the greatest common divisor of a, b, and c by $\gcd(a, b, c)$.

a. Show that $\gcd(a, b, c) = \gcd(\gcd(a, b), c)$.
b. Show that $\gcd(a, b, c)$ is the smallest positive integer d for which there exist integers x, y, and z such that $d = ax + by + cz$.
c. If $\gcd(a, b, c) = 1$, does it follow that a, b, and c are pairwise relatively prime?
d. Give a formula for $\gcd(a, b, c)$ in terms of the canonical factorizations of a, b, and c where $2 \leq a \leq b \leq c$.

21. Let x be any integer and let $A = \{0, 1, \ldots, 6\}$.

a. Show that $\{x \text{ MOD } 7, (x + 3) \text{ MOD } 7, \ldots, (x + 3^6) \text{ MOD } 7\} = A$.
b. Show that $\{x \text{ MOD } 7, (x + 2) \text{ MOD } 7, \ldots, (x + 2^6) \text{ MOD } 7\} \neq A$.

22. Prove: If $ac \text{ MOD } n = bc \text{ MOD } n$ and $d = \gcd(c, n)$, then $a \text{ MOD } (n/d) = b \text{ MOD } (n/d)$. (Hence if c and n are relatively prime, then $a \text{ MOD } n = b \text{ MOD } n$.)

23. Let a and b be integers and let p be prime.

a. Prove: If a^2 MOD $p = b^2$ MOD p, then either a MOD $p = b$ MOD p or a MOD $p = -b$ MOD p.

b. Give an example to show the necessity of the requirement that p be prime in part a.

24. Prove: If a MOD $n = b$ MOD n and $m \mid n$, then a MOD $m = b$ MOD m.

25. Prove: If a MOD $n = b$ MOD n and $d > 0$ divides each of a, b, and n, then

$$\frac{a}{d} \text{ MOD } \frac{n}{d} = \frac{b}{d} \text{ MOD } \frac{n}{d}$$

26. Let $a = p_1^{a_1}p_2^{a_2} \cdots p_m^{a_m}$ and $b = q_1^{b_1}q_2^{b_2} \cdots q_n^{b_n}$ be the canonical factorizations of a and b. What conditions must be satisfied by the exponents a_i and b_i if

a. a is a perfect square?

b. b is a perfect cube?

c. $a \mid b$?

27. Prove by induction that the following statements hold for all $n \in \mathbb{N}$.

a. $1(1!) + 2(2!) + \cdots + n(n!) = (n + 1)! - 1$

b. 3 divides $2^{2n} - 1$

c. $1^2 - 2^2 + 3^2 - 4^2 + \cdots + (-1)^{n+1}n^2 = \dfrac{(-1)^{n+1}n(n + 1)}{2}$

d. $\dfrac{1}{1 \cdot 2} + \dfrac{1}{2 \cdot 3} + \cdots + \dfrac{1}{n(n + 1)} = \dfrac{n}{n + 1}$

e. $1 \cdot 2 + 2 \cdot 3 + \cdots + n(n + 1) = \dfrac{n(n + 1)(n + 2)}{3}$

f. $\dfrac{1}{\sqrt{1}} + \dfrac{1}{\sqrt{2}} + \cdots + \dfrac{1}{\sqrt{n}} \leq 2\sqrt{n} - 1$

28. Use induction to prove DeMorgan's laws for sets $A_1, A_2, \ldots, A_n$:

a. $\left(\bigcup_{i=1}^{n} A_i\right)' = \bigcap_{i=1}^{n} A_i'$ **b.** $\left(\bigcap_{i=1}^{n} A_i\right)' = \bigcup_{i=1}^{n} A_i'$

29. Let $A_1, A_2, \ldots, A_m$ be finite, nonempty subsets of a universal set U. Use induction to prove the extended form of the multiplication principle:

$$n(A_1 \times A_2 \times \cdots \times A_m) = n(A_1)n(A_2) \cdots n(A_m)$$

30. Use induction to prove the extended version of Euclid's lemma: If a_1, $a_2, \ldots, a_n$ are integers and p is a prime such that $p \mid (a_1a_2 \cdots a_n)$, then $p \mid a_i$ for some i, $1 \leq i \leq n$.

31. Let a and d be real numbers. Use induction to prove that

$$a + (a + d) + (a + 2d) + \cdots + (a + nd) = \frac{(n + 1)(2a + nd)}{2}$$

32. Use induction to prove that the number of primes is infinite. (Hint: for each positive integer n, let $P(n)$ be the statement, "There exist at least n primes.")

33. The following argument purports to prove that for any $n \in \mathbb{N}$ and for any set S of n cars, all cars in S get the same average number of miles per gallon of fuel. What is the flaw in the argument?

1. Clearly, the result holds when $n = 1$.
2. Assume the result holds when $n = k$ for some arbitrary $k \geq 1$. In other words, all cars in any set of k cars get the same average mpg.
3. Show the result holds when $n = k + 1$, that is, show all cars in any set of $k + 1$ cars get the same average mpg. Let S be a set of $k + 1$ cars. Choose subsets A and B of S such that $n(A) = n(B) = k$, $A \cap B \neq \phi$, and $A \cup B = S$. Then, by the induction hypothesis, all cars in A get the same average mpg and all cars in B get the same average mpg. Since there is at least one car in $A \cap B$, it follows that all cars in $S = A \cup B$ must get the same average mpg. Thus the result holds for $n = k + 1$ and therefore, by the PMI, the result holds for every $n \in \mathbb{N}$.

Four Relations

4.1 INTRODUCTION

At this stage in our development the student should have achieved some familiarity with the rudiments of set theory and the divisibility properties of integers. In this chapter we will make use of the language of set theory to define the notion of a relation between two sets.

A rather nice example of a relation is provided by the idea of divisibility for positive integers. Given two positive integers a and b, we can ask whether a divides b; if so, we say that "a is related to b." It will turn out that "divides" is a relation from $\mathbb{N}$ to $\mathbb{N}$. Notice that it is possible that a is related to b but that b is not related to a; this will happen, in particular, if $a = 2$ and $b = 6$. Because of this, it makes good sense to single out the ordered pair (a,b) if $a \mid b$. Thus the relation determines a unique set of ordered pairs, namely, $R = \{(a,b) \mid a, b \in \mathbb{N} \wedge a \mid b\}$.

As another example, consider the sets P and L of all points and lines, respectively, in a given plane. Given a point p and a line l, we can ask whether p is incident with l (p is on l). Then incidence will turn out to be a relation from P to L; in this case the relation determines the set $\{(p,l) \mid p \in P \wedge l \in L \wedge p \in l\}$.

What should be clear from these examples is that a relation from a set A to a set B determines a unique subset of the Cartesian product $A \times B$. This discussion indicates, in a certain sense, the general definition of the term "relation."

DEFINITION 4.1

A *relation* from a set A to a set B is defined to be any subset of $A \times B$. If R is a relation from A to B and $(a,b) \in R$, then we say that *a is related to b* and we write $a \ R \ b$.

Since the preceding definition involves two sets A and B, we sometimes refer to such a relation as a *binary relation*.

Example 4.1 Let

$$A = \{\text{Brinkerhoff, Chan, McKenna, Slonneger, Will, Yellen}\}$$

be the set of computer science instructors at a small college, and let

$$B = \{\text{CS105, CS260, CS261, CS360, CS450, CS460}\}$$

be the set of computer science courses offered next semester at that college. Then $A \times B$ gives all possible pairings of instructors and courses. Let the relation R from A to B be given by

$$\begin{aligned} R = \{&(\text{Brinkerhoff,CS105}), (\text{Brinkerhoff,CS360}),\\ &(\text{Chan,CS260}), (\text{Chan,CS360}), (\text{Chan,CS460}),\\ &(\text{McKenna,CS105}), (\text{McKenna,CS260}),\\ &(\text{Slonneger,CS260}), (\text{Slonneger,CS261}), (\text{Will,CS105}),\\ &(\text{Yellen,CS261}), (\text{Yellen,CS450})\}\end{aligned}$$

Then R might tell us, for example, which instructors are assigned to teach which courses. ■

Example 4.2 Let P be the set of primes. Define a relation R from P to $\mathbb{N}$ by

$$p \; R \; n \leftrightarrow p \mid n$$

Find (a) all primes p such that $p \; R \; 126$; (b) all n such that $3 \; R \; n$.

Solution Since $126 = (2)(3)(3)(7)$, for (a) we have that $2 \; R \; 126$, $3 \; R \; 126$, and $7 \; R \; 126$. For (b), if $3 \; R \; n$ then $3 \mid n$; hence n is a multiple of 3. Conversely, if $3 \mid n$, then $3 \; R \; n$. Thus the set of positive integers related to 3 under R is $\{3, 6, 9, \ldots\}$. ■

Example 4.3 Define a relation F from $\mathbb{R}$ to $\mathbb{R}$ by $x \; F \; y \leftrightarrow x^2 + y^2 \leq 4$. Then F consists of all points (x,y) in the coordinate plane such that $x^2 + y^2 \leq 4$. Notice that F consists of all points that lie on or within the circle of radius 2, centered at the origin, as shown in Figure 4.1.

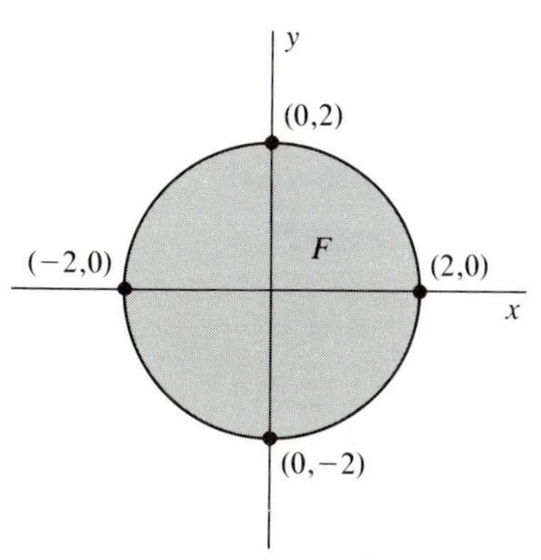

FIGURE 4.1 The relation $F = \{(x,y) \mid x^2 + y^2 \leq 4\}$ ■

Suppose that A and B are finite sets, with $A = \{a_1, a_2, \ldots, a_m\}$ and $B = \{b_1, b_2, \ldots, b_n\}$. Let R be a relation from A to B. One way to represent R is to form a table having m rows and n columns, a row for each element of A and a column for each element of B. Label the rows a_1 through a_m and the columns b_1 through b_n. If $a_i \, R \, b_j$, then place a "1" in the position of the table corresponding to row a_i and column b_j; if it is not the case that $a_i \, R \, b_j$, then place a "0" in that position. For example, the table representing the relation of Example 4.1 is shown in Figure 4.2.

	CS105	*CS260*	*CS261*	*CS360*	*CS450*	*CS460*
Brinkerhoff	1	0	0	1	0	0
Chan	0	1	0	1	0	1
McKenna	1	1	0	0	0	0
Slonneger	0	1	1	0	0	0
Will	1	0	0	0	0	0
Yellen	0	0	1	0	1	0

FIGURE 4.2 Table representation of the relation of Example 4.1

Another way to represent a relation R from a finite set A to a finite set B makes use of a very fertile and popular area of mathematics called *graph theory*. We shall have more to say about this subject in Chapter 7; however, for the moment we wish to make use of what is called a *bipartite graph*. Let the sets A and B be as just mentioned. Start with two rows of points, one consisting of m points that are labelled by the elements of A, the other consisting of n points labelled by the elements of B. We refer to these points as *vertices*. Next, if two elements a_i and b_j are related, then the corresponding vertices are joined by a line segment, called an *edge*. As an example, in Figure 4.3 we show the bipartite graph that represents the relation R of Example 4.1.

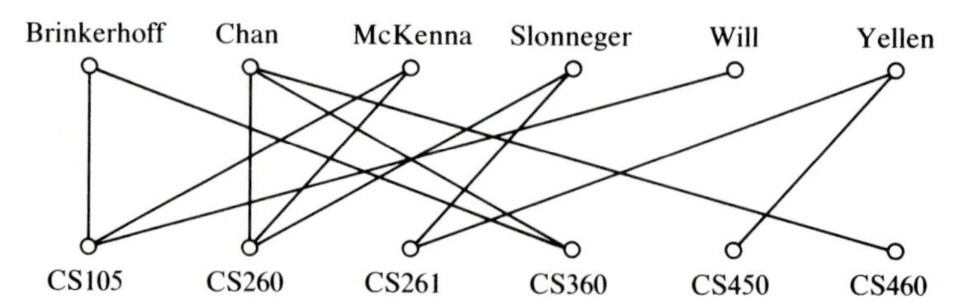

FIGURE 4.3 Bipartite graph representation of the relation of Example 4.1.

If R is a relation from a set A to a set B, then we know that R is some subset of $A \times B$. Thus R determines a subset of A, namely, the set of all first coordinates of the ordered pairs of R. Similarly, a subset of B is determined by R. These subsets, formally defined in Definition 4.2, will be discussed in some detail in Chapter 5.

DEFINITION 4.2

Let R be a relation from the set A to the set B. The *domain* of R is the set *dom R* defined by

$$\text{dom } R = \{a \in A \mid (a,b) \in R \text{ for some } b \in B\}$$

The *image* (or *range*) of R is the set *im R* defined by

$$\text{im } R = \{b \in B \mid (a,b) \in R \text{ for some } a \in A\}$$

Example 4.4 In Example 4.1, dom $R = A$ and im $R = B$. In Example 4.2, dom $R = P$ and im $R = \mathbb{N} - \{1\}$. As another example, let A be the set of undergraduates enrolled this semester at your school and let B be the set of undergraduate mathematics courses that your school offers. Define R from A to B by $a\ R\ b \leftrightarrow$ student a is taking course b this semester. Then dom R consists only of those students in A who are taking a mathematics course this semester. Since it is almost certainly the case that some students are not taking any mathematics courses, it is probably true that dom R is a proper subset of A. Similarly, im R consists only of those mathematics courses being offered this semester. If it is the case that not every undergraduate mathematics course is offered in a given semester, then im R is a proper subset of B. ■

Exercises 4.1

1. Each of the following parts defines a relation F from $\mathbb{R}$ to $\mathbb{R}$. Picture each relation as a subset of $\mathbb{R} \times \mathbb{R}$, using a coordinate plane.
   a. $F = \{(x,y) \mid y = 2x + 1\}$ **b.** $F = \{(x,y) \mid y \leq 2x + 1\}$
   c. $F = \{(x,y) \mid y = x^2\}$ **d.** $F = \{(x,y) \mid 4 > y > x^2\}$
   e. $F = \{(x,y) \mid |x| + |y| = 4\}$ **f.** $F = \{(x,y) \mid x > 1 \wedge y < 2\}$
2. Let C be the set of classes you are taking this semester and let D be the

set of days on which classes are held at your school. Define a relation R from C to D by $c\ R\ d \leftrightarrow$ class c meets on day d.

a. List the ordered pairs in R.
b. Give the table representation of R.
c. Represent R as a bipartite graph.
d. Find dom R and im R.

3. Let $A = \{a_1, a_2, \ldots, a_m\}$ and $B = \{b_1, b_2, \ldots, b_n\}$ be finite sets. A relation R from A to B can be represented in Pascal as a two-dimensional Boolean array; such an array can be declared as follows:

```
CONST M = 3; N = 4;   (*or any positive integers*)
TYPE RELATION = ARRAY[1..M,1..N] OF BOOLEAN;
VAR R : RELATION;
```

The idea is that R[I,J] = TRUE if and only if $a_i\ R\ b_j$.

Suppose that $A = \{1, 2, 3\}$ and $B = \{1, 2, 3, 4\}$. Give the representation of R if

a. $R = \{(1,1), (1,3), (2,4), (3,1), (3,2), (3,4)\}$
b. $R = \phi$, that is, R is the *empty relation*
c. $R = A \times B$
d. $R = \{(1,1), (2,2), (3,3)\}$
e. $R = \{(x,y) \mid x \in A \wedge y \in B \wedge x \neq y\}$
f. $R = \{(x,y) \mid x \in A \wedge y \in B \wedge x < y\}$

4.2 EQUIVALENCE RELATIONS

One type of relation that is encountered frequently in mathematics is a relation from a set A to itself. We have already seen an example of such a relation, namely, the relation divides from $\mathbb{N}$ to $\mathbb{N}$. Since many important and well-used relations are of this type, they are referred to in a special way. We present this in the following definition, along with several key properties that these relations may possess.

DEFINITION 4.3

Given a set A, a relation R from A to A will be termed a *relation on* A. If R is a relation on A, then

1. R is *reflexive* if $a\ R\ a$ for all $a \in A$.
2. R is *symmetric* if $a\ R\ b \rightarrow b\ R\ a$ for all $a, b \in A$.
3. R is *antisymmetric* if $(a\ R\ b \wedge b\ R\ a) \rightarrow a = b$ for all $a, b \in A$.
4. R is *transitive* if $(a\ R\ b \wedge b\ R\ c) \rightarrow a\ R\ c$ for all $a, b, c \in A$.

In terms of ordered pairs, it should be noted that a relation R on A is

1. reflexive $\leftrightarrow (a,a) \in R$ for all $a \in A$;
2. symmetric $\leftrightarrow (a,b) \in R \rightarrow (b,a) \in R$ for all $a,b \in A$;
3. antisymmetric $\leftrightarrow ((a,b) \in R \wedge (b,a) \in R) \rightarrow a = b$ for all $a,b \in A$;
4. transitive $\leftrightarrow ((a,b) \in R \wedge (b,c) \in R) \rightarrow (a,c) \in R$ for all $a,b,c \in A$.

Example 4.5 Define the relation R on $\mathbb{Z}$ by

$$a \; R \; b \leftrightarrow 3 \mid (a - b)$$

Show that R is reflexive, symmetric, and transitive.

Solution Since $3 \mid (a - a)$ for any integer a, the relation R is reflexive. For all $a,b \in \mathbb{Z}$, if $3 \mid (a - b)$, then $3 \mid (b - a)$. It follows that R is symmetric. Also, by Theorem 3.2, part 3, if $3 \mid (a - b)$ and $3 \mid (b - c)$, then 3 divides $(a - b) + (b - c) = a - c$. This shows that R is transitive. The relation R is not antisymmetric; for instance, $6 \; R \; 9$ and $9 \; R \; 6$, but $6 \neq 9$. ■

Example 4.6 Let Y be a nonempty set and define the relation R on the power set of Y by the following rule:

$$\text{For } A, B \in \mathcal{P}(Y), \quad A \; R \; B \leftrightarrow A \subseteq B.$$

Show that R is reflexive, antisymmetric, and transitive, but not symmetric.

Solution Since $A \subseteq A$ for every $A \in \mathcal{P}(Y)$, the relation R is reflexive. For all $A, B \in \mathcal{P}(Y)$, if $A \subseteq B$ and $B \subseteq A$, then $A = B$. This shows that R is antisymmetric. Also, by Theorem 2.1, if $A, B, C \in \mathcal{P}(Y)$ with $A \subseteq B$ and $B \subseteq C$, then $A \subseteq C$. It follows that R is transitive. The relation R is not symmetric, for if $A, B \in \mathcal{P}(Y)$ and A is a proper subset of B, then A is related to B but B is not related to A. ■

Example 4.7 Consider the relation R defined from A = {Brinkerhoff, Chan, McKenna, Slonneger, Will, Yellen} to B = {CS105, CS260, CS261, CS360, CS450, CS460} defined in Example 4.1. Define a relation T on A by agreeing that two instructors who teach different sections of the same course will be related. The relation T is symmetric but, in this case, is not transitive. This is because Brinkerhoff is related to Chan (both teach CS360) and Chan is related to Slonneger (both teach CS260), but Brinkerhoff is not related to Slonneger. ■

A relation R on a finite set A can be represented geometrically by using the following scheme. Each element of A is represented by a point in the plane. Further, if $a_1 \; R \; a_2$ for some $a_1, a_2 \in A$, where $a_1 \neq a_2$, then a directed line segment is drawn from a_1 to a_2, as illustrated in Figure 4.4(a). If $a \; R \; a$ for some $a \in A$, then a directed simple closed curve,

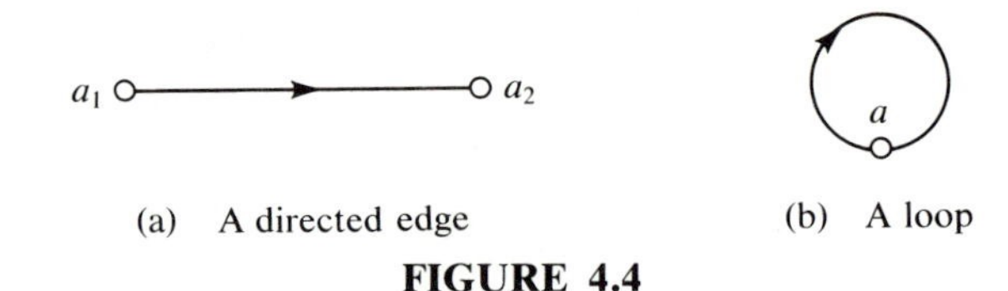

(a) A directed edge (b) A loop

FIGURE 4.4

called a *loop,* is drawn from a to a, as shown in Figure 4.4(b). The resulting structure is referred to as a *directed graph;* the points are called *vertices* and the directed line segments and loops are called *directed edges* (or *arcs*).

Example 4.8 Consider the relation R defined on $\mathcal{P}(Y)$, where $Y = \{1, 2\}$, by $A\ R\ B \leftrightarrow A \subseteq B$. This is just a special case of the relation considered in Example 4.6. The directed graph for this relation is shown in Figure 4.5(a). Notice that since R is reflexive, there is a loop at each vertex. When the relation under consideration is understood to be reflexive, it is customary to draw its directed graph without loops, as shown in Figure 4.5(b).

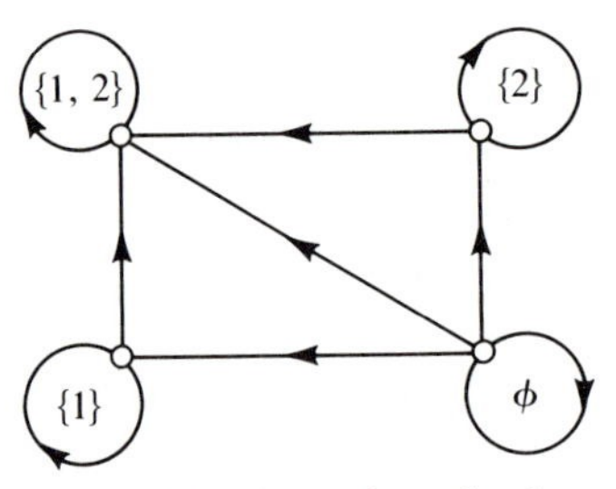

(a) The directed graph of the relation R of Example 4.8

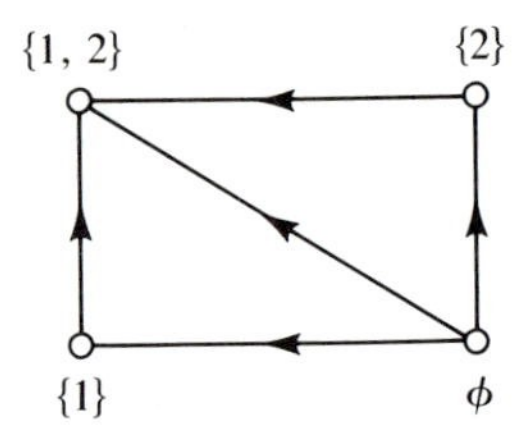

(b) The directed graph of R, drawn with the understanding that R is reflexive

FIGURE 4.5 ■

If a relation R on a finite set A is symmetric, then for each pair of distinct, related elements a_1 and a_2 of A, the associated directed graph contains both a directed edge from a_1 to a_2 and one from a_2 to a_1. In this case it is customary to replace these two directed edges by a single (undirected) edge, as shown in Figure 4.6. The resulting structure is called a *graph*. The concepts of a graph and a directed graph will be more formally defined in a later chapter.

a_1 ——— a_2

FIGURE 4.6 An (undirected) edge

Example 4.9 Let $X = \{1, 2, 3, 4, 5, 6\}$ and let R be the relation on X defined by

$$x_1 \, R \, x_2 \leftrightarrow \gcd(x_1, x_2) = 1$$

Then R is symmetric and the graph of R is given in Figure 4.7.

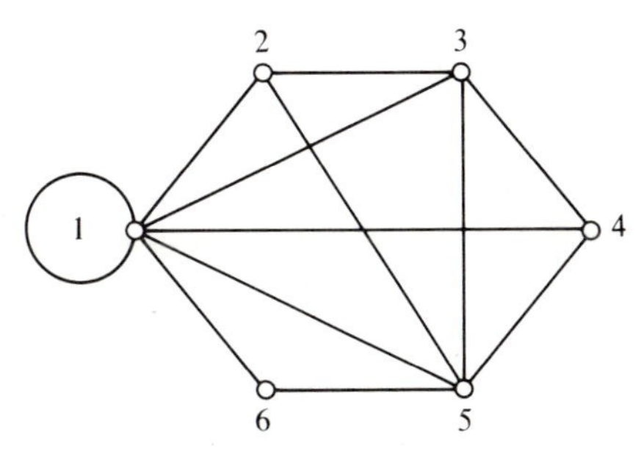

FIGURE 4.7 Representing a relation using a graph ■

A relation R on a set A may have one or more of the properties of being reflexive, symmetric, antisymmetric, or transitive. Certain combinations of these properties lead to some important kinds of relations for which a general theory has been developed. For example, each of the following relations is reflexive, symmetric, and transitive:

1. "equals" on $\mathbb{Z}$
2. "having the same birthday" on a set of people
3. "congruence" on the set of triangles in a plane

DEFINITION 4.4

A relation R on a set A is called an *equivalence relation* if it is reflexive, symmetric, and transitive.

Example 4.10 Let C be the set of currently enrolled undergraduate computer science majors at a certain university. Suppose that two students are related if they have the same faculty advisor. Then it is easy to verify that this relation is reflexive, symmetric, and transitive and hence is an equivalence relation. ■

Example 4.11 Let $S = \{1, 2, 3\}$. For each of the following relations on S, determine whether it is an equivalence relation.

(a) $R_1 = \{(1,2), (2,3), (1,3), (3,2), (2,1), (3,1)\}$
(b) $R_2 = \{(1,1), (2,2), (3,3), (1,2), (2,1), (2,3), (3,2)\}$
(c) $R_3 = \{(1,1), (2,2), (3,3), (1,2), (2,3), (1,3)\}$
(d) $R_4 = \{(1,1), (2,2), (3,3), (1,2), (2,1)\}$

Solution (a) It is clear that R_1 is symmetric but not reflexive. It is also not transitive. For let $a = 1$, $b = 2$, and $c = 1$ in the definition. Then we have $a\ R_1\ b$ and $b\ R_1\ c$, but a is not related to c under R_1.
(b) It is evident that R_2 is both reflexive and symmetric. However, R_2 is not transitive, for 1 is related to 2 and 2 is related to 3, but 1 is not related to 3.
(c) The relation R_3 is reflexive and transitive, but it is not symmetric.
(d) The relation R_4 is reflexive, symmetric, and transitive. Hence, of these four relations, only R_4 is an equivalence relation on S. ■

Consider the relation R of Example 4.5, where for integers a and b, we had $a\ R\ b \leftrightarrow 3 \mid (a - b)$. If $a\ R\ b$, then there is some integer n such that $3n = a - b$. Now suppose that b MOD 3 $= r$ (recall that $r \in \{0, 1, 2\}$). Then there is some integer q such that $b = 3q + r$. Since $b = a - 3n$,

$$a - 3n = 3q + r$$

so

$$\begin{aligned} a &= 3q + 3n + r \\ &= 3(q + n) + r \end{aligned}$$

This shows that a MOD 3 $= r$. Hence, if $a\ R\ b$, then a MOD 3 $= b$ MOD 3. In other words, a and b yield the same remainder upon division by 3. We shall say that "a is congruent to b modulo 3" and, for this particular relation, denote $a\ R\ b$ by the notation $a \equiv b(\text{mod } 3)$. This relation is called "congruence modulo 3." There is nothing special about the use of the integer 3 in this example; we can just as well use any positive integer m.

DEFINITION 4.5

Let m be any positive integer. Given integers a and b, we say that a *is congruent to b modulo m*, denoted $a \equiv b(\text{mod } m)$, provided $m \mid (a - b)$. This relation on $\mathbb{Z}$ is referred to as *congruence modulo m*.

THEOREM 4.1 For any $m \in \mathbb{N}$, the relation of congruence modulo m is an equivalence relation on $\mathbb{Z}$. ■

We leave the proof of this theorem to Exercise 2. The student should note that if a MOD $m = r$, then $a \equiv r(\text{mod } m)$, but not conversely. The relation of congruence modulo m gives rise to some interesting results in number theory and provides a number of very useful and fundamental examples in the area of modern algebra.

Consider once again the relation of congruence modulo 3 on $\mathbb{Z}$. The very use of the word "relation" suggests asking for the set of all "relatives" of a given fixed integer t; that is, we ask for the set of all those integers a for which $a \equiv t(\text{mod } 3)$. For any integer a, it has already been noted that a MOD 3 is exactly one of 0, 1, or 2, so it seems reasonable to consider the set of relatives of each of 0, 1, and 2. In particular, let

$$[0] = \{a \in \mathbb{Z} \mid a \equiv 0(\text{mod } 3)\}$$
$$[1] = \{a \in \mathbb{Z} \mid a \equiv 1(\text{mod } 3)\}$$

and

$$[2] = \{a \in \mathbb{Z} \mid a \equiv 2(\text{mod } 3)\}$$

What are the elements of [0]? This is easily determined in the following string:

$$a \in [0] \leftrightarrow a \equiv 0(\text{mod } 3) \leftrightarrow 3 \mid a \leftrightarrow a = 3k \text{ for some } k \in \mathbb{Z}$$

Thus we see that $[0] = \{3k \mid k \in \mathbb{Z}\}$. Similarly, we can obtain

$$[1] = \{3k + 1 \mid k \in \mathbb{Z}\} \qquad \text{and} \qquad [2] = \{3k + 2 \mid k \in \mathbb{Z}\}$$

There are some relevant facts that should be observed about the sets [0], [1], and [2]:

1. [0], [1], and [2] are each nonempty.
2. [0], [1], and [2] are pairwise disjoint.
3. $[0] \cup [1] \cup [2] = \mathbb{Z}$.

Facts 2 and 3 follow from the division algorithm, since each $a \in \mathbb{Z}$ is uniquely expressible in the form $a = 3k + r$, where $r = 0$, 1, or 2.

The preceding discussion can be applied to any equivalence relation.

DEFINITION 4.6

Let R be an equivalence relation on a set A. For each $a \in A$, the *equivalence class* of a is the set

$$[a] = \{x \in A \mid x \mathrel{R} a\}$$

Example 4.12 Consider again the equivalence relation of Example 4.10. Let F be the set of computer science faculty advisors, and for $f \in F$, let C_f denote the set of students in C who have f as an advisor. If $c \in C$ is a student, then $[c]$ is the set of students in C who have the same advisor as c; thus if c has f for an advisor, then $[c] = C_f$. It follows that the sets C_f, $f \in F$, are the equivalence classes of this relation. Note that, since each student has a unique advisor, the sets C_f are pairwise disjoint and

$$\bigcup_{f \in F} C_f = C$$ ■

Example 4.13 (a) For the equivalence relation R_4 of Example 4.11, $[1] = [2] = \{1, 2\}$, and $[3] = \{3\}$.
(b) Define the relation R on $S = \{1, 2, 3, 4, 5\}$ by

$$S = \{(1,1), (2,2), (3,3), (4,4), (5,5), (1,2), (2,1), (1,3), (3,1), (2,3), (3,2), (4,5), (5,4)\}$$

Then it can be verified that R is an equivalence relation on S, where $[1] = [2] = [3] = \{1, 2, 3\}$ and $[4] = [5] = \{4, 5\}$.
(c) Define the relation R on $\mathbb{Z} - \{0\}$ by $a\ R\ b \leftrightarrow ab > 0$. Then R is an equivalence relation, $[1] = [2] = \cdots = \mathbb{N}$, and $[-1] = [-2] = \cdots = \{-1, -2, -3, \ldots\}$. ■

In each part of Example 4.13, we are given an equivalence relation R on a set A and we obtain (from the equivalence classes) a pairwise disjoint collection of nonempty subsets of A whose union is A. Such a collection is given a special name.

DEFINITION 4.7

A set $\mathcal{P}$ of subsets of a set A is called a *partition* of A if the following conditions hold:

1. $B \in \mathcal{P} \rightarrow B \neq \phi$
2. $B, C \in \mathcal{P} \rightarrow (B = C \vee B \cap C = \phi)$
3. $\bigcup_{B \in \mathcal{P}} B = A$

Example 4.14 (a) The relation congruence modulo 3 on $\mathbb{Z}$ yields the partition $\mathcal{P} = \{[0], [1], [2]\}$ of $\mathbb{Z}$, where $[r] = \{3k + r \mid k \in \mathbb{Z}\}$.
(b) The relation R of Example 4.13(b) yields the partition $\mathcal{P} = \{\{1, 2, 3\}, \{4, 5\}\}$ of S.
(c) The relation R of Example 4.13(c) yields the partition of $\mathbb{Z} - \{0\}$ into the set of positive integers and the set of negative integers. ■

Example 4.15 The set of computer science courses offered at a certain college can be partitioned according to the level of the course as follows:

A = the set of 100-level courses = {CS105, CS125}
B = the set of 200-level courses = {CS205, CS260, CS261, CS265}
C = the set of 300-level courses = {CS340, CS350, CS360, CS361, CS380}
D = the set of 400-level courses = {CS400, CS450, CS460, CS480}

We can define a relation R on the set of computer science courses by saying that two courses at the same level are related. Then it can easily be verified that R is an equivalence relation and that A, B, C, and D are the equivalence classes of R. ■

As defined, the notions of equivalence relation and partition seem to be quite different, but as we have seen in the preceding examples, they are very much related. The next result reveals this relationship.

THEOREM 4.2 **Fundamental theorem on equivalence relations** If R is an equivalence relation on a set A, then the set

$$\mathcal{P} = \{[a] \mid a \in A\}$$

is a partition of A. Conversely, if $\mathcal{P}$ is a partition of A, then the relation R, defined on A by

$$a\ R\ b \leftrightarrow \exists C \in \mathcal{P}(a \in C \wedge b \in C)$$

is an equivalence relation on A.

Proof We first prove that if R is an equivalence relation on A, then the collection $\mathcal{P}$ as defined in the statement of the theorem is a partition of A. First, it is clear that $[a]$ is nonempty for every $a \in A$; in particular, $a \in [a]$ since R is reflexive. Next, suppose that $[a]$, $[b] \in \mathcal{P}$ and that $[a]$ and $[b]$ are not disjoint; then we must show that $[a] = [b]$. This is accomplished by showing both $[a] \subseteq [b]$ and $[b] \subseteq [a]$, and since the proofs of both inclusions are the same, we consider only the proof that $[a] \subseteq [b]$. Let $x \in [a]$. Since $[a] \cap [b] \neq \phi$, there is some element $c \in A$ such that $c \in [a]$ and $c \in [b]$. Since R is an equivalence relation, it follows that $x\ R\ a$, $a\ R\ c$, and $c\ R\ b$. But $x\ R\ a$ and $a\ R\ c$ imply that $x\ R\ c$, and then $x\ R\ c$ and $c\ R\ b$ imply that $x\ R\ b$. Therefore $x \in [b]$, showing that $[a] \subseteq [b]$. So $\mathcal{P}$ is a pairwise disjoint collection. Lastly, if $a \in A$ then $a \in [a]$, so that $A \subseteq \bigcup_{c \in A} [c]$, and it follows that $A = \bigcup_{c \in A} [c]$. Thus $\mathcal{P}$ is a partition of A.

Next, suppose $\mathcal{P}$ is a partition of A and the relation R is defined as in the statement of the theorem. It must be shown that R is an equivalence relation. To see that R is reflexive, let $a \in A$. Since $\mathcal{P}$ is a partition of A, there is some $C \in \mathcal{P}$ such that $a \in C$. Thus $a\ R\ a$. For the symmetric property, let $a, b \in A$ and assume $a\ R\ b$. Then there is some $C \in \mathcal{P}$ such that $a \in C$ and $b \in C$. But this is also the condition for $b\ R\ a$, so R is symmetric. Finally, suppose for some $a, b, c \in A$, that $a\ R\ b$ and $b\ R\ c$.

Then there are sets $D, E \in \mathcal{P}$ such that $a, b \in D$ and $b, c \in E$. Then $b \in D \cap E$, so D and E are not disjoint. Since $\mathcal{P}$ is a partition, it follows that $D = E$ and hence that $a\ R\ c$. This shows that R is transitive and, therefore, that R is an equivalence relation. ∎

To repeat, if R is an equivalence relation, then the set

$$\mathcal{P} = \{[a] \mid a \in A\}$$

of equivalence classes of R is a partition of A. Conversely, given a partition $\mathcal{C}$ of a set A, the partition $\mathcal{C}$ determines an equivalence relation R on A, where $a\ R\ b$ if and only if a and b belong to the same set in $\mathcal{C}$. We say that $\mathcal{C}$ *induces* an equivalence relation on A. Thus, in a sense, the notions of equivalence relation and partition are the same.

Example 4.16 Each part gives a partition of the set $A = \{1, 2, 3, 4, 5, 6\}$. Draw the graph of the equivalence relation on A induced by the partition. (Since an equivalence relation is understood to be reflexive, omit the loop at each vertex.)

(a) $\mathcal{P}_1 = \{\{1, 2\}, \{3, 4\}, \{5, 6\}\}$
(b) $\mathcal{P}_2 = \{\{1\}, \{2\}, \{3, 4, 5, 6\}\}$
(c) $\mathcal{P}_3 = \{\{1, 2, 3\}, \{4, 5, 6\}\}$

Solution The graphs of $\mathcal{P}_1$, $\mathcal{P}_2$, and $\mathcal{P}_3$ are shown in Figure 4.8 (a), (b), and (c), respectively.

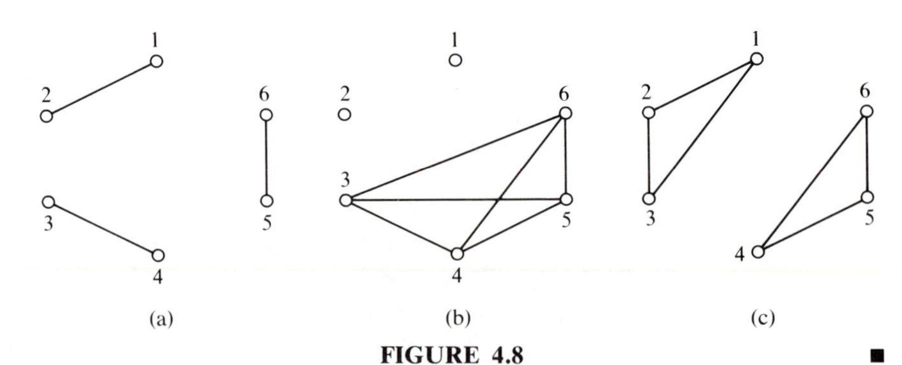

FIGURE 4.8 ∎

In the special case of congruence modulo m on $\mathbb{Z}$, the implied partition is $\{[a] \mid a \in \mathbb{Z}\}$, where $[a] = \{c \in \mathbb{Z} \mid c \equiv a(\text{mod } m)\}$. We shall call $[a]$ the *residue class* of a: here residue means remainder. One obvious question that arises is, What are the residue classes of congruence modulo m? For any $a \in \mathbb{Z}$, we know that a MOD $m = r$, where $0 \leq r \leq m - 1$, so $a \in [r]$. Thus every integer belongs to exactly one of the residue classes $[0], [1], \ldots, [m - 1]$, and these residue classes are distinct (hence pairwise disjoint). Thus $\{[0], [1], \ldots, [m - 1]\}$ is a partition of $\mathbb{Z}$.

DEFINITION 4.8

The set

$$\mathbb{Z}_m = \{[0], [1], \ldots, [m-1]\}$$

is called the *set of residue classes modulo m*.

Given $r \in \mathbb{Z}$, $0 \leq r \leq m - 1$, what do the elements of the residue class $[r]$ look like? We proceed as we did earlier for the special case $m = 3$:

$$\begin{aligned} c \in [r] &\leftrightarrow c \equiv r(\text{mod } m) \\ &\leftrightarrow m \mid (c - r) \\ &\leftrightarrow c - r = mq \text{ for some } q \in \mathbb{Z} \\ &\leftrightarrow c = mq + r \text{ for some } q \in \mathbb{Z} \end{aligned}$$

Hence $[r] = \{mq + r \mid q \in \mathbb{Z}\}$. For example, if $m = 4$, then the residue classes modulo 4 are

$$\begin{aligned} [0] &= \{4q \mid q \in \mathbb{Z}\} = \{\ldots, -8, -4, 0, 4, 8, \ldots\} \\ [1] &= \{4q + 1 \mid q \in \mathbb{Z}\} = \{\ldots, -7, -3, 1, 5, 9, \ldots\} \\ [2] &= \{4q + 2 \mid q \in \mathbb{Z}\} = \{\ldots, -6, -2, 2, 6, 10, \ldots\} \\ [3] &= \{4q + 3 \mid q \in \mathbb{Z}\} = \{\ldots, -5, -1, 3, 7, 11, \ldots\} \end{aligned}$$

With regard to congruence modulo m, it is very important to keep in mind that $[a] = [b] \leftrightarrow m \mid (a - b)$. Thus when $m = 4$, for instance, we see that

$$\cdots = [-6] = [-2] = [2] = [6] = [10] = \cdots$$

Each of the numbers -6, -2, 2, 6, 10, etc., is called a *representative* of the residue class [2]. This is because we may use any one of these integers to refer to the class. In some situation it might be desirable to use $[-6]$ instead of [2]. In general, with respect to congruence modulo m, each of the integers $mq + r$, $q \in \mathbb{Z}$, is a representative of the residue class $[r]$.

Exercises 4.2

1. For each of the following relations R on $\{1, 2, 3, 4, 5\}$, determine whether R is (i) reflexive, (ii) symmetric, (iii) transitive, and (iv) antisymmetric. If R is symmetric, draw the graph that represents R; otherwise draw the directed graph that represents R.

a. $R = \{(1,1), (2,2), (2,3), (3,2), (3,3), (3,4), (4,3), (4,4), (5,5)\}$

b. $R = \{(1,2), (1,4), (1,5), (2,4), (2,5), (3,4), (3,5), (4,5)\}$

c. $R = \{(1,3), (1,5), (2,4), (3,1), (3,5), (4,2), (5,1), (5,3)\}$

d. $R = \{(1,1), (1,3), (1,5), (2,2), (2,4), (3,1), (3,3), (3,5), (4,2), (4,4), (5,1), (5,3), (5,5)\}$

2. Prove Theorem 4.1.

3. For each of the following relations, determine whether it is (i) reflexive, (ii) symmetric, and (iii) transitive.

a. Define R on $\mathbb{N}$ by $a\ R\ b \leftrightarrow \gcd(a, b) = 1$.

b. Define R on $\mathbb{N} - \{1\}$ by $a\ R\ b \leftrightarrow$ there is a prime p that is a factor of both a and b.

c. Define R on the power set of a nonempty set X by $A\ R\ B \leftrightarrow A \cap B = \phi$.

4. Let m be a positive integer. Show that integers a and b are congruent modulo m if and only if a MOD $m = b$ MOD m.

5. a. Define the relation R on $X = \{2, 3, 4, 6, 8, 12, 24\}$ by $a\ R\ b \leftrightarrow a \mid b$. Draw the directed graph that represents R.

b. Let $X = \{1, 2, 3\}$. Draw the graph of the relation R defined in Exercise 3c.

c. Draw the directed graph of the relation R defined on $\mathcal{P}(\{1, 2, 3\})$ by $A\ R\ B \leftrightarrow A \subseteq B$.

d. Draw the graph of the relation defined in Example 4.13(b).

6. Each part below gives a set S and a relation $\sim$ on S ("$a \sim b$" means "a is related to b"). Verify that $\sim$ is an equivalence relation and describe the equivalence classes into which $\sim$ partitions S.

a. S is the set of all polygons: for $A, B \in S$, $A \sim B \leftrightarrow A$ and B have the same number of sides.

b. $S = \mathbb{Z}$ and for $a,b \in S$, $a \sim b \leftrightarrow |a - 2| = |b - 2|$.

c. S is the set of all alumni of your college; for $a,b \in S$, $a \sim b \leftrightarrow a$ and b both graduated the same year.

7. Let A be a nonempty set and let B be a fixed subset of A. Define a relation $\sim$ on $\mathcal{P}(A)$ by

$$\text{For } C, D \in \mathcal{P}(A), \quad C \sim D \leftrightarrow C \cap B = D \cap B$$

a. Show that $\sim$ is an equivalence relation on $\mathcal{P}(A)$.

b. For the particular case where $A = \{1, 2, 3, 4, 5\}$, $B = \{1, 2, 5\}$, and $C = \{2, 4, 5\}$, find $[C]$.

8. Which of the residue classes $[3], [6], [8], [39], [44], [-2], [-3], [-6]$, and $[-12]$ are equal in $\mathbb{Z}_9$?

9. Let S be the set of all points in the xy-plane, excluding the origin. For $r \in (0, \infty)$, let C_r denote the set of all points on the circle of radius r, centered at the origin. Let $\mathscr{C} = \{C_r \mid r \in (0, \infty)\}$.

a. Show that $\mathscr{C}$ is a partition of S.

Let $\sim$ denote the equivalence relation on S induced by $\mathscr{C}$.

b. Is $(2,4) \sim (4,-2)$? **c.** Is $(3,4) \sim (0,-5)$? **d.** Is $(2,3) \sim (1,4)$?

e. State a nice condition under which $(a,b) \sim (c,d)$.

10. How many distinct residue classes are there in $\mathbb{Z}_m$?

4.3 PARTIAL-ORDER RELATIONS

Consider the relation "is a subset of" or simply "$\subseteq$" on the power set $\mathcal{P}(U)$, where $U = \{1, 2, \ldots, 10\}$. This relation is reflexive and transitive, but it is clearly not symmetric. It is antisymmetric since, for any subsets A and B of U, $(A \subseteq B \wedge B \subseteq A) \rightarrow A = B$. Relations that are reflexive, antisymmetric, and transitive play an important role in mathematics and are addressed in this section.

DEFINITION 4.9

A relation R on a set A is called a *partial-order relation* if R is reflexive, antisymmetric, and transitive. We also refer to R as a *partial ordering* of A and call (A, R) a *partially ordered set* (or *poset*).

Example 4.17 Consider the relation divides on $\mathbb{N}$. Show that divides is a partial ordering of $\mathbb{N}$. Is divides a partial ordering of $\mathbb{Z}$?

Solution We must verify that divides on $\mathbb{N}$ is reflexive, antisymmetric, and transitive. The reflexive property is obvious, and antisymmetry and transitivity follow from Theorem 3.2, parts 4 and 2, respectively. The relation divides is not a partial ordering of $\mathbb{Z}$, because it is not antisymmetric; for example, $-2 \mid 2$ and $2 \mid -2$, but $-2 \neq 2$. ■

Perhaps the partial ordering most familiar to the student is the standard ordering "less than or equal to" or "$\leq$" on the set of real numbers. It is the prototype of a partial-order relation and, for this reason, it is common to use the symbol "$\lesssim$" to denote a general partial ordering.

Some authors use the term "ordering" of a set A to mean a relation $\lesssim$ on A that is transitive. If such a relation is also antisymmetric, then given $a \lesssim b$ and $a \neq b$, we shall write "$a < b$" and say *a is less than b* (or *a precedes b*). The term "partial" is used to describe an ordering in which not every two distinct elements are necessarily related.

In Section 4.2 we discussed the use of a directed graph as a means of representing a relation on a finite set A. In case $(A, \lesssim)$ is a poset, this method of representation can be simplified. First of all, since a partial ordering is reflexive, we omit the loop at each vertex. Second, we adopt the convention of omitting any directed edge that is implied by transitivity; in other words, there will be a directed edge from a to b if and only if $a < b$ and there is no $c \in A$ with $a < c$ and $c < b$. Finally, since A is a poset, it is possible to construct its directed graph so that if $a < b$, then the vertex corresponding to b lies above the vertex corresponding to a. For this reason we simply agree to use edges in place of directed edges, understanding that the orientation of all edges is from bottom to top. The resulting representation is then a graph and is called the *Hasse diagram* of the poset.

Example 4.18 (a) The Hasse diagram of the poset $(\mathcal{P}(\{1, 2, 3\}), \subseteq)$ is shown in Figure 4.9(a). Note in this case that the Hasse diagram is the same as the subset diagram of $\mathcal{P}(\{1, 2, 3\})$.
(b) Let A denote the set of positive divisors of 24 and let "|" denote the relation divides. Then the Hasse diagram of $(A, |)$ is given in Figure 4.9(b).

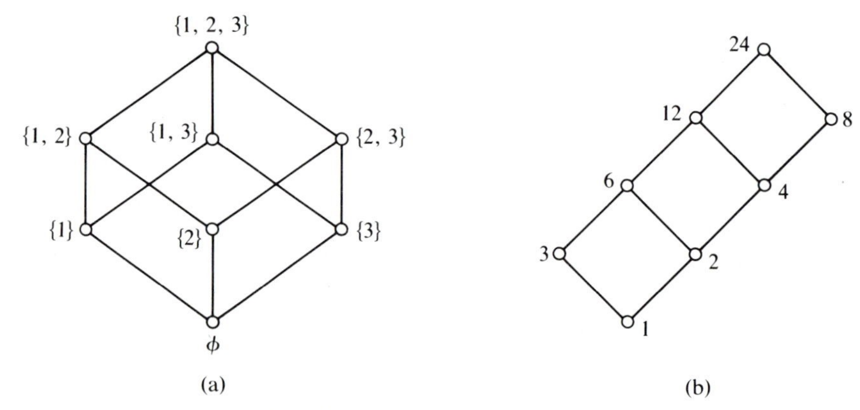

FIGURE 4.9 Two Hasse diagrams ■

DEFINITION 4.10

Let $(A, \lesssim)$ be a poset and let $B \subseteq A$.

1. An element $x \in A$ is called a *lower bound* for B if $x \lesssim y$ for all $y \in B$. An element $z \in A$ is called the *greatest lower bound* for B if the following hold:
   (a) z is a lower bound for B,
   (b) if x is any lower bound for B, then $x \lesssim z$.
   We write $z = \text{glb}(B)$.
2. An element $x \in$ A is called an *upper bound* for B if $y \lesssim x$ for all $y \in B$. An element $z \in A$ is called the *least upper bound* for B if the following hold:
   (a) z is an upper bound for B,
   (b) if x is any upper bound for B, then $z \lesssim x$.
   We write $z = \text{lub}(B)$.

It should be noted, in the context of Definition 4.10, that glb(B) and lub(B), should they exist, are uniquely determined. This is suggested by the language used in the definition. For example, suppose that both z_1 and z_2 satisfy the conditions of glb(B). Then both are lower bounds for B and, by (b), we have both $z_1 \lesssim z_2$ and $z_2 \lesssim z_1$. Since $\lesssim$ is antisymmetric, it follows that $z_1 = z_2$. So if glb(B) exists, it is unique.

Example 4.19 Consider the poset whose Hasse diagram is given in Figure 4.10. Let $B = \{g, h, i\}$, $C = \{g, h\}$, and $D = \{e, f\}$. Find (if it exists) (a) lub(B), (b) lub(C), (c) lub(D), (d) glb(D).

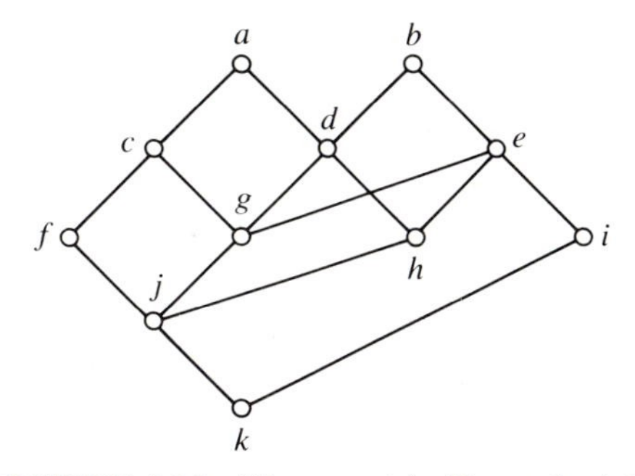

FIGURE 4.10 The poset in Example 4.19

Solution (a) The set of upper bounds for B is $\{b, e\}$, and since $e < b$, we have that $e = \text{lub}(B)$.
(b) The set of upper bounds for C is $\{a, b, d, e\}$; but lub(C) does not exist, since d and e are not comparable.
(c) The set D has no upper bounds, so lub(D) does not exist.
(d) The elements j and k are both lower bounds for D; since $j > k$, it follows that $j = \text{glb}(D)$. ■

Example 4.20 Consider again the poset $(\mathcal{P}(A), \subseteq)$ and let $\mathcal{B}$ be a nonempty subset of $\mathcal{P}(A)$. Find (a) glb($\mathcal{B}$), (b) lub($\mathcal{B}$).

Solution (a) If C is a lower bound for $\mathcal{B}$, then $C \subseteq B$ for every $B \in \mathcal{B}$. But then $C \subseteq S$, where

$$S = \bigcap_{B \in \mathcal{B}} B$$

It is also the case that S is a lower bound for $\mathcal{B}$. Therefore, $\text{glb}(\mathcal{B}) = S$. In words, the greatest lower bound for the collection $\mathcal{B}$ is the intersection of all the sets in $\mathcal{B}$.

(b) Similarly, we leave it to the reader to show (in Exercise 4) that the least upper bound for the collection $\mathcal{B}$ is the union of all the sets in $\mathcal{B}$:

$$\text{lub}(B) \quad \bigcup_{B \in \mathcal{B}} B$$

■

Notice that in the poset $(\mathcal{P}(A), \subseteq)$, every subset $\mathcal{B}$ of $\mathcal{P}(A)$ has both a least upper bound and a greatest lower bound. Posets in which every pair of distinct elements has both a greatest lower bound and a least upper bound are designated by a special term.

DEFINITION 4.11

If every two elements of a poset $(A, \preceq)$ possess both a greatest lower bound and a least upper bound, then the poset is called a *lattice*. Given x and y in A, we call $\text{glb}(\{x, y\})$ the *meet* of x and y and denote it by $x \wedge y$. We call $\text{lub}(\{x, y\})$ the *join* of x and y and denote it by $x \vee y$.

Thus it follows that the poset $(\mathcal{P}(A), \subseteq)$ is a lattice; for $B, C \in \mathcal{P}(A)$, $B \wedge C = B \cap C$ and $B \vee C = B \cup C$.

Example 4.21 Consider the set $\mathbb{N}$ of positive integers under the relation divides. If e is a lower bound of a and b, then $e \mid a$ and $e \mid b$, so e is a common divisor of a and b. Hence the meet of a and b is the greatest common divisor of a and b; that is, $a \wedge b = \text{gcd}(a, b)$. Similarly, if c is an upper bound of a and b, then $a \mid c$ and $b \mid c$, so c is a common multiple of a and b. It follows that the join of a and b is the least common multiple of a and b; that is, $a \vee b = \text{lcm}(a, b)$. Since, for any two positive integers a and b, both $\text{gcd}(a, b)$ and $\text{lcm}(a, b)$ exist, we may conclude that the poset $(\mathbb{N}, \mid)$ is a lattice. ■

As mentioned before, not every two elements in a poset are necessarily comparable. In $(\mathbb{N}, \mid)$, for example, neither of the integers 3 nor 8 divides the other; hence 3 and 8 are not comparable in this poset. However, in $(\mathbb{N}, \leq)$ (here $\leq$ has the usual interpretation "less than or equal to"), if $a, b \in \mathbb{N}$, then $a \leq b$ or $b \leq a$. This brings us to the next definition.

DEFINITION 4.12

A partial order relation $\lesssim$ on a set A is called a *total ordering* if, for any two elements $a, b \in A$, either $a \lesssim b$ or $b \lesssim a$.

We remark that if $\lesssim$ is a total ordering on the set A, then the condition given in Definition 4.12 can be restated as follows:

> For any two elements $a, b \in A$, precisely one of $a = b$, $a < b$, or $b < a$ holds.

This is sometimes called the *law of trichotomy*. A set A, together with a total ordering $\lesssim$, is called a *totally ordered set*. In some settings a total ordering is called a *linear, simple, complete,* or *full ordering*.

Example 4.22 (a) Let A be the set of positive divisors of 24. The poset $(A, |)$, as shown in Figure 4.9(b), is not totally ordered; for example, 4 and 6 are not comparable.
(b) Let B be the set of positive divisors of 81. The poset $(B, |)$, as shown in Figure 4.11(a), is a totally ordered set.
(c) As shown in Figure 4.9(a), the poset $(\mathcal{P}(\{1, 2, 3\}), \subseteq)$ is not a totally ordered set.
(d) Let $\mathcal{B} = \{\phi, \{1\}, \{1, 2\}, \{1, 2, 3\}\}$. Then $(\mathcal{B}, \subseteq)$ is a totally ordered set; its Hasse diagram is shown in Figure 4.11(b).

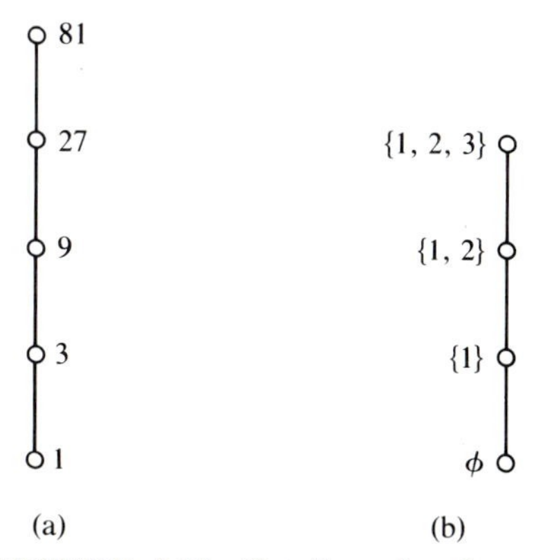

FIGURE 4.11 Totally ordered sets ■

Exercises 4.3

1. Each part gives a partial ordering $\lesssim$ on $X = \{a, b, c, d, e, f\}$. Draw the Hasse diagram of each poset $(X, \lesssim)$ and determine whether each poset is a lattice.
   a. $f \lesssim d \lesssim b \lesssim a, \quad f \lesssim e \lesssim c \lesssim a$
   b. $f \lesssim d \lesssim c \lesssim a, \quad e \lesssim d \lesssim c \lesssim b$
   c. $f \lesssim d \lesssim b \lesssim a, \quad f \lesssim e \lesssim c \lesssim a, \quad e \lesssim b, \quad d \lesssim c$
   d. $f \lesssim e \lesssim c \lesssim b \lesssim a, \quad e \lesssim d \lesssim b$
2. If $(A, \lesssim)$ is a poset and $B \subseteq A$, show that $(B, \lesssim)$ is a poset. We call B a *subposet of* A.
3. Define the relation $\lesssim$ on $\mathbb{Z} \times \mathbb{Z}$ by $(a,b) \lesssim (c,d) \leftrightarrow (a \leq c \wedge b \leq d)$.
   a. Show that $\lesssim$ is a partial ordering on $\mathbb{Z} \times \mathbb{Z}$.
   b. For $A = \{(3,2), (0,-3), (-1,4)\}$, find lub($A$) and glb($A$).
   c. Generalize part b. For $A = \{(a_1,b_1), (a_2,b_2), \ldots, (a_n,b_n)\}$, describe lub($A$) and glb($A$).
   d. Is $(\mathbb{Z} \times \mathbb{Z}, \lesssim)$ a lattice?
4. Consider the poset $(\mathcal{P}(A), \subseteq)$, and let $\mathcal{B}$ be a nonempty subset of $\mathcal{P}(A)$. Show that
$$\text{lub}(\mathcal{B}) = \bigcup_{B \in \mathcal{B}} B$$
5. Let $C = \{1, 2, 3\}$ and consider the poset $(C \times C, \lesssim)$, where the relation $\lesssim$ is defined in Exercise 3. Draw the Hasse diagram of $(C \times C, \lesssim)$.
6. Define the relation $<=$ on $\mathbb{Z} \times \mathbb{Z}$ by
$$(a,b) <= (c,d) \leftrightarrow ((a < c) \vee (a = c \wedge b \leq d))$$
   a. Show that $<=$ is a total ordering on $\mathbb{Z} \times \mathbb{Z}$.
   b. For $C = \{1, 2, 3\}$, draw the Hasse diagram of $(C \times C, <=)$.

4.4 *n*-ARY RELATIONS

We end this chapter with a very brief discussion of an extension of the definition of relation. Recall that a relation is also called a binary relation; this terminology emphasizes the fact that two sets are involved in the definition of relation. There is a natural extension of this definition to any finite number of sets.

DEFINITION 4.13

Let $A_1, A_2, \ldots, A_n$ be sets, $n \geq 2$. An *n-ary relation* among these sets is defined to be any subset of the Cartesian product $A_1 \times A_2 \times \cdots \times A_n$.

A relation among three sets is most often referred to as a *ternary relation,* while a relation among four sets is called a *quaternary relation.* If, in Definition 4.13, $A_1 = A_2 = \cdots = A_n = A$, then the relation is called an *n-ary relation on A.*

Example 4.23 The chairperson of the mathematics department at your college may be interested in the following sets:

- A the set of mathematics instructors
- B the set of mathematics courses to be offered next semester
- C the set of classrooms used for mathematics
- D the set of time slots during which courses meet (for example, MWF 2:00–3:00 or TTh 11:00–12:30)

In making up the teaching schedule, the chairperson might first come up with a binary relation R from A to B that assigns instructors to courses. Next, (s)he might decide when and where each course will meet; this will yield a ternary relation S among B, C, and D, where $(b,c,d) \in S$ if course b meets in location c during time slot d. The relations R and S could then be "joined" to yield the quaternary relation T among A, B, C, and D, where $(a,b,c,d) \in T$ means that instructor a will teach course b in location c during time slot d. Notice that

$$(a,b,c,d) \in T \leftrightarrow ((a,b) \in R \wedge (b,c,d) \in S)$$ ■

Example 4.24 Define a ternary relation R on $\mathbb{N}$ by

$$(a,b,c) \in R \leftrightarrow c^2 = a^2 + b^2$$

In other words, a, b, and c are related provided they form a Pythagorean triple. (See Chapter Problem 19 in Chapter 3.) Then, for instance, $(3,4,5) \in R$ and $(5,12,13) \in R$. ■

In the area of computer science known as database management, the main concern is the efficient manipulation of vast amounts of data. For example, suppose that a major university must maintain information concerning each student's registration. This could consist of several n-ary relations for various values of n. Here are several possibilities:

1. PDATA is a 5-ary relation whose 5-tuples have the form

   ```
   (NAME, HOME_ADDRESS, BIRTHDATE, SEX, IDNUMBER)
   ```

2. DEPT is a 4-ary relation with elements of the form

   ```
   (IDNUMBER, MAJOR, EARNED_HOURS, GPA)
   ```

3. ENROLLMENT is a ternary relation with elements of the form

   ```
   (NAME, IDNUMBER, COURSEID)
   ```

4. SCHEDULE is a 5-ary relation whose elements have the form

   ```
   (COURSEID, COURSENAME, TIME, LOCATION, INSTRUCTOR)
   ```

Thus, for the relation PDATA, let N be the set of all student names, A the set of all home addresses, B the set of all calendar dates, S the set {M, F}, and I the set of all student identification numbers. Then PDATA is a 5-ary relation among N, A, B, S, and I; that is, PDATA is a subset of $N \times A \times B \times S \times I$.

In the language of computer science, the set

$$D = \{\text{PDATA, DEPT, ENROLLMENT, SCHEDULE}\}$$

is called a *relational database,* and the relations in D are commonly referred to as *tables*. Indeed, it is easy to imagine the relations in D being displayed in tabular form; for example, PDATA would be a table with headings NAME, HOME_ADDRESS, BIRTHDATE, SEX, and IDNUMBER. The rows of this table would contain the various 5-tuples of PDATA, with the value of each coordinate listed in a column under the appropriate heading. These columns are called *attributes* of the particular relation. Notice, for instance, that the relations PDATA and ENROLLMENT have the attributes NAME and IDNUMBER in common.

In working with a relational database, there are three basic operations that are commonly used. In the discussion that follows, we shall describe these operations and demonstrate how they might be put to good use. To make life easy, we shall deal with two relations, R_1 and R_2, which are shown in Figures 4.12 and 4.13, respectively.

COURSEID	COURSENAME	TIME	LOCATION	INSTRUCTOR
CS105	Computing I	8	GEM 174	Straat
CS105	Computing I	9	GEM 174	Jones
CS105	Computing I	10	GEM 175	Moses
CS260	Data Structures	9	GEM 175	Straat
CS260	Data Structures	10	GEM 174	Lewis
CS340	Software Design	2	GEM 180	Lewis
MA122	Calculus I	11	GEM 180	Jones
MA122	Calculus I	1	GEM 180	Lewis
MA331	Algebra	10	GEM 179	Straat
MA350	Statistics	1	GEM 174	Moses

FIGURE 4.12 The relation R_1

COURSEID	TEXTAUTHOR	PREREQ	CREDITS
CS105	Allenton	CS102	3
CS260	Paston	CS105	3
CS340	Strong	CS261	3
MA122	Calvin	MA106	4
MA331	Seelo	MA231	3
MA350	Normal	MA231	3

FIGURE 4.13 The relation R_2

Now suppose that we would like to obtain a list of all courses being taught by Professor Straat, along with the text being used and the prerequisite for each course. Observe that R_1 contains the information about which courses Professor Straat is teaching (R_2 does not), while R_2 contains the text and prerequisite data (R_1 does not). Thus we must somehow make use of both R_1 and R_2 to obtain the desired information. We shall do this by applying the *relational database operators* described below.

1. *SELECT* is an operation that derives a new relation S from a given relation R, where S consists of all elements of R that satisfy specified conditions. For example, the instruction

   SELECT FROM R_1 WHERE INSTRUCTOR = 'STRAAT'

   will produce the relation S consisting of all elements of R_1 whose INSTRUCTOR coordinate is Straat. The relation S is shown in Figure 4.14.

COURSEID	COURSENAME	TIME	LOCATION	INSTRUCTOR
CS105	Computing I	8	GEM 174	Straat
CS260	Data Structures	9	GEM 175	Straat
MA331	Algebra	10	GEM 179	Straat

FIGURE 4.14 The relation S

2. *PROJECT* forms a new relation P from a given relation R by extracting specified attributes (or columns) of R and, at the same time, removing any duplicate elements (rows) that may result from this extraction. For example, the instruction

   PROJECT R_1 OVER LOCATION AND INSTRUCTOR

yields the relation P shown in Figure 4.15. (Note that the element (GEM 180, Lewis) is not listed twice.)

LOCATION	INSTRUCTOR
GEM 174	Straat
GEM 174	Jones
GEM 175	Moses
GEM 175	Straat
GEM 174	Lewis
GEM 180	Lewis
GEM 180	Jones
GEM 179	Straat
GEM 174	Moses

FIGURE 4.15 The relation P

3. *JOIN* forms a new relation as follows. Given an m-ary relation R and an n-ary relation S with common attribute A, the JOIN of R and S over A is an $(m + n - 1)$-ary relation J. The relation J consists of all $(m + n - 1)$-tuples obtained by taking elements $r \in R$ and $s \in S$ with common A-coordinate (the A-coordinates of r and s are equal), forming the $(n - 1)$-tuple s' by deleting the A-coordinate of s, and then concatenating r with s'. This operation can also be applied over several common attributes of R and S. For example, the instruction

   JOIN R_1 AND R_2 OVER COURSEID

   results in the relation J shown in Figure 4.16.

COURSEID	COURSENAME	TIME	LOCATION	INSTRUCTOR	TEXTAUTHOR	PREREQ	CREDITS
CS105	Computing I	8	GEM 174	Straat	Allenton	CS102	3
CS105	Computing I	9	GEM 174	Jones	Allenton	CS102	3
CS105	Computing I	10	GEM 175	Moses	Allenton	CS102	3
CS260	Data Structures	9	GEM 175	Straat	Paston	CS105	3
CS260	Data Structures	10	GEM 174	Lewis	Paston	CS105	3
CS340	Software Design	2	GEM 180	Lewis	Strong	CS261	3
MA122	Calculus I	11	GEM 180	Jones	Calvin	MA106	4
MA122	Calculus I	1	GEM 180	Lewis	Calvin	MA106	4
MA331	Algebra	10	GEM 179	Straat	Seelo	MA231	3
MA350	Statistics	1	GEM 174	Moses	Normal	MA231	3

FIGURE 4.16 The relation J

Let us go back to the problem posed earlier, that of determining all courses being taught by Professor Straat, along with the text and prerequisites for each course. We can get this information from the database by using the SELECT operator on the relation *J* of Figure 4.16. Simply use the instruction

```
SELECT FROM J WHERE INSTRUCTOR = 'STRAAT'
```

to obtain the relation *T* of Figure 4.17.

COURSEID	COURSENAME	TIME	LOCATION	INSTRUCTOR	TEXTAUTHOR	PREREQ	CREDITS
CS105	Computing I	8	GEM 174	Straat	Allenton	CS102	3
CS260	Data Structures	9	GEM 175	Straat	Paston	CS105	3
MA331	Algebra	10	GEM 179	Straat	Seelo	MA231	3

FIGURE 4.17 The relation *T*

Then, to obtain the desired information, employ the instruction

```
PROJECT T OVER COURSEID AND TEXTAUTHOR AND PREREQ
```

The result is the table of Figure 4.18.

COURSEID	TEXTAUTHOR	PREREQ
CS105	Allenton	CS102
CS260	Paston	CS105
MA331	Seelo	MA231

FIGURE 4.18

A relational database is one kind of "database management system." Such systems are used by many large corporations (such as airline companies, insurance companies, and banks) for organizing, storing, manipulating, and retrieving large amounts of data within a computing environment. In this context, relational database management systems are often relatively efficient and easy to use.

Exercises 4.4

1. For the relations R_1 and R_2 of Figures 4.12 and 4.13, determine the relations produced by the following operations.
   a. SELECT FROM R_1 WHERE TIME = 10
   b. PROJECT R_2 OVER COURSEID AND TEXTAUTHOR

c. `SELECT FROM R₂ WHERE CREDITS = 3`
d. `PROJECT R₁ OVER COURSEID AND COURSENAME AND TIME`
e. `SELECT FROM R₁ WHERE TIME ≠ 8 AND (INSTRUCTOR = 'STRAAT' OR INSTRUCTOR = 'MOSES')`

2. What might the following instruction produce, given the relations R_1 and R_2 of Figures 4.12 and 4.13?

```
SELECT FROM (PROJECT R₁ OVER COURSEID AND INSTRUCTOR)
WHERE INSTRUCTOR = 'STRAAT'
```

3. Suppose the relation R_3, with attributes COURSEID, TIME, LOCATION, and ENROLLMENT, gives the enrollment (number of students) in each of the courses listed in Figure 4.12. Discuss how the operations SELECT, PROJECT, and JOIN can be used (with R_1 and R_3) to obtain the following information.
   a. A relation Q_1 with attributes TIME and ENROLLMENT
   b. A relation Q_2 giving the enrollment in each section of CS105
   c. A ternary relation Q_3 with attributes COURSEID, INSTRUCTOR, and ENROLLMENT
   d. A relation Q_4 giving the enrollment in each course taught by Professor Lewis
4. Indicate how the relation of Figure 4.18 can be obtained from R_1 and R_2 by applying PROJECT, JOIN, and SELECT, in that order. Are there advantages or disadvantages to doing it this way, as opposed to the way it was done in the text?

CHAPTER PROBLEMS

1. Can a relation R on a nonempty set A be both symmetric and antisymmetric?
2. Let R be a symmetric and transitive relation on a nonempty set A. Under what condition will R be reflexive?
3. Let R be a relation on a nonempty set A. Complete each of the following statements:
   a. R is not reflexive if and only if ______.
   b. R is not symmetric if and only if ______.
   c. R is not antisymmetric if and only if ______.
   d. R is not transitive if and only if ______.
4. Let R be a reflexive relation on a set A. Prove that R is an equivalence relation if and only if, given any elements a, b, and c of A (not necessarily distinct), the following condition holds:

$$(a\ R\ b \wedge a\ R\ c) \rightarrow b\ R\ c$$

5. Determine whether each of the following relations $\sim$ is an equivalence relation on $\mathbb{R} \times \mathbb{R}$. For those that are, describe (geometrically) the equivalence class containing the point (a,b).

a. $(a,b) \sim (c,d) \leftrightarrow a + d = b + c$

b. $(a,b) \sim (c,d) \leftrightarrow (a - 1)^2 + b^2 = (c - 1)^2 + d^2$

c. $(a,b) \sim (c,d) \leftrightarrow (a - c)(b - d) = 0$

d. $(a,b) \sim (c,d) \leftrightarrow |a| + |b| = |c| + |d|$

e. $(a,b) \sim (c,d) \leftrightarrow ab = cd$

6. For each of the following relations $\sim$ defined on $\mathbb{R}$, verify that $\sim$ is an equivalence relation and describe the partition of $\mathbb{R}$ induced by $\sim$.

a. $x \sim y \leftrightarrow \text{int}(x) = \text{int}(y)$, where $\text{int}(x)$ is the largest integer n such that $n \leq x$

b. $x \sim y \leftrightarrow \text{round}(x) = \text{round}(y)$, where $\text{round}(x)$ is the integer nearest x (and, for example, $\text{round}(2.5) = 3$)

c. $x \sim y \leftrightarrow |x| = |y|$

d. $x \sim y \leftrightarrow x - y \in \mathbb{Z}$

e. $x \sim y \leftrightarrow x - y \in \mathbb{Q}$

7. Define the relation $\sim$ on $\mathbb{Z} \times \mathbb{N}$ by $(a,b) \sim (c,d) \leftrightarrow ad = bc$.

a. Show that $\sim$ is an equivalence relation.

b. Describe the equivalence class $[(a,b)]$ in a nice way. (Hint: think of (a,b) as the fraction a/b.)

8. Each of the following parts gives a set X, a relation $\lesssim$ on X, and a subset Y of X. (i) Verify that $(X, \lesssim)$ is a poset and (ii) draw the Hasse diagram of the (sub)poset $(Y, \lesssim)$.

a. $X = \mathbb{R}, x \lesssim y \leftrightarrow (x = y \vee |x| < |y|), Y = \{-3, -2, -1, 0, 1, 2, 3\}$

b. $X = \mathbb{Z}, x \lesssim y \leftrightarrow (x = y \vee x \text{ MOD } 4 < y \text{ MOD } 4)$, $Y = \{-4, -3, -2, -1, 0, 1, 2, 3, 4\}$

c. $X = \mathbb{Z} \times \mathbb{Z}, (a,b) \lesssim (c,d) \leftrightarrow ((a,b) = (c,d) \vee a^2 + b^2 < c^2 + d^2)$, $Y = \{0, 1, 2\} \times \{0, 1, 2\}$

9. Let $A = \{(a,b) \mid a \in \mathbb{N} \wedge b \in \mathbb{N} \wedge \gcd(a,b) = 1\}$. Define a relation $<=$ on A by $(a,b) <= (c,d) \leftrightarrow ad \leq bc$. Show that $<=$ defines a partial ordering of A.

10. Define a relation $\sim$ on $\mathbb{Z}$ by $a \sim b \leftrightarrow 3 \mid (a + 2b)$.

a. Show that $\sim$ is an equivalence relation on $\mathbb{Z}$.

b. Determine the partition of $\mathbb{Z}$ induced by $\sim$.

11. Each part gives a set X and a relation R on X. Determine whether R is (i) reflexive, (ii) symmetric, (iii) antisymmetric, and (iv) transitive.

a. $X = \mathcal{P}(\{1, 2, 3, 4\}), A \; R \; B \leftrightarrow A \subseteq B \cup \{1\}$

b. $X = \mathbb{Z}_7, [x] \; R \; [y] \leftrightarrow [x - y] = [1]$ or $[x - y] = [6]$

c. $X = \mathbb{Z}, x \; R \; y \leftrightarrow |x - y| > 2$

d. $X = \mathbb{R}, x \; R \; y \leftrightarrow xy \geq 0$

e. $X = (0, 1), x \; R \; y \leftrightarrow xy \in \mathbb{Q}$

12. Consider the poset whose Hasse diagram is shown in Figure 4.19(a). Let $A = \{c, d, e\}$, $B = \{f, d\}$, and $C = \{f, e\}$. Find, if it exists,

a. glb(A) **b.** lub(A) **c.** glb(B)
d. lub(B) **e.** glb(C) **f.** lub(C)

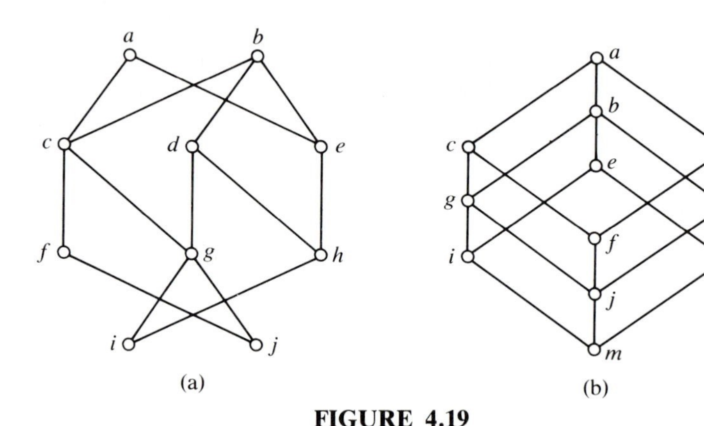

FIGURE 4.19

13. Consider the lattice whose Hasse diagram is shown in Figure 4.19(b). Find

a. $i \vee a$ **b.** $i \wedge c$ **c.** $g \vee k$ **d.** $b \wedge d$
e. $e \wedge f$ **f.** $e \vee f$ **g.** $d \wedge g$ **h.** $a \wedge m$

14. Draw the Hasse diagram for the lattice $(A, |)$, where A is the set of positive divisors of

a. 42 **b.** 54 **c.** 100 **d.** 126

15. A relation R is *irreflexive* if

$$\forall a, b \in A(a\ R\ b \rightarrow a \neq b)$$

In other words, A is irreflexive if no element of A is related to itself.

a. Give an example of a relation on $\mathbb{Z}$ that is irreflexive and symmetric but not transitive.

b. Give an example of a relation on $\mathbb{N}$ that is irreflexive and transitive but not symmetric.

c. Prove that a nonempty relation R on a set A cannot be irreflexive, symmetric, and transitive.

16. Let R be a relation from A to B. The *converse* of R is the relation R^c from B to A defined by

$$\text{For } a \in A, b \in B, \quad b\ R^c\ a \leftrightarrow a\ R\ b$$

a. For finite sets A and B, how are the table representations of R and R^c related?

b. For a finite set A, let R be a relation on A. How are the directed graphs of R and R^c related?
c. Let R be a relation on A. Under what condition does $R = R^c$?
d. Prove: If R is a partial-order relation on A, then so is R^c.

17. Let R be a relation on A. Define the relation R^r on A by

$$a\ R^r\ b \leftrightarrow (a\ R\ b \vee a = b)$$

a. Show that the relation R^r is reflexive.
b. Note that $R \subseteq R^r$. Under what condition does $R = R^r$?
c. Prove that if R' is a reflexive relation on A and $R \subseteq R'$, then $R^r \subseteq R'$.
The relation R^r is called the *reflexive closure* of R.
d. Find the reflexive closure of the relation $<$ on $\mathbb{Z}$.

18. Let R be a relation on A. Define the relation R^s on A by

$$a\ R^s\ b \leftrightarrow (a\ R\ b \vee b\ R\ a)$$

a. Show that the relation R^s is symmetric.
b. Note that $R \subseteq R^s$. Under what condition does $R = R^s$?
c. Prove that if R' is a symmetric relation on A and $R \subseteq R'$, then $R^s \subseteq R'$.
d. Show that $R^s = R \cup R^c$ (see Problem 16).
The relation R^s is called the *symmetric closure* of R.
e. Find the symmetric closure of the relation $<$ on $\mathbb{Z}$.

19. Let R be a relation on a set A. The *transitive closure* of R is the relation R^t on A that has the following properties: (i) R^t is transitive; (ii) $R \subseteq R^t$; (iii) if R' is a transitive relation on A and $R \subseteq R'$, then $R^t \subseteq R'$. Informally, R^t is the smallest transitive relation on A that contains R.
a. Find the transitive closure of the relation R on $\mathbb{Z}$ defined by $a\ R\ b \leftrightarrow b = a + 1$.
b. Find the transitive closure of the relation "is the father of" on your family tree.
c. Define the relation R^2 on A by $a\ R^2\ b \leftrightarrow (a\ R\ b \vee \exists c(a\ R\ c \wedge c\ R\ b))$. In view of Problems 17 and 18, it is tempting to define R^t in the way we have defined R^2. However, show that R^2 is not transitive, in general.

20. Let R be a relation on A. Prove each of the following:
a. If R is symmetric and transitive, then so is R^r.
b. If R is reflexive and transitive, then R^s is reflexive but may not be transitive.
c. If R is reflexive and symmetric, then so is R^t.
d. $((R^r)^s)^t$ is an equivalence relation on A.

21. Let R be a relation on A and let S be a relation on B. Define the relations T_1 and T_2 on $A \times B$ as follows:

$$(a_1,b_1)\ T_1\ (a_2,b_2) \leftrightarrow (a_1\ R\ a_2 \wedge b_1\ S\ b_2)$$
$$(a_1,b_1)\ T_2\ (a_2,b_2) \leftrightarrow ((a_1 \neq a_2 \wedge a_1\ R\ a_2) \vee (a_1 = a_2 \wedge b_1\ S\ b_2))$$

a. Consider, one at a time, each of the properties (i) reflexive, (ii) symmetric, (iii) antisymmetric, and (iv) transitive. What can be said about T_1 and T_2 if it is known that both R and S possess a given property?

b. What can be said about T_1 and T_2 if both R and S are (i) equivalence relations? (ii) partial orderings?

22. Let R and S be relations on a set A. Suppose R is an equivalence relation and S is a partial ordering. What can be said about the relation $R \cap S$?

23. An element m in a poset $(A, \preceq)$ is called a *minimal element* if

$$\forall a \in A(a \preceq m \rightarrow a = m)$$

In other words, m is a minimal element if there does not exist an element $a \in A$ such that $a < m$. Similarly, m is called a *maximal element* if $\forall a \in A(m \preceq a \rightarrow a = m)$. For example, in the poset of Figure 4.10, a and b are maximal elements and k is a minimal element.

a. What are the maximal and minimal elements in the poset of Figure 4.19(a)?

b. What are the maximal and minimal elements of $(\mathcal{P}(X), \subseteq)$?

c. What are the minimal elements of $(\mathbb{N} - \{1\}, |)$?

d. Show that every nonempty, finite poset A contains a minimal (and a maximal) element. Given $A = \{a_1, a_2, \ldots, a_n\}$, describe an algorithm for finding a minimal element of A.

e. Give an example of a poset that has no maximal elements.

24. Given a nonempty finite poset $(A, \preceq)$, we sometimes wish to *embed* the partial ordering $\preceq$ in a total order $<=$. That is, we wish to define a total ordering $<=$ on A such that if a and b are elements of A and $a \preceq b$, then $a <= b$. We can then write $A = \{a_1, a_2, \ldots, a_n\}$, where a_1 precedes a_2, a_2 precedes a_3, and so on, in the total ordering. The process of finding such a total ordering is called *sorting*. One method of sorting is called the *topological sort*. This method makes use of the fact (Problem 23) that every nonempty finite poset contains a minimal element. Choose a minimal element m of A and set $a_1 = m$. Next, for a_2, choose a minimal element of the poset $A - \{a_1\}$. Continue this process until the sort is completed.

a. Describe more formally the algorithm for the topological sort, giving a justification for each step.

b. Apply the algorithm to the poset of Figure 4.10.

25. Let $(A, \preceq)$ be a poset. A *chain* in $(A, \preceq)$ is a nonempty subset C of A

such that $(C, \precsim)$ is totally ordered. If C is finite we may write $C = \{c_1, c_2, \ldots, c_n\}$ where $c_1 < c_2 < \cdots c_n$; the number n is called the *length* of the chain C. As an example, in the poset of Figure 4.10, $j < g < e < b$ is a chain of length 4.

a. Find all chains of length 5 in the poset of Figure 4.10.

b. Find the length of a longest chain in the poset of Figure 4.19(a).

c. Find the length of a longest chain in the poset $(\mathcal{P}(X), \subseteq)$, where $n(X) = m$.

d. Prove: If $a_1 < a_2 < \cdots < a_n$ is a longest chain in $(A, \precsim)$, then a_n is maximal in $(A, \precsim)$ (and a_1 is minimal).

e. Let the length of a longest chain in $(A, \precsim)$ be $n \geq 2$. Prove: If M is the set of maximal elements in A, then the length of a longest chain in $(A - M, \precsim)$ is $n - 1$.

26. An *antichain* in a poset $(A, \precsim)$ is a nonempty subset B of A in which no two distinct elements are comparable. Symbolically, B is an antichain if

$$\forall a,b \in B(a \precsim b \to a = b)$$

For example, $B = \{d, e, f\}$ is an antichain in the poset of Figure 4.10. For the poset of Figure 4.10,

a. Find an antichain having 4 elements.

b. Find the largest antichain that contains the element j.

c. Show that no antichain has 5 elements.

Let $(A, \precsim)$ be a nonempty poset.

d. Show that the set of minimal (or maximal) elements of A is an antichain.

e. Let $a_1 < a_2 < \cdots < a_n$ be a longest chain in A, and for $1 \leq i \leq n - 1$, define $B_i = \{b \in A \mid a_i \precsim b < a_{i+1}\}$. Show that each B_i is an antichain.

27. A lattice L is said to be *distributive* if the distributive laws hold: $\forall a,b,c \in L$,

$$(1) \quad a \vee (b \wedge c) = (a \vee b) \wedge (a \vee c)$$
$$(2) \quad a \wedge (b \vee c) = (a \wedge b) \vee (a \wedge c)$$

Verify that each of the following lattices is a distributive lattice.

a. $(\mathcal{P}(X), \subseteq)$ ★**b.** $(\mathbb{N}, \mid)$

28. Let P denote the Pascal character set. Then P contains the letters A, B, . . . , Z, the digits 0, 1, . . . , 9, and other special characters such as), +, =, etc. A *character string* is a sequence of characters from P enclosed in single quotes; for example, 'FREDONIA' and '93 MAIN ST.' are character strings. The *length* of a character string is the number of characters it contains (not counting the enclosing quotes). Thus the length of 'FREDONIA' is 8 and the length of '93 MAIN ST.'

is 11 (blanks are counted). The unique string of length zero is called the *empty string* and is denoted ''.

Comparison of two character strings is made possible by the *collating sequence* of P, which totally orders P. Let us denote this ordering by $\leq$. In Pascal, 'A' < 'B' < · · · < 'Z' and '0' < '1' < · · · < '9'. However, whether '9' < 'A' or 'Z' < '0' varies from system to system. In this problem we consider two possible ways of extending the ordering of the Pascal characters to an ordering of all character strings.

The first method is called *lexicographic ordering* and will be denoted $\leq_l$. In this ordering the empty string precedes any nonempty string, while for two nonempty strings, s_1 = '$a_1a_2 \cdots a_m$' and s_2 = '$b_1b_2 \cdots b_n$', $s_1 \leq_l s_2$ provided either $m \leq n$ and $a_i = b_i$, $1 \leq i \leq m$, or if, proceeding from left to right, i is the first value where $a_i \neq b_i$ and $a_i < b_i$.

a. Verify that $\leq_l$ is a total ordering.

b. Find character strings $s_1, s_2, s_3, \ldots$, such that $s_1 \geq_l s_2 \geq_l s_3 \geq_l \cdots$.

The second method of ordering is called *standard ordering* and will be denoted $\leq_s$. For two strings s_1 and s_2, if the length of s_1 is less than the length of s_2, then $s_1 \leq_s s_2$. If s_1 and s_2 have the same length, then we apply lexicographic ordering.

c. Verify that $\leq_s$ is a total ordering.

d. Given any nonempty set of character strings S, show that S has a smallest element. Compare with part b.

29. Let $\mathcal{P}_1$ and $\mathcal{P}_2$ be partitions of a nonempty set A. We say that $\mathcal{P}_2$ *refines* $\mathcal{P}_1$ if for every set B_2 in $\mathcal{P}_2$ there is a set B_1 in $\mathcal{P}_1$ such that $B_2 \subseteq B_1$. For example, with $A = \{1, 2, 3, 4\}$, the partition $\mathcal{P}_2 = \{\{1, 2\}, \{3\}, \{4\}\}$ refines $\mathcal{P}_1 = \{\{1, 2, 3\}, \{4\}\}$.

a. Show that the collection of all partitions of A under the relation "refines" is a poset.

b. It can be shown that this poset is a lattice. For two partitions $\mathcal{P}_1$ and $\mathcal{P}_2$, describe $\mathcal{P}_1 \wedge \mathcal{P}_2$ and $\mathcal{P}_1 \vee \mathcal{P}_2$.

30. Prove: If the length of a longest chain in the finite poset $(A, \preceq)$ is n, then A can be partitioned into n antichains. (Hint: Use induction on n. Let M be the set of maximal elements in A and consider $A - M$. See Problems 25 and 26.)

Five Functions

5.1 INTRODUCTION

Recall that a relation R from a set A to a set B is any subset of $A \times B$. The domain of R is the set

$$\text{dom } R = \{a \in A \mid (a,b) \in R \text{ for some } b \in B\}$$

and the image of R is the set

$$\text{im } R = \{b \in B \mid (a,b) \in R \text{ for some } a \in A\}$$

We have already encountered two very special types of relations, namely, equivalence relations and partial-order relations. These were well motivated by some rather classical examples: "equals" is an equivalence relation on $\mathbb{N}$ and "is less than or equal to" is a partial ordering of $\mathbb{N}$.

In this chapter we study what is probably the most fundamental type of relation used in mathematics.

DEFINITION 5.1

A relation f from a set A to a set B is called a *function from A to B* if the following conditions hold.

1. $\text{dom } f = A$
2. No two distinct ordered pairs in f have the same first coordinate.

We denote the fact that f is a function from A to B by writing $f: A \to B$. The set A is called the *domain* of the function f, and we call B the *codomain* of f. If $A = B$, then we call f a *function on A*.

We remark that condition 2 of the definition of a function can be stated symbolically as follows:

$$((x,y) \in f \land (x,z) \in f) \to y = z$$

Example 5.1 Determine which of the following relations are functions and find the image of those that are.

(a) The relation f_1 on $\{1, 2, 3\}$ given by $f_1 = \{(1,2), (2,1), (3,2)\}$.
(b) The relation f_2 on $\{1, 2, 3\}$ given by $f_2 = \{(1,1), (1,3), (2,3), (3,1)\}$.
(c) The relation f_3 on $\mathbb{R}$ defined by $f_3 = \{(x,y) \mid x^2 + y^2 = 4\}$.
(d) The relation f_4 on $\mathbb{Z}$ defined by $f_4 = \{(m,n) \mid n = 2m + 1\}$.
(e) The relation f_5 defined on your family tree by

$$(x,y) \in f_5 \leftrightarrow y \text{ is the (biological) father of } x$$

Solution (a) The relation f_1 is a function; $\operatorname{im} f_1 = \{1, 2\}$.
(b) The relation f_2 is not a function because $(1,1)$ and $(1,3)$ both belong to f_2.
(c) The relation f_3 is not a function because, for example, $(0,2) \in f_3$ and $(0,-2) \in f_3$.
(d) This relation is a function; $\operatorname{im} f_4$ is precisely the set of odd integers.
(e) This relation is a function since each person has a unique father. The image of f_5 consists of those people in your family who are (or were) fathers. ■

Suppose now that f is a function from A to B. Then conditions 1 and 2 in Definition 5.1 can be replaced by the following single condition:

Given $x \in A$, there is a unique $y \in B$ such that $(x,y) \in f$.

We call y the *image of x under f* and we write $y = f(x)$; this equation is read, "y equals f of x." Thus $(x,y) \in f$ if and only if $y = f(x)$, and it should then be observed that

$$f = \{(x,y) \mid x \in A \text{ and } y = f(x)\} = \{(x,f(x)) \mid x \in A\}$$

It is common to refer to the equation $y = f(x)$ as the "defining equation" of the function f. Indeed, it is common to define a function by writing, "Define $f: A \to B$ by $y = f(x)$." In some sense this equation can be viewed as a rule indicating how to find the image of a given $x \in A$. In fact, it is not uncommon to see the following statement as a definition of a function $f: A \to B$:

3. A function from A to B is a rule f that associates with each $x \in A$ a unique element $y \in B$.

This statement is, at best, imprecise. The reason for this lies in the lack of a precise meaning for the term "rule." Condition 3 simply relies on the reader's intuition as to what the term "rule" means.

In the foregoing discussion it is clear that, given $f: A \to B$, the uniqueness of the image of each element of A under f is a key feature in the definition of a function. Consider the following attempt at defining a function from $\mathbb{Z}_3$ to $\mathbb{Z}_6$. Define $g: \mathbb{Z}_3 \to \mathbb{Z}_6$ by $g([a]_3) = [a]_6$. Here $[a]_3$ denotes an arbitrary element of $\mathbb{Z}_3$ and $[a]_6$ an arbitrary element of $\mathbb{Z}_6$. The question is, do we really have a function? In particular, is each $[a]_3$ in $\mathbb{Z}_3$ associated with exactly one element of $\mathbb{Z}_6$? The problem is that there are infinitely many ways to represent $[a]_3$; for example, $[5]_3 = [2]_3$. In this particular case we see that $g([5]_3) = [5]_6$ and $g([2]_3) = [2]_6$, but $[5]_6 \neq [2]_6$. Thus we are forced to conclude that g associates with $[2]_3$ at least two different elements of $\mathbb{Z}_6$. Therefore, g is not a function.

In general terms, what went wrong with the previous example is that an attempt was made to define a function g from a set A to a set B, where the elements of A can be represented by more than one "name." For instance, $[2]_3$ and $[5]_3$ are different names for the same element of $\mathbb{Z}_3$. If g is to be a function, then the image of each element of A must be independent of the name chosen to represent it. If this happens, then we shall emphasize the property by stating that g is *well-defined.*

Example 5.2 Define $g: \mathbb{Z}_7 \to \mathbb{Z}_7$ by $g([a]) = [4a]$. Show that g is well-defined.

Solution Suppose that $[a] = [b]$ in $\mathbb{Z}_7$. We must show that $g([a]) = g([b])$. Since $[a] = [b]$, it follows that $7 \mid (a - b)$; that is, there exists an integer q such that $a - b = 7q$. Thus

$$g([a]) = [4a] = [4(b + 7q)] = [4b + 28q] = [4b] = g([b])$$

(Note that $[4b + 28q] = [4b]$, since $7 \mid 28q$.) Thus g is well-defined. ■

If f is a function whose domain and range are both subsets of $\mathbb{R}$, then associated with each ordered pair $(x,y) \in f$ there is a uniquely determined point (x,y) in the xy-coordinate plane. In Chapter 4, the set of all points so determined was called the graph of f. Here we shall also speak of the graph of such a function f. We assume the reader has had considerable experience graphing functions.

Example 5.3 Sketch the graphs of the following functions.

(a) $f: \mathbb{R} \to \mathbb{R}$; $f(x) = 4 - 2x$

(b) $f: (0, \infty) \to (0, \infty)$; $f(x) = 1/x$

(c) $f: \mathbb{R} \to [-4, \infty)$; $f(x) = x^2 - 2x - 3$

Solution The graphs are shown in Figure 5.1(a), (b), and (c). ■

If X and Y are subsets of $\mathbb{R}$ and $f: X \to Y$ is given, then the property

$$((x,y_1) \in f \wedge (x,y_2) \in f) \to y_1 = y_2$$

can be interpreted geometrically. It implies that each vertical line intersects the graph of f in at most one point. Thus, if we are given a set of points in a coordinate plane, then we can determine whether the associated set of ordered pairs is a function by applying this "vertical line test."

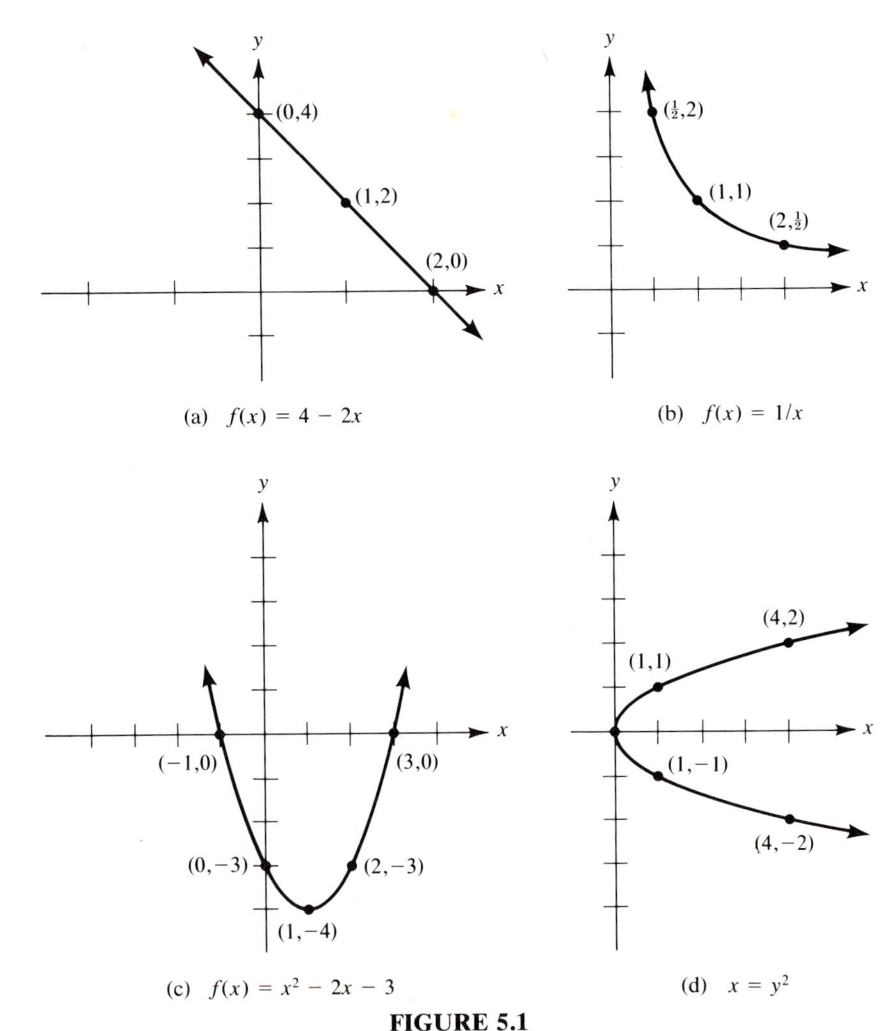

FIGURE 5.1

Example 5.4 It is easy to apply the vertical line test to each of the graphs in Figure 5.1. Those in (a) through (c) are seen to determine functions, which of course we already know from the previous example. Figure 5.1(d) is the graph of the relation $f = \{(x,y) \mid y \in \mathbb{R} \wedge x = y^2\}$. Note that any vertical line $x = a^2$, where $a \neq 0$, intersects the graph of f in two points, namely, (a^2,a) and $(a^2,-a)$. This shows that this relation is not a function. ■

Exercises 5.1

1. Let $S = \{1, 2, 3, 4\}$. Determine whether each of the following relations f on S is a function.

a. $f = \{(1,2), (3,4), (4,1)\}$

b. $f = \{(1,3), (2,3), (3,3), (4,3)\}$
c. $f = \{(1,1), (2,2), (3,3), (3,4), (4,4)\}$
d. $f = \{(1,3), (2,4), (3,1), (4,2)\}$

2. Which of the following equations determine functions on $\mathbb{Z}_6$, that is, which are well-defined?
a. $g([a]) = [a + 1]$ **b.** $g([a]) = [2a]$
c. $g([a]) = [a \text{ DIV } 2]$ **d.** $g([a]) = [-a]$

3. Determine which of the following relations are functions and find the image of those that are.
a. The relation f on $\mathbb{R}$; $(x,y) \in f \leftrightarrow y = x^2$.
b. The relation g on $\mathbb{R}$; $(x,y) \in g \leftrightarrow |x - y| = 2$.
c. The relation h from $\mathbb{N} - \{1\}$ to the set P of primes; $(n,p) \in h \leftrightarrow p$ is the smallest prime factor of n.
d. The relation f on your family tree; $(x,y) \in f \leftrightarrow y$ is an uncle of x.
e. The relation g on your family tree; $(x,y) \in f \leftrightarrow y$ is the eldest son of x's (paternal) grandfather.

5.2 ONE-TO-ONE AND ONTO FUNCTIONS

Given a function $f: A \to B$, it should be emphasized that for each $a \in A$, there is exactly one $b \in B$ such that $(a,b) \in f$. However, it need not be the case that for each $d \in B$ there is exactly one $c \in A$ such that $(c,d) \in f$. In fact, for some $d \in B$ there may be no $c \in A$ with $(c,d) \in f$, or there may be several elements $c \in A$ with $(c,d) \in f$. For example, consider the function $f: \mathbb{R} \to \mathbb{R}$, where $f(x) = x^2 + 1$. It is clear that $f(x) \geq 1$ for all $x \in \mathbb{R}$, so, for instance, there is no $x \in \mathbb{R}$ for which $f(x) = 0$. On the other hand, it is readily observed that $f(1) = f(-1) = 2$; actually, $f(a) = f(-a) = a^2 + 1$ for all $a \in \mathbb{R}$.

Some functions $f: A \to B$ satisfy the property that for each $b \in B$ there is at most one $a \in A$ such that $f(a) = b$. This condition may be rephrased as follows: For $a_1, a_2 \in A$, if $f(a_1) = f(a_2)$ then $a_1 = a_2$. For example, consider the function $f: \mathbb{R} \to \mathbb{R}$ defined by $f(x) = 3x + 5$. To show that the aforementioned condition holds, suppose that $f(a_1) = f(a_2)$ for some a_1, $a_2 \in \mathbb{R}$. Then $3a_1 + 5 = 3a_2 + 5$, from which it easily follows that $a_1 = a_2$.

DEFINITION 5.2

A function $f: A \to B$ is called *one-to-one* if, for all $a_1, a_2 \in A$, the following condition holds:

$$f(a_1) = f(a_2) \to a_1 = a_2$$

As just indicated, the condition stated in the definition that a function be one-to-one is a workable one. It is sometimes convenient to use the condition in this form:

$$\text{For all } a_1, a_2 \in A, \quad a_1 \neq a_2 \rightarrow f(a_1) \neq f(a_2).$$

This is obtained by taking the contrapositive of the implication.

When is the function $f: A \rightarrow B$ not one-to-one? Taking the negation of the condition in Definition 5.2, we obtain

$$f \text{ is not one-to-one} \leftrightarrow \text{for some } a_1, a_2 \in A, \quad a_1 \neq a_2 \wedge f(a_1) = f(a_2).$$

Example 5.5 Determine which of the following functions are one-to-one.

(a) $f:\{1, 2, 3\} \rightarrow \{1, 2, 3\}; \quad f(1) = 2, f(2) = 2, f(3) = 1$
(b) $f:\{1, 2, 3\} \rightarrow \{1, 2, 3\}; \quad f(1) = 3, f(2) = 2, f(3) = 1$
(c) $f:\mathbb{Z} \rightarrow \mathbb{Z}; \quad f(m) = m - 1$
(d) $g:\mathbb{Z} \rightarrow \mathbb{Z}; \quad g(m) = 3m + 1$
(e) $h:\mathbb{Z} \rightarrow \mathbb{N}; \quad h(m) = |m| + 1$
(f) $p:\mathbb{Q} - \{1\} \rightarrow \mathbb{Q}; \quad p(r) = r/(1 - r)$

Solution (a) Here f is not one-to-one since $f(1) = f(2) = 2$.

(b) This f is one-to-one since no two elements of $\{1, 2, 3\}$ have the same image under f.

(c) This function f maps each integer to its predecessor, and it is easily seen to be one-to-one.

(d) If $g(m_1) = g(m_2)$, where $m_1, m_2 \in \mathbb{Z}$, then $3m_1 + 1 = 3m_2 + 1$, which implies that $m_1 = m_2$. So g is one-to-one.

(e) If $h(m_1) = h(m_2)$, then $|m_1| + 1 = |m_2| + 1$, which implies that $|m_1| = |m_2|$. But this does not imply that $m_1 = m_2$, which leads us to suspect that h is not one-to-one. Indeed, if we let $m_1 = -1$ and $m_2 = 1$, then we see that $h(-1) = h(1) = 2$, which shows that h is not one-to-one.

(f) Let $r_1 \neq 1$ and $r_2 \neq 1$ be rational numbers. Then

$$\begin{aligned} p(r_1) = p(r_2) &\rightarrow r_1/(1 - r_1) = r_2/(1 - r_2) \\ &\rightarrow r_1(1 - r_2) = r_2(1 - r_1) \\ &\rightarrow r_1 - r_1r_2 = r_2 - r_1r_2 \\ &\rightarrow r_1 = r_2 \end{aligned}$$

This shows that the function p is one-to-one. ■

Suppose we have a function $f: X \rightarrow Y$, where both X and Y are subsets of $\mathbb{R}$. How can we determine from the graph of f whether f is one-to-one? Well, if f is not one-to-one, then there exist $x_1 \neq x_2$ such that $f(x_1) = f(x_2)$. Letting $y_1 = f(x_1)$, we have the two distinct points (x_1, y_1) and (x_2, y_1) that are on the graph of f, and that are also on the horizontal line $y = y_1$. Conversely, if some horizontal line intersects the graph of f in more than one point, then f is not one-to-one. This yields the "horizontal line test":

the function f is one-to-one if and only if every horizontal line intersects the graph of f in at most one point.

Example 5.6 Apply the horizontal line test to determine whether the functions defined in Example 5.3 are one-to-one. (See Figure 5.1.)

Solution (a) The function $f:\mathbb{R} \to \mathbb{R}$ defined by $f(x) = 4 - 2x$ is one-to-one by the test.

(b) The function $f:(0, \infty) \to (0, \infty)$ defined by $f(x) = 1/x$ is also seen to be one-to-one.

(c) The function $f:\mathbb{R} \to [-4, \infty)$ defined by $f(x) = x^2 - 2x - 3$ fails the test. For example, the line $y = 0$ (x-axis) intersects the graph in the points $(-1,0)$ and $(3,0)$. Thus f is not one-to-one. ■

Given a function $f:A \to B$, what can be said about $\operatorname{im} f$? Knowing nothing else, all that can be said is that $\operatorname{im} f \subseteq B$. One extreme possibility is provided by choosing a fixed element $b \in B$ and defining $g:A \to B$ by $g(a) = b$ for all $a \in A$. In this case $\operatorname{im} g = \{b\}$ and we call g a *constant function* (the value of g is constant at b). The other extreme is the case of a function $f:A \to B$ for which $\operatorname{im} f = B$.

DEFINITION 5.3

A function $f:A \to B$ is called *onto* if $\operatorname{im} f = B$.

Observe that a function $f:A \to B$ is onto provided, for each $b \in B$, there exists an $a \in A$ such that $f(a) = b$. This condition provides a very common method for proving that a given function $f:A \to B$ is onto. Choose an arbitrary element $b \in B$, set $f(a) = b$, and then attempt to solve the equation for a in terms of b. If such a solution exists and is in A, then f is onto. On the other hand, if for some $b \in B$ there is no solution in A to the equation $f(a) = b$, then f is not onto.

Example 5.7 Determine which of the following functions are onto.

(a) $f:\{1, 2, 3\} \to \{1, 2, 3\};\quad f(1) = 2, f(2) = 2, f(3) = 1$

(b) $f:\{1, 2, 3\} \to \{1, 2, 3\};\quad f(1) = 3, f(2) = 2, f(3) = 1$

(c) $f:\mathbb{Z} \to \mathbb{Z};\quad f(m) = m - 1$

(d) $g:\mathbb{Z} \to \mathbb{Z};\quad g(m) = 3m + 1$

(e) $h:\mathbb{Z} \to \mathbb{N};\quad h(m) = |m| + 1$

(f) $p:\mathbb{Q} - \{1\} \to \mathbb{Q};\quad p(r) = r/(1 - r)$

(g) $g:\mathbb{Q} \to \mathbb{Q};\quad g(r) = 3r + 1$

Solution (a) Here f is not onto since $\operatorname{im} f = \{1, 2\} \neq \{1, 2, 3\}$.

(b) This f is onto.

(c) This function f is onto since, for any $n \in \mathbb{Z}$, $f(n+1) = (n+1) - 1 = n$.
(d) Let us attempt to show that g is onto. Let $n \in \mathbb{Z}$. We wish to find $m \in \mathbb{Z}$ such that $g(m) = n$. Now, $g(m) = n \leftrightarrow 3m + 1 = n$. This shows that $n \in \text{im } g \leftrightarrow n \equiv 1(\text{mod } 3)$. In other words, im $g = \{\ldots, -5, -2, 1, 4, 7, \ldots\}$. Thus g is not onto.
(e) For $n \in \mathbb{N}$,

$$\begin{aligned} h(m) = n &\leftrightarrow |m| + 1 = n \\ &\leftrightarrow |m| = n - 1 \\ &\leftrightarrow m = n - 1 \quad \text{or} \quad m = 1 - n \end{aligned}$$

Thus h is onto; in fact, each $n \in \mathbb{N}$, other than $n = 1$, is the image of two integers: $n - 1$ and $1 - n$.
(f) For $s \in \mathbb{Q}$,

$$\begin{aligned} p(r) = s &\leftrightarrow r/(1 - r) = s \\ &\leftrightarrow r = s - rs \\ &\leftrightarrow r + rs = s \\ &\leftrightarrow r = s/(s + 1) \end{aligned}$$

Thus if $s \neq -1$, then $r = s/(s+1) \in \mathbb{Q} - \{1\}$ and $p(r) = s$. However, there does not exist $r \in \mathbb{Q} - \{1\}$ such that $p(r) = -1$. Therefore, im $p = \mathbb{Q} - \{-1\}$, and the function p just misses being onto.
(g) Note that this function g has the same rule as the function of part (d), but the domain and codomain have been changed from $\mathbb{Z}$ to $\mathbb{Q}$. Let's see what happens. Let $s \in \mathbb{Q}$; we wish to find $r \in \mathbb{Q}$ such that $g(r) = s$. Now, $g(r) = s \leftrightarrow 3r + 1 = s \leftrightarrow r = (s - 1)/3$. This shows that g is onto; for each rational number s, the image of the rational number $r = (s - 1)/3$ under g is s. This example illustrates the important point that "ontoness" for a given function depends not only on the rule for the function, but on the domain and codomain as well. ■

We have seen examples of functions that are one-to-one and not onto, and the reverse possibility, functions that are onto but not one-to-one. Under what conditions does the existence of one condition imply the other? One very important case is supplied by the following theorem.

THEOREM 5.1 Let A and B be finite sets with $n(A) = n(B)$ and let f be a function from A to B. Then f is one-to-one if and only if f is onto.

Proof Let $n(A) = n(B) = m$ and suppose $A = \{a_1, a_2, \ldots, a_m\}$.

We first assume that f is one-to-one and show that f is onto. The image of f is the set

$$\begin{aligned} \text{im } f &= \{f(a) \mid a \in A\} \\ &= \{f(a_1), f(a_2), \ldots, f(a_m)\} \end{aligned}$$

We know that $\text{im } f \subseteq B$. If we can show that $n(\text{im } f) = m$, then we'll have both $\text{im } f \subseteq B$ and $n(\text{im } f) = n(B)$. We may then conclude that $\text{im } f = B$. To show that $n(\text{im } f) = m$, it suffices to show that $f(a_1), f(a_2), \ldots, f(a_m)$ are distinct. Suppose $f(a_i) = f(a_j)$ for some i and j. Since f is one-to-one, $f(a_i) = f(a_j)$ implies that $a_i = a_j$ and hence that $i = j$. This shows that $f(a_1)$, $f(a_2), \ldots, f(a_m)$ are distinct, and hence we conclude that $\text{im } f = B$.

Next we assume that f is onto and show that f is one-to-one. Since f is onto, $\text{im } f = B$. Then $\{f(a_1), f(a_2), \ldots, f(a_m)\} = B$ and $n(B) = m$, so it must be that $f(a_1), f(a_2), \ldots, f(a_m)$ are distinct. Hence $a_i \neq a_j$ implies that $f(a_i) \neq f(a_j)$, which shows that f is one-to-one. ■

Theorem 5.1 can be nicely applied in various situations. For example, consider the function $f: \mathbb{Z}_{30} \to \mathbb{Z}_{30}$ defined by $f([a]) = [7a]$. (We leave it to the reader to show that f is well-defined.) The following steps show that f is one-to-one:

$$
\begin{aligned}
f([a]) = f([b]) &\to [7a] = [7b] \\
&\to 7a \equiv 7b (\text{mod } 30) \\
&\to 30 \mid (7a - 7b) \\
&\to 30 \mid 7(a - b) \\
&\to 30 \mid (a - b) \qquad (\text{since } \gcd(7, 30) = 1) \\
&\to [a] = [b]
\end{aligned}
$$

Thus f is one-to-one. We now get that f is onto with no additional work; just apply Theorem 5.1!

We inject a word of caution with regard to the proper application of Theorem 5.1. In order to apply this theorem to a function $f: A \to B$, it must be made certain that A and B are finite sets and that $n(A) = n(B)$.

Let $A = \{a_1, a_2, \ldots, a_m\}$ and suppose that $f: A \to A$ is a one-to-one function. It then follows that the ordered m-tuple $(f(a_1), f(a_2), \ldots, f(a_m))$ is simply an ordered arrangement, indeed, a permutation of the set A. Accordingly, the function f is also called a permutation of A. In general, if A is any set and $f: A \to A$ is both one-to-one and onto, then the function f is also referred to as a *permutation* of A.

Example 5.8 As just shown, the function $f: \mathbb{Z}_{30} \to \mathbb{Z}_{30}$ defined by $f([a]) = [7a]$ is a permutation of $\mathbb{Z}_{30}$. From Examples 5.5(c) and 5.7(c), it follows that the function $f: \mathbb{Z} \to \mathbb{Z}$ defined by $f(m) = m - 1$ is a permutation of $\mathbb{Z}$. In Example 5.7(g), we showed that the function $g: \mathbb{Q} \to \mathbb{Q}$ defined by $g(r) = 3r + 1$ is onto. It is not hard to show that this function is also one-to-one. Therefore, g is a permutation of $\mathbb{Q}$. ■

Suppose now that we are given $f: A \to B$; then we know that $\text{im } f =$

$\{f(a) \mid a \in A\}$. It seems natural to write $\operatorname{im} f = f(A)$. More generally, if $C \subseteq A$, then we define the *image of C under f* to be the set

$$f(C) = \{f(c) \mid c \in C\}$$

Turn the situation around: if $D \subseteq B$, then how should we refer to those elements $a \in A$ for which $f(a) \in D$? In a sense, seeking such elements in A has the effect of reversing the direction of f. In other words, the action proceeds from D to A. Formally, we define the *inverse image of D under f* to be the set

$$f^{-1}(D) = \{a \in A \mid f(a) \in D\}$$

Example 5.9 Let g be the permutation of $\mathbb{Q}$ defined by $g(r) = 3r + 1$. Find the following sets:

(a) $g(\mathbb{Z})$
(b) $g(E)$, where E is the set of even integers
(c) $g^{-1}(\mathbb{N})$
(d) $g^{-1}(D)$, where D is the set of odd integers

Solution (a) If $m \in \mathbb{Z}$, then $g(m) = 3m + 1$. Thus $g(\mathbb{Z}) = \{3m + 1 \mid m \in \mathbb{Z}\}$.
(b) We have

$$\begin{aligned} k \in g(E) &\leftrightarrow g(2m) = k \quad \text{for some } m \in \mathbb{Z} \\ &\leftrightarrow 3(2m) + 1 = k \quad \text{for some } m \in \mathbb{Z} \\ &\leftrightarrow 6m + 1 = k \quad \text{for some } m \in \mathbb{Z} \end{aligned}$$

Thus $g(E) = \{6m + 1 \mid m \in \mathbb{Z}\}$.
(c) We have

$$\begin{aligned} r \in g^{-1}(\mathbb{N}) &\leftrightarrow 3r + 1 = n \quad \text{for some } n \in \mathbb{N} \\ &\leftrightarrow r = \frac{n-1}{3} \quad \text{for some } n \in \mathbb{N} \end{aligned}$$

Hence $g^{-1}(\mathbb{N}) = \{(n-1)/3 \mid n \in \mathbb{N}\} = \{0, \frac{1}{3}, \frac{2}{3}, 1, \frac{4}{3}, \frac{5}{3}, \ldots\}$.
(d) Similarly,

$$\begin{aligned} r \in g^{-1}(D) &\leftrightarrow 3r + 1 = 2m + 1 \quad \text{for some } m \in \mathbb{Z} \\ &\leftrightarrow r = \frac{2m}{3} \quad \text{for some } m \in \mathbb{Z} \end{aligned}$$

Hence $g^{-1}(D) = \{2m/3 \mid m \in \mathbb{Z}\} = \{\ldots, -\frac{4}{3}, -\frac{2}{3}, 0, \frac{2}{3}, \frac{4}{3}, \ldots\}$. ■

THEOREM 5.2 Given $f: A \to B$, let C and D be subsets of A and let E and F be subsets of B. Then the following relationships hold.

1. (a) $f(C \cup D) = f(C) \cup f(D)$
   (b) $f^{-1}(E \cup F) = f^{-1}(E) \cup f^{-1}(F)$

2. (a) $f(C \cap D) \subseteq f(C) \cap f(D)$
   (b) $f^{-1}(E \cap F) = f^{-1}(E) \cap f^{-1}(F)$
3. (a) $f(C) - f(D) \subseteq f(C - D)$
   (b) $f^{-1}(E) - f^{-1}(F) = f^{-1}(E - F)$
4. (a) If $C \subseteq D$, then $f(C) \subseteq f(D)$
   (b) If $E \subseteq F$, then $f^{-1}(E) \subseteq f^{-1}(F)$

Proof We prove 1(a) and 3(b). The proofs of the remaining parts are left to Exercise 2.

The proof that $f(C \cup D) = f(C) \cup f(D)$ proceeds as follows:

$$\begin{aligned} y \in f(C \cup D) &\leftrightarrow \exists x \in C \cup D(y = f(x)) \\ &\leftrightarrow (\exists x \in C(y = f(x))) \vee (\exists x \in D(y = f(x))) \\ &\leftrightarrow y \in f(C) \vee y \in f(D) \\ &\leftrightarrow y \in f(C) \cup f(D) \end{aligned}$$

Thus $f(C \cup D) = f(C) \cup f(D)$.

The proof that $f^{-1}(E) - f^{-1}(F) = f^{-1}(E - F)$ proceeds along similar lines:

$$\begin{aligned} x \in (f^{-1}(E) - f^{-1}(F)) &\leftrightarrow x \in f^{-1}(E) \wedge x \notin f^{-1}(F) \\ &\leftrightarrow f(x) \in E \wedge f(x) \notin F \\ &\leftrightarrow f(x) \in E - F \\ &\leftrightarrow x \in f^{-1}(E - F) \end{aligned}$$

Hence $f^{-1}(E) - f^{-1}(F) = f^{-1}(E - F)$. ■

There are several additional properties involving the notions of image and inverse image. Some of these will be addressed in the exercises.

Exercises 5.2

1. Verify whether each of the following functions is (i) one-to-one and (ii) onto.

   a. $f: \mathbb{Z} \rightarrow \mathbb{N}$; $f(x) = x^2 + 1$ **b.** $g: \mathbb{R} \rightarrow \mathbb{R}$; $g(x) = x^3$

   c. $h: \mathbb{R} \rightarrow \mathbb{R}$; $h(x) = x^3 - x$ **d.** $p: \mathbb{Q} \rightarrow \mathbb{R}$; $p(x) = 2^x$

   e. The "cardinality function" n from $\mathcal{P}(A)$, where A is a finite nonempty set, to $\{0, 1, \ldots, n(A)\}$: for $B \in \mathcal{P}(A)$, $n(B)$ = the number of elements in B

   f. The "complement function" c on $\mathcal{P}(A)$: for $B \in \mathcal{P}(A)$, $c(B) = A - B$

2. Prove the remaining parts of Theorem 5.2: part 1(b), parts 2(a) and 2(b), part 3(a), and parts 4(a) and 4(b).

3. Each of the following parts gives sets A and B and a function $f: A \rightarrow B$.

Determine (i) whether f is one-to-one, (ii) whether f is onto, and (iii) if $A = B$, whether f is a permutation.

a. $A = \{1, 2, 3, 4\}$, $B = \{1, 2, 3\}$, $f(1) = 2, f(2) = 3, f(3) = 1, f(4) = 1$
b. $A = \{1, 2, 3\}$, $B = \{1, 2, 3, 4\}$, $f(1) = 3, f(2) = 2, f(3) = 1$
c. $A = B = \{1, 2, 3, 4\}$, $f(1) = 2, f(2) = 1, f(3) = 2, f(4) = 1$
d. $A = B = \{1, 2, 3, 4\}$, $f(1) = 3, f(2) = 4, f(3) = 1, f(4) = 2$
e. $A = B = \mathbb{Z}$, $f(n) = -n$
f. $A = B = \mathbb{Z}$, if $n \geq 0, f(n) = n + 1$, if $n < 0, f(n) = n - 1$
g. $A = \{1, 2, 3\}$, $B = \{0, 1, 2\}$, $f(1) = 0, f(2) = 1, f(3) = 2$

4. Let A and B be finite nonempty sets with $n(A) = s$ and $n(B) = t$. What can be said about the possibility that a function $f\colon A \to B$ is one-to-one or onto if

a. $s < t$? **b.** $t < s$?

5. For each of the following functions $f\colon \mathbb{Z}_{12} \to \mathbb{Z}_{12}$, (i) determine whether f is a permutation, (ii) find $f(C)$ where $C = \{[1], [5], [7], [11]\}$, and (iii) find $f^{-1}(D)$ where $D = \{[4], [8]\}$.

a. $f([n]) = [2n]$ **b.** $f([n]) = [4n]$ **c.** $f([n]) = [5n]$

6. Give an example of a function $f\colon \mathbb{N} \to \mathbb{N}$ that is

a. neither one-to-one nor onto;
b. one-to-one but not onto;
c. onto but not one-to-one;
d. both one-to-one and onto.

7. Each of the following parts refers to the corresponding function $f\colon A \to B$ defined in Exercise 3. For $C \subseteq A$ and $D \subseteq B$, find $f(C)$ and $f^{-1}(D)$.

a. $C = \{1, 2\} = D$
b. $C = \{1, 3\}$, $D = \{2, 4\}$
c. $C = \{1, 3\}$, $D = \{1\}$
d. $C = \{2\} = D$
e. $C = \{2m \mid m \in \mathbb{Z}\}$, $D = \mathbb{N}$
f. $C = \mathbb{N}$, $D = \{2m \mid m \in \mathbb{Z}\}$
g. $C = \{1\} = D$

5.3 INVERSE FUNCTIONS AND COMPOSITION

Suppose that $f\colon A \to B$ is both one-to-one and onto. Since f is onto, given any $b \in B$, there is an element $a \in A$ such that $f(a) = b$. Moreover, since f is one-to-one, this element a is uniquely determined. Thus for each $b \in B$ there is exactly one $a \in A$ such that $b = f(a)$. We can then define a new function $g\colon B \to A$ as follows: for each $b \in B$,

$$g(b) = a \leftrightarrow b = f(a)$$

In other words, $g(b)$ is that unique element $a \in A$ for which $f(a) = b$.

DEFINITION 5.4

Let $f:A \to B$ be one-to-one and onto. The function $g:B \to A$ defined by

$$g(b) = a \leftrightarrow b = f(a)$$

for each $b \in B$ is called the *inverse function* of f and is denoted by f^{-1}.

The situation of a function $f:A \to B$ and its inverse $f^{-1}:B \to A$ is depicted in Figure 5.2, where $f(a) = b$.

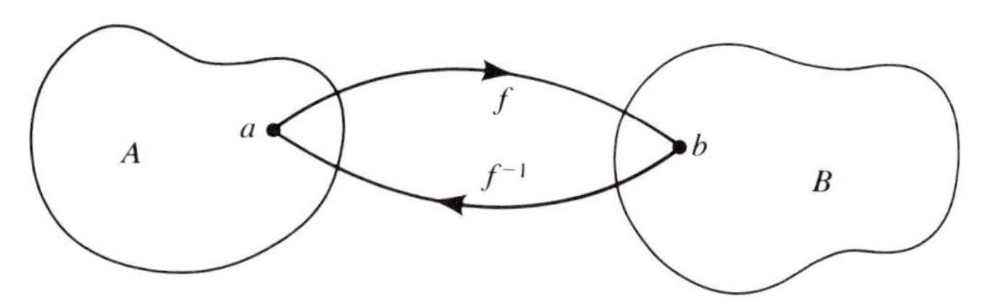

FIGURE 5.2

Let $f:A \to B$ be a one-to-one and onto function and let $D \subseteq B$. At this point we have two interpretations for the notation $f^{-1}(D)$: one is the inverse image of D under the function f, and the other is the image of D under the function f^{-1}. In Exercise 8, the reader will be asked to show that these two sets are equal. It should be pointed out, however, that if f is not both one-to-one and onto, then only the former interpretation for $f^{-1}(D)$ applies.

THEOREM 5.3 If $f:A \to B$ is both one-to-one and onto, then the inverse function $f^{-1}:B \to A$ is also one-to-one and onto.

Proof We first show that f^{-1} is one-to-one. Suppose $f^{-1}(b_1) = f^{-1}(b_2) = a$ for some $b_1, b_2 \in B$ and $a \in A$. Then, by definition, $b_1 = f(a)$ and $b_2 = f(a)$. Since f is a function, it follows that $b_1 = b_2$. Thus f^{-1} is one-to-one.

Next we show that f^{-1} is onto. Let $a \in A$, and suppose $f(a) = b$. Then, again by definition, $a = f^{-1}(b)$, so f^{-1} is onto. ∎

If $f:A \to B$ is one-to-one and onto, how do we find the inverse function $f^{-1}:B \to A$? For example, given $f:\mathbb{R} \to \mathbb{R}$ defined by $f(x) = 3x - 7$, how can we find f^{-1}? Given $r \in \mathbb{R}$, we want to find $x \in \mathbb{R}$ such that $f^{-1}(r) = x$. This means that $r = f(x)$ or, in particular, that $r = 3x - 7$. Elementary algebra shows that $x = (r + 7)/3$, hence $f^{-1}(r) = (r + 7)/3$.

Example 5.10 In Examples 5.5 and 5.7, it was shown that the function $p:\mathbb{Q} - \{1\} \to \mathbb{Q} - \{-1\}$ defined by $p(r) = r/(1 - r)$ is one-to-one and onto. Find p^{-1}.

Solution If $x \in \mathbb{Q} - \{-1\}$ and $p^{-1}(x) = r$, where $r \in \mathbb{Q} - \{1\}$, then $p(r) = x$. Thus $r/(1 - r) = x$, or $r = x(1 - r)$. Solving for r, we obtain $r = x/(x + 1)$. Therefore, $p^{-1}:\mathbb{Q} - \{-1\} \to \mathbb{Q} - \{1\}$ is given by $p^{-1}(x) = x/(x + 1)$. ■

Example 5.11 In the last section it was shown that the function $f:\mathbb{Z}_{30} \to \mathbb{Z}_{30}$, defined by $f([a]) = [7a]$, is a permutation of $\mathbb{Z}_{30}$. If X is a set and $f:X \to X$ is a permutation of X, then it follows from Theorem 5.3 that f^{-1} is also a permutation of X. Find f^{-1} for the function $f:\mathbb{Z}_{30} \to \mathbb{Z}_{30}$ just given.

Solution We have that

$$
\begin{aligned}
f^{-1}([b]) = [a] &\leftrightarrow f([a]) = [b] \\
&\leftrightarrow [7a] = [b] \\
&\leftrightarrow b = 30q + 7a \quad \text{for some } q \in \mathbb{Z} \\
&\leftrightarrow 13b = 13(30q + 7a) \quad \text{for some } q \in \mathbb{Z} \\
&\leftrightarrow 13b = 30(13q + 3a) + a \quad \text{for some } q \in \mathbb{Z} \\
&\leftrightarrow 13b - a = 30(13q + 3a) \quad \text{for some } q \in \mathbb{Z} \\
&\leftrightarrow [a] = [13b]
\end{aligned}
$$

Therefore, $f^{-1}([b]) = [13b]$. ■

There are various ways in which two functions may be combined to produce a third function. One of the most common and important of these is called "composition." Suppose $f:A \to B$ and $g:B \to C$ are given. Then, for any $a \in A$, there is a unique $b \in B$ such that $f(a) = b$. For this element b, there is a unique $c \in C$ such that $c = g(b) = g(f(a))$. Hence with each $a \in A$ there is associated a unique element $c \in C$, where $c = g(f(a))$. This association allows us to define a new function $h:A \to C$ by $h(a) = g(f(a))$. This situation is depicted in Figure 5.3.

DEFINITION 5.5

Given $f:A \to B$ and $g:B \to C$, we define the *composite function* $g \circ f:A \to C$ as follows: for each $a \in A$,

$$(g \circ f)(a) = g(f(a))$$

Example 5.12 (a) Let E be the set of even integers, let $f:\mathbb{Z} \to E$ be defined by $f(m) = 2m$, and let $g:E \to \mathbb{N}$ be defined by $g(n) = \frac{|n|}{2} + 1$. Then $g \circ f:\mathbb{Z} \to \mathbb{N}$ is defined by

$$(g \circ f)(x) = g(f(x)) = g(2x) = \frac{|2x|}{2} + 1 = |x| + 1$$

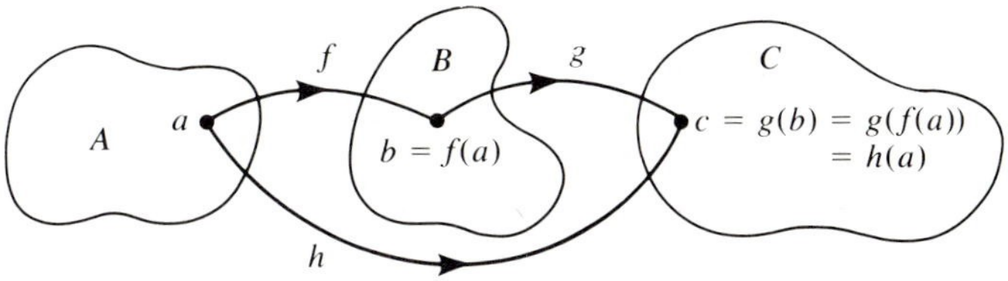

FIGURE 5.3

(b) Let $f:\mathbb{R} - \{0\} \to \mathbb{R} - \{1\}$ and $g:\mathbb{R} - \{1\} \to \mathbb{R} - \{2\}$ be given by $f(x) = (x + 1)/x$ and $g(x) = 3x - 1$. Then $g \circ f:\mathbb{R} - \{0\} \to \mathbb{R} - \{2\}$ is defined by

$$\begin{aligned}(g \circ f)(x) = g(f(x)) &= g((x + 1)/x) \\ &= \frac{3(x+1)}{x} - 1 \\ &= \frac{3(x+1)}{x} - \frac{x}{x} \\ &= \frac{2x+3}{x}\end{aligned}$$

■

Example 5.13 Let $f:\mathbb{Z} \to \mathbb{Z}$ and $g:\mathbb{Z} \to \mathbb{Z}$ be defined by $f(x) = x + 3$ and $g(x) = -x$. Find (a) $f \circ g:\mathbb{Z} \to \mathbb{Z}$, and (b) $g \circ f:\mathbb{Z} \to \mathbb{Z}$.

Solution (a) $(f \circ g)(x) = f(g(x)) = f(-x) = -x + 3$
(b) $(g \circ f)(x) = g(f(x)) = g(x + 3) = -(x + 3) = -x - 3$
It should be noted that $f \circ g \neq g \circ f$, and that f, g, $f \circ g$, and $g \circ f$ are all permutations of $\mathbb{Z}$. ■

In what follows, we shall be interested in knowing when certain functions are "equal." What does this mean? After all, if f and g are functions, then each is a set of ordered pairs, and perhaps equality should simply mean that these sets are equal. This condition would certainly imply that the two functions have the same domain and image, although not necessarily the same codomain. In fact, we shall insist that the codomains be the same as well.

DEFINITION 5.6

The functions $f:A \to B$ and $g:C \to D$ are *equal* if and only if $A = C$, $B = D$, and $f(a) = g(a)$ for all $a \in A$.

Thus, for instance, the functions $f_1:\mathbb{R} \to \mathbb{R}$ and $f_2:\mathbb{R} \to [0,\infty)$ defined by $f_1(x) = f_2(x) = x^2$ are not equal by our definition.

There are several interesting results that involve composition of functions and the properties onto and one-to-one. For example, suppose $f:A \to B$ is both one-to-one and onto; thus $f^{-1}:B \to A$ exists. Given any $a \in A$, if $f(a) = b$, then we know that $f^{-1}(b) = a$ and hence that $f^{-1}(f(a)) = a$. So the composite function $f^{-1} \circ f:A \to A$ satisfies the property $(f^{-1} \circ f)(a) = a$ for all $a \in A$. In a similar fashion, we can determine that $f \circ f^{-1}:B \to B$ satisfies the condition $(f \circ f^{-1})(b) = b$ for every $b \in B$. Note that each of $f \circ f^{-1}$ and $f^{-1} \circ f$ is a function of the type $h:X \to X$, where $h(x) = x$ for all $x \in X$.

DEFINITION 5.7

For any nonempty set A, the function $I_A:A \to A$ defined by $I_A(a) = a$ for all $a \in A$ is called the *identity function on A*.

In view of the preceding discussion, if $f:A \to B$ is one-to-one and onto, then $f^{-1} \circ f = I_A$ and $f \circ f^{-1} = I_B$. A very basic and easily verified property of identity functions is contained in the following theorem.

THEOREM 5.4 Given any function $f:A \to B$, $I_B \circ f = f$ and $f \circ I_A = f$.

Proof See Exercise 2. ■

THEOREM 5.5 Given $f:A \to B$ and $g:B \to C$, the following hold:

1. If f and g are both one-to-one, then $g \circ f$ is one-to-one.
2. If f and g are both onto, then $g \circ f$ is onto.

Proof First consider the proof of 1. Suppose, for $a_1, a_2 \in A$, that $(g \circ f)(a_1) = (g \circ f)(a_2)$. Then, by the definition of composition, $g(f(a_1)) = g(f(a_2))$. Since g is one-to-one, we have $f(a_1) = f(a_2)$. Then, since f is also one-to-one, $a_1 = a_2$. Hence $g \circ f$ is one-to-one.

Next, consider the proof of 2. To prove that $g \circ f$ is onto, begin with an arbitrary element $c \in C$. Since g is onto, there is an element $b \in B$ for which $g(b) = c$. Since $b \in B$ and since f is onto, there is an element $a \in A$ for which $f(a) = b$. Thus $(g \circ f)(a) = g(f(a)) = g(b) = c$ and it follows that $g \circ f$ is onto. ■

COROLLARY 5.5 If f and g are permutations of the set A, then $f \circ g$ is also a permutation of A. ■

For each of the statements 1 and 2 in Theorem 5.5, the converse is false (see Exercise 4). However, a partial converse does hold.

THEOREM 5.6 Given $f:A \to B$ and $g:B \to C$, the following hold:

1. If $g \circ f$ is one-to-one, then f is one-to-one.
2. If $g \circ f$ is onto, then g is onto.

Proof We prove only part 1. Suppose then that $f(a_1) = f(a_2)$ for some $a_1, a_2 \in A$. Then since $f(a_1) \in B$, we have that $g(f(a_1)) = g(f(a_2))$ or $(g \circ f)(a_1) = (g \circ f)(a_2)$. Since $g \circ f$ is given to be one-to-one, we may conclude that $a_1 = a_2$. Thus the function f is one-to-one. ■

Given functions $f:A \to B$, $g:B \to C$, and $h:C \to D$, we see that $h \circ g$ is a function from B to D and $g \circ f$ is a function from A to C. Thus $(h \circ g) \circ f$ and $h \circ (g \circ f)$ are both functions from A to D. In fact, they are equal!

THEOREM 5.7 Given functions $f:A \to B$, $g:B \to C$, and $h:C \to D$,

$$(h \circ g) \circ f = h \circ (g \circ f)$$

Proof Both $(h \circ g) \circ f$ and $h \circ (g \circ f)$ have domain A and codomain D. If $a \in A$, then

$$\begin{aligned}(h \circ (g \circ f))(a) &= h((g \circ f)(a)) \\ &= h(g(f(a))) \\ &= (h \circ g)(f(a)) \\ &= ((h \circ g) \circ f)(a)\end{aligned}$$

Thus, by the definition of equality of functions, $(h \circ g) \circ f = h \circ (g \circ f)$. ■

Exercises 5.3

1. For each of the following functions f, find f^{-1}.

a. $f:\mathbb{Q} \to \mathbb{Q}$; $f(x) = 4x + 2$

b. $f:\mathbb{R} - \{1\} \to \mathbb{R} - \{1\}$; $f(x) = x/(x - 1)$

c. $f:\mathbb{Z}_{12} \to \mathbb{Z}_{12}$; $f([x]) = [5x]$

d. $f:\mathbb{Z} \to \mathbb{N}$; $f(x) = \begin{cases} 2x + 1, & x \geq 0 \\ -2x, & x < 0 \end{cases}$

e. $f:\{1, 2, 3, 4\} \to \{1, 2, 3, 4\}$; $f(1) = 4, f(2) = 1, f(3) = 2, f(4) = 3$

f. $f:\{1, 2, 3, 4\} \to \{1, 2, 3, 4\}$; $f(1) = 3, f(2) = 4, f(3) = 1, f(4) = 2$

2. Prove Theorem 5.4.

3. Find $g \circ f$.

a. $f:\mathbb{Z} \to \mathbb{Z}_5$; $f(n) = [n]$, $g:\mathbb{Z}_5 \to \mathbb{Z}_5$; $g([n]) = [-n]$

b. $f:\mathbb{R} \to (0, 1)$; $f(x) = 1/(x^2 + 1)$, $g:(0, 1) \to (0, 1)$; $g(x) = 1 - x$

c. $f:\mathbb{R} - \{2\} \to \mathbb{R} - \{0\}$; $f(x) = 1/(x - 2)$, $g:\mathbb{R} - \{0\} \to \mathbb{R} - \{0\}$; $g(x) = 1/x$

d. $f:\mathbb{R} \to [1, \infty)$; $f(x) = x^2 + 1$, $g:[1, \infty) \to [0, \infty)$; $g(x) = \sqrt{x - 1}$

e. $f:\{1, 2, 3, 4\} \to \{1, 2, 3, 4\}$; $f(1) = 4, f(2) = 1, f(3) = 2, f(4) = 3$,
$g:\{1, 2, 3, 4\} \to \{1, 2, 3, 4\}$; $g(1) = 3$, $g(2) = 4$, $g(3) = 1$, $g(4) = 2$

4. Give examples of functions $f:A \to B$ and $g:B \to C$ such that

a. $g \circ f$ and f are both one-to-one but g is not;

b. $g \circ f$ and g are both onto but f is not.

5. Find f^{-1}, g^{-1}, $f \circ g$, $g \circ f$, $(f \circ g)^{-1}$, and $g^{-1} \circ f^{-1}$, where f and g are the following permutations:

a. $f:\mathbb{Z} \to \mathbb{Z}$; $f(x) = x + 1$, $g:\mathbb{Z} \to \mathbb{Z}$; $g(x) = 2 - x$

b. $f:\mathbb{Z}_7 \to \mathbb{Z}_7$; $f([x]) = [x + 3]$, $g:\mathbb{Z}_7 \to \mathbb{Z}_7$; $g([x]) = [2x]$

c. $f:\{1, 2, 3, 4\} \to \{1, 2, 3, 4\}$; $f(1) = 4, f(2) = 1, f(3) = 2, f(4) = 3$,
$g:\{1, 2, 3, 4\} \to \{1, 2, 3, 4\}$; $g(1) = 3$, $g(2) = 4$, $g(3) = 1$, $g(4) = 2$

6. Define functions f and g on your family tree by

$$f(x) = \text{the father of } x$$

$$g(x) = \text{the eldest child of the father of } x$$

Describe $f \circ f$, $f \circ g$, $g \circ f$, and $g \circ g$.

7. Prove: If $f:A \to B$ and $g:B \to C$ are both one-to-one and onto, then $(g \circ f)^{-1} = f^{-1} \circ g^{-1}$.

8. Let $f:A \to B$ be a one-to-one and onto function and let $D \subseteq B$. Let C_1 be the inverse image of D under f and let C_2 be the image of D under f^{-1}. Show that $C_1 = C_2$.

9. Let $f:A \to B$ and $g:B \to A$ both be one-to-one and onto functions. Prove: If $f \circ g = I_B$ and $g \circ f = I_A$, then $g = f^{-1}$.

5.4 RECURSION AND ITERATION

Suppose that we are given three pegs, labelled peg 1, peg 2, and peg 3, planted vertically on a board. In addition, we are given n circular disks of different diameters, each disk having a hole through its center. The disks are labelled $d_1, d_2, \ldots, d_n$ in order of decreasing size (d_1 the largest and d_n the smallest), and then placed on peg 1 in the order $d_1, d_2, \ldots, d_n$ (see Figure 5.4).

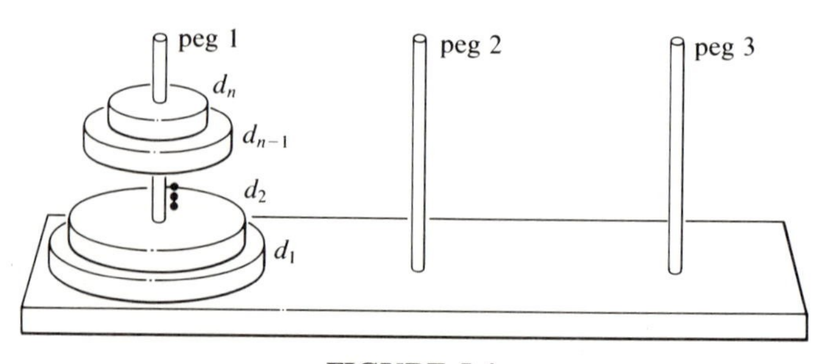

FIGURE 5.4

We wish to consider the problem of transferring the n disks from peg 1 to some other peg, say peg 3, subject to the following rules:

1. Each move consists of transferring the top disk from one peg to another peg.
2. At no time can a larger disk be placed atop a smaller one.

How many moves are required to accomplish this?

This problem is known as the *Towers of Hanoi problem*. We begin our attempt at a solution by letting $h(n)$ denote the number of moves required for n disks, where $n \geq 1$. Clearly, $h(1) = 1$. Suppose $n > 1$; then in order to transfer d_1 to a disk-free peg 3, we must first transfer $d_2, d_3, \ldots,$ d_n from peg 1 to peg 2. According to our notational convention, this will require $h(n - 1)$ moves. If this has been accomplished, then d_1 can be transferred from peg 1 to peg 3 in one move. At this point we have the situation depicted in Figure 5.5, with $h(n - 1) + 1$ moves having taken place. Finally, $h(n - 1)$ moves are required to transfer $d_2, d_3, \ldots, d_n$ from peg 2 to peg 3, and the desired transfer can be accomplished in $(h(n - 1) + 1) + h(n - 1) = 2h(n - 1) + 1$ moves.

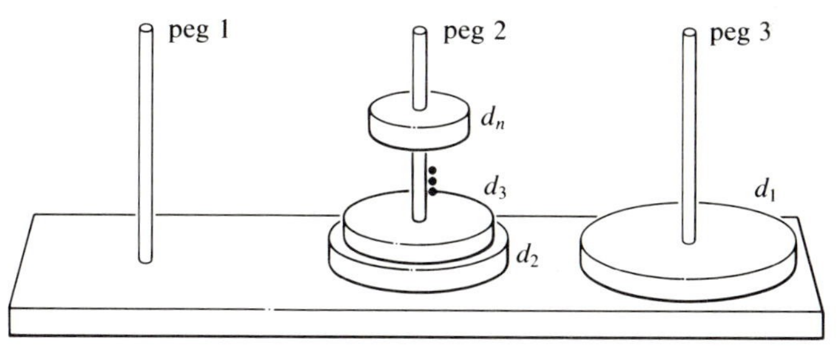

FIGURE 5.5

Thus we have the following information about $h(n)$:

1. $h(1) = 1$
2. $h(n) = 2h(n - 1) + 1$ for $n \geq 2$

We seek an explicit formula for $h(n)$ in terms of n, for each $n \geq 1$. What we have obtained so far is the value of $h(1)$ and a description of $h(n)$ in terms of $h(n - 1)$ for $n \geq 2$. This is an example of a "recursively defined" function.

DEFINITION 5.8

A function $f:\mathbb{N} \to \mathbb{N}$ is said to be *recursively* (or *inductively*) *defined* if, for some positive integer n_0, the following hold:

1. $f(1), f(2), \ldots, f(n_0)$ are known.
2. For $n > n_0$, $f(n)$ is defined in terms of $f(k)$ for some values of k, $1 \le k \le n - 1$.

We call this description of f a *recurrence relation*. We refer to $f(1)$, $f(2), \ldots, f(n_0)$ as *initial values*.

Example 5.14 Recall that the numbers 1, 1, 2, 3, 5, 8, 13, . . . are known as the Fibonacci numbers. Letting $f(n)$ denote the nth Fibonacci number, $n \ge 1$, the function $f(n)$ is defined recursively as follows:

1. $f(1) = f(2) = 1$
2. For $n \ge 3$, $f(n) = f(n-2) + f(n-1)$

Note that the recurrence relation states that each Fibonacci number, from the third on, is the sum of the previous two Fibonacci numbers. ■

Example 5.15 Let $g(n)$ be the number of k-element subsets of an n-element set, where k is a fixed positive integer and $n \ge k$. (Then $g(n) = C(n,k)$.) We note that $g(k) = 1$. Find a recurrence relation for $g(n)$, $n > k$.

Solution Let A be a set with n elements, $n > k$. We begin by considering the k-element subsets of $A - \{x\}$, where x is an arbitrary (but fixed) element of A. Since $A - \{x\}$ has $n - 1$ elements, it will have $g(n-1)$ k-element subsets. Summing now over all n elements that can be x gives a total of $n \cdot g(n-1)$ k-element subsets. (We illustrate the case $n = 5$ and $k = 3$, where $A = \{1, 2, 3, 4, 5\}$, at the end of this example.) However, these $n \cdot g(n-1)$ subsets are not distinct. So we must ask ourselves, How many times does each k-element subset appear? Well, a given k-element subset B will appear as a subset of $A - \{x\}$ whenever $x \in A - B$, and there are $n - k$ such x's. (As is illustrated for $n = 5$ and $k = 3$, the subset $B = \{1, 2, 3\}$ appears twice in the list, for $x = 4$ and for $x = 5$.) Therefore, the number of distinct k-element subsets of A is $(n \cdot g(n-1))/(n-k)$, which yields the recurrence relation

$$g(n) = \frac{n \cdot g(n-1)}{n-k}, \qquad n > k$$

The 3-element subsets of $\{1, 2, 3, 4, 5\} - \{x\}$ are listed in tabular form. Note that each subset appears twice.

$x = 1$	$x = 2$	$x = 3$	$x = 4$	$x = 5$
$\{2, 3, 4\}$	$\{1, 3, 4\}$	$\{1, 2, 4\}$	$\{1, 2, 3\}$	$\{1, 2, 3\}$
$\{2, 3, 5\}$	$\{1, 3, 5\}$	$\{1, 2, 5\}$	$\{1, 2, 5\}$	$\{1, 2, 4\}$
$\{2, 4, 5\}$	$\{1, 4, 5\}$	$\{1, 4, 5\}$	$\{1, 3, 5\}$	$\{1, 3, 4\}$
$\{3, 4, 5\}$	$\{3, 4, 5\}$	$\{2, 4, 5\}$	$\{2, 3, 5\}$	$\{2, 3, 4\}$

■

Example 5.16 Consider character strings of length $n \geq 1$ consisting entirely of the letters 'A' and 'B'. Let $l(n)$ be the number of such strings that do not contain two consecutive letters 'A'. Then $l(1) = 2$ ('A' and 'B') and $l(2) = 3$ ('AB', 'BA', and 'BB'). Find a recurrence relation for $l(n)$, $n \geq 3$.

Solution Let x denote a string of length n having no consecutive letters 'A'. If the first character of x is 'B', then x may be completed in $l(n - 1)$ ways, with any substring of length $n - 1$ having no consecutive letters 'A'. On the other hand, if the first character of x is 'A', then necessarily the second character of x is 'B'; now x may be completed in $l(n - 2)$ ways. It follows that

$$l(n) = l(n - 1) + l(n - 2), \qquad n \geq 3$$

Note that the recurrence relation for l is the same as that for the Fibonacci numbers; only the initial values differ. ■

We now return to the problem of the Towers of Hanoi and attempt to find an explicit formula for the function $h(n)$. At this point it is known that $h(1) = 1$ and, for $n \geq 2$, $h(n) = 2h(n - 1) + 1$. Thus it is readily seen that $h(2) = 2h(1) + 1 = 2 \cdot 1 + 1 = 3$. In fact, we can proceed as follows to obtain successive values of h:

$$\begin{aligned} h(3) &= 2h(2) + 1 = 2 \cdot 3 + 1 = 7 \\ h(4) &= 2h(3) + 1 = 2 \cdot 7 + 1 = 15 \\ h(5) &= 2h(4) + 1 = 2 \cdot 15 + 1 = 31 \end{aligned}$$

and so on.

This method of obtaining successive values of a recursively defined function is called *iteration*. Iteration is easily implemented in most programming languages using a counter-controlled loop; with a computer, the values of a function like $h(n)$ can be found for many values of n. To illustrate, the following Pascal segment uses iteration (a FOR loop) to output values of $h(n)$ for $n = 1, 2, \ldots, 25$.

```
BEGIN
  WRITELN('VALUE OF N      VALUE OF H(N)');
  N := 1;  (*initialize N*)
  H := 1;  (*initialize H(N)*)
  WRITELN(N:10,H:17);
  FOR N := 2 TO 25 DO
     BEGIN
       H := 2*H + 1;  (*use the recurrence relation to
                        compute next value of H(N)*)
       WRITELN(N:10,H:17)
     END
END.
```

Upon inspection of several values of $h(n)$, perhaps the student can guess a formula for $h(n)$ in terms of n. It certainly appears that $h(n) = 2^n - 1$ is a good guess. Can we in fact prove such a formula holds for each $n \in \mathbb{N}$? What method of proof should be employed? Induction, of course!

As usual, let S be the set of all positive integers n for which $h(n) = 2^n - 1$.

1. Show that $1 \in S$. We have already observed that $h(1) = 1 = 2^1 - 1$, so $1 \in S$.
2. Assume $k \in S$ for some arbitrary $k \geq 1$. The induction hypothesis (IHOP) is that $h(k) = 2^k - 1$.
3. Show that $k + 1 \in S$. We must show that $h(k + 1) = 2^{k+1} - 1$. Observe that

$$\begin{aligned} h(k+1) &= 2h(k) + 1 \qquad \text{(by the recurrence relation)} \\ &= 2(2^k - 1) + 1 \qquad \text{(by IHOP)} \\ &= 2^{k+1} - 1 \end{aligned}$$

   Hence $k + 1 \in S$.

It follows that $S = \mathbb{N}$, namely, that $h(n) = 2^n - 1$ for all $n \in \mathbb{N}$. Thus, in the Towers of Hanoi problem, $2^n - 1$ moves are required to transfer the n disks from peg 1 to peg 3.

Example 5.17 Find a formula for the function $g(n)$, where $g(1) = 1$ and, for $n \geq 2$, $g(n) = g(n - 1) + 2n - 1$.

Solution The first several values of $g(n)$ are as follows:

$$\begin{aligned} g(2) &= g(1) + 2(2) - 1 = 4 \\ g(3) &= g(2) + 2(3) - 1 = 9 \\ g(4) &= g(3) + 2(4) - 1 = 16 \end{aligned}$$

It seems reasonable at this stage to conjecture that $g(n) = n^2$.

It is easy to check that the function $g(n) = n^2$ satisfies the recurrence relation $g(n) = g(n - 1) + 2n - 1$ and the initial condition $g(1) = 1$. What we would like to show is that the function $g(n) = n^2$ is *the only function* from $\mathbb{N}$ to $\mathbb{N}$ that satisfies the given recurrence relation and initial condition. In other words, we must prove that if $g:\mathbb{N} \to \mathbb{N}$ has the properties

1. $g(1) = 1$
2. $g(n) = g(n-1) + 2n - 1 \quad \text{for all } n \geq 2$

then $g(n) = n^2$. The proof is by induction and the details are left to the reader. ■

It should be mentioned that general methods exist for solving certain kinds of recurrence relations, but it is not our aim to go into them here. Such methods are normally covered in a course in combinatorics, or in a more advanced course in discrete methods. Our aim is to emphasize the process of finding a recurrence relation for a particular function (which is quite often the hardest part). Once found, a program can then be written that uses the recurrence relation and the initial values to compute values of the function.

It has already been observed how an iterative program can be written to find successive values of a recursively defined function. What is especially intriguing is that many high-level languages, like Pascal, allow the writing of subprograms that call themselves. Note the analogy with the way in which a recursive function is defined in terms of itself. As a result, such programs are said to be *recursive*.

As an illustration, we present a recursive Pascal function to evaluate n factorial. Note that $n!$ can be defined recursively as follows:

$$0! = 1$$
$$n! = n \cdot (n-1)! \quad \text{for } n \geq 1$$

The Pascal program is based on this formula.

```
FUNCTION FACTORIAL(N:INTEGER):INTEGER;
  VAR M,T : INTEGER;
  BEGIN
    IF N = 0 THEN
      FACTORIAL := 1 (*if N = 0, then return 1 as the
                        value of FACTORIAL*)
    ELSE
      BEGIN
        M := N - 1;   (*else assign the value N - 1 to M*)
        T := FACTORIAL(M);  (*call FACTORIAL recursively
                               to find the value M!*)
        FACTORIAL := N * T  (*and then return N·M! as the
                               value of FACTORIAL*)
      END
  END;
```

Suppose now that, in a main program that has defined FACTORIAL, a reference is made to FACTORIAL(3). Then calls to the function FACTORIAL are made and assignment statements are executed as indicated by the following outline:

1. A call is made to FACTORIAL with N = 3:

```
M := N - 1;  (*M is assigned the value 2*)
T := FACTORIAL(M);  (*Before it can be assigned to T, the value
                      of FACTORIAL(2) must be computed*)
```

2. A call is made to FACTORIAL with N = 2:

```
M := N - 1;  (*M is assigned the value 1*)
T := FACTORIAL(M);  (*The value of FACTORIAL(1) is needed*)
```

3. A call is made to FACTORIAL with N = 1:

```
M := N - 1;  (*M is assigned the value 0*)
T := FACTORIAL(M);  (*The value of FACTORIAL(0) is needed*)
```

4. A call is made to FACTORIAL with N = 0:

```
FACTORIAL := 1;  (*The value of FACTORIAL(0) is 1*)
```

5. Control now returns to FACTORIAL with N = 1:

```
T := FACTORIAL(M);  (*T is assigned the value of FACTORIAL(0) = 1*)
FACTORIAL := N * T;  (*The value of FACTORIAL(1) is 1·1 = 1*)
```

6. Control returns to FACTORIAL with N = 2:

```
T := FACTORIAL(M);  (*T is assigned the value of FACTORIAL(1) = 1*)
FACTORIAL := N * T;  (*The value of FACTORIAL(2) is 2·1 = 2*)
```

7. Control returns to FACTORIAL with N = 3:

```
T := FACTORIAL(M);  (*T is assigned the value of FACTORIAL(2) = 2*)
FACTORIAL := N * T;  (*The value of FACTORIAL(3) is 3·2 = 6*)
```

8. Return control to the main program; the value of FACTORIAL(3) is 6.

Perhaps it is instructive to consider an iterative function for evaluating n factorial. The reader is urged to trace the following function, say with

N = 3, just to see how it contrasts with the recursive function FACTORIAL.

```
FUNCTION IFACTORIAL(N:INTEGER):INTEGER;
  VAR PRODUCT,FACTOR : INTEGER;
  BEGIN
    PRODUCT := 1;
    FOR FACTOR := 1 TO N DO
      PRODUCT := PRODUCT * FACTOR;
    IFACTORIAL := PRODUCT
  END;
```

We shall not attempt a formal comparison of the two methods; to do so we need to define formally the notion of "computational complexity," whereby an algorithm is analyzed as to its memory requirements and execution time. We leave it to the interested student to experiment with the two methods. For example, each program could be run for various values of N, perhaps making a graph of execution time versus N.

Exercises 5.4

1. For a fixed positive integer n and for $0 \leq m \leq n$, let $k(m)$ be the number of m-element subsets of an n-element set. Then $k(0) = 1$. Find a recurrence relation for $k(m)$ in terms of $k(m - 1)$, $1 \leq m \leq n$.

2. Use the recurrence relation obtained in Exercise 1 to write a recursive Pascal function COMBO that returns the number of M-element subsets of an N-element set, where M and N are integers with $0 \leq M \leq N$.

3. Consider character strings of length $n \geq 1$ consisting only of the letters 'A', 'B', and 'C'. Find initial values and a recurrence relation for

a. the number $h_1(n)$ of such strings not containing the substring 'AA';

b. the number $h_2(n)$ of such strings not containing either of the substrings 'AA' or 'AB';

c. the number $h_3(n)$ of such strings not containing the substring 'BBB';

d. the number $h_4(n)$ of such strings not containing the substring 'AB'. (Careful!)

4. Let the function $g(n)$ be recursively defined by $g(1) = 2$ and, for $n \geq 2$,

$$g(n) = g(n - 1) + 2n$$

Guess a formula for $g(n)$ and then use induction to try to prove that your guess is correct.

5. Let $t(n)$ be the number of ways in which $2n$ tennis players can be paired to play n matches. (For example, if $n = 3$ and the players are denoted A, B, C, D, E, and F, then one possible pairing is to have A play B, C play D, and E play F; another possibility is to have A play F, B play E, and C play D.) Find initial values and a recurrence relation for $t(n)$.

CHAPTER PROBLEMS

1. Give an example of a function $f:[0, 1] \to [0, 1]$ that is
   a. both one-to-one and onto;
   b. one-to-one but not onto;
   c. onto but not one-to-one;
   d. neither one-to-one nor onto.
2. Do Problem 1 for $f:E \to \mathbb{N}$, where E is the set of even integers.
3. Given $f:X \to Y$, prove that f is one-to-one if and only if

$$\forall A \subseteq X, \forall B \subseteq X(f(A \cap B) = f(A) \cap f(B))$$

4. Let $f:X \to Y$, where X and Y are subsets of $\mathbb{R}$. The function f is said to be *increasing* if the condition

$$x_1 < x_2 \to f(x_1) < f(x_2)$$

holds for all $x_1, x_2 \in X$. Similarly, the function f is said to be *decreasing* if the condition

$$x_1 < x_2 \to f(x_1) > f(x_2)$$

holds for all $x_1, x_2 \in X$. If f is either increasing or decreasing, then we say that f is *monotonic*.
   a. Prove that a monotonic function is one-to-one.
   b. Apply the result of part a to show that the function $f:\mathbb{R} \to \mathbb{R}$ defined by $f(x) = x^3 + x - 2$ is one-to-one.
5. Determine $f \circ g$ and $g \circ f$ for the following functions.
   a. $f:\mathbb{R} \to \mathbb{R}$; $f(x) = x^2 + x$, $g:\mathbb{R} \to \mathbb{R}$; $g(x) = 3x + 4$
   b. $f:(0, \infty) \to (0, \infty)$; $f(x) = x/(x^2 + 1)$, $g:(0, \infty) \to (0, \infty)$; $g(x) = 1/x$
6. Give an example of functions $f:\mathbb{R} \to \mathbb{R}$ and $g:\mathbb{R} \to \mathbb{R}$ such that $f \neq g$, $f \neq I_{\mathbb{R}}$, $g \neq I_{\mathbb{R}}$, $f \circ g \neq I_{\mathbb{R}}$, and $f \circ g = g \circ f$.
7. For each function, determine whether it is one-to-one and whether it is onto. If the function is both one-to-one and onto, find its inverse.
   a. $f:\mathbb{R} \to \mathbb{R}$; $f(x) = x^3 + 1$
   b. $g:\mathbb{R} - \{\frac{1}{2}\} \to \mathbb{R} - \{\frac{3}{2}\}$; $g(x) = 3x/(2x - 1)$
   c. $h:\mathbb{Z} \to \mathbb{N}$; $h(x) = |x - 4| + 1$
   d. $p:\mathbb{N} \to \mathbb{N}$; $p(x) = x^2$
   e. $q:[0, \infty) \to [0, \infty)$; $q(x) = x^2$
8. Let f, g, and h be the functions defined on $\mathbb{Z}$ as follows:

$$f(m) = m + 1, \qquad g(m) = 2m, \qquad h(m) = \begin{cases} 0 & \text{if } m \text{ is even} \\ 1 & \text{if } m \text{ is odd} \end{cases}$$

Find the following composite functions.
   a. $f \circ g$ b. $g \circ f$ c. $f \circ h$ d. $h \circ f$
   e. $g \circ h$ f. $h \circ g$ g. $g \circ g$ h. $h \circ f \circ g$

9. Let P denote the Pascal character set, $\mathcal{S}$ the set of Pascal character strings, and $\mathcal{S}^+$ the set of nonempty strings. Also, for $x, y \in \mathcal{S}$, let $x \# y$ denote the concatenation of x with y. Define the following functions.

(i) For a fixed $\alpha \in P$, the functions $\rho_\alpha : \mathcal{S} \to \mathcal{S}$ and $s_\alpha : \mathcal{S} \to \mathcal{S}$ are defined by

$$\rho_\alpha(x) = \text{'}\alpha\text{'} \# x \qquad \text{and} \qquad s_\alpha(x) = x \# \text{'}\alpha\text{'}.$$

(ii) The function $r : \mathcal{S} \to \mathcal{S}$ is defined as follows: $r(x)$ is the string obtained by reversing the characters in x.

(iii) The function $t : \mathcal{S}^+ \to \mathcal{S}$ is defined as follows: $t(x)$ is the string obtained by deleting the first character of x.

For example, if α = 'S' and x = 'TAB', then $\rho_\alpha(x)$ = 'STAB', $s_\alpha(x)$ = 'TABS', $r(x)$ = 'BAT', and $t(x)$ = 'AB'.

a. Determine which of these functions are one-to-one.

b. Find the image of each of the given functions and determine which are onto.

c. Describe those strings x for which $r(x) = x$.

d. Show that $r \circ \rho_\alpha = s_\alpha \circ r$.

e. Describe $\rho_\alpha \circ t$ and $t \circ \rho_\alpha$.

10. Given a set X, let $\mathcal{R}(X)$ denote the set of real-valued functions on X:

$$\mathcal{R}(X) = \{f \mid f : X \to \mathbb{R}\}$$

For $f, g \in \mathcal{R}(X)$, we define functions $h, k \in \mathcal{R}(X)$ as follows.

(i) $h : X \to \mathbb{R}$ is defined by $h(x) = f(x) + g(x)$.

(ii) $k : X \to \mathbb{R}$ is defined by $k(x) = f(x) \cdot g(x)$.

Determine the functions h and k if

a. $X = \mathbb{N}$, $f(x) = x^2$, and $g(x) = 2x + 1$;

b. $X = \mathbb{R} - \{0\}$, $f(x) = (x^2 + 1)/x$, and $g(x) = x/(x^2 + 1)$;

c. $X = \mathbb{R}$, $f(x) = x^2 - 2x + 3$, and $g(x) = -x^2 + 2x - 3$.

11. Let A be a nonempty set. For $B \subseteq A$, define the *characteristic function* χ_B of B by $\chi_B : A \to \{0, 1\}$, where

$$\chi_B(a) = \begin{cases} 0 & \text{if } a \notin B \\ 1 & \text{if } a \in B \end{cases}$$

For $B \subseteq A$, $C \subseteq A$, and $a \in A$, prove the following.

a. $\chi_{B \cap C}(a) = \chi_B(a) \cdot \chi_C(a)$

b. $\chi_{B \cup C}(a) = \chi_B(a) + \chi_C(a) - (\chi_B(a) \cdot \chi_C(a))$

c. $\chi_A(a) = 1$

d. $\chi_{A-B}(a) = 1 - \chi_B(a)$

e. $\chi_{B-C}(a) = \chi_B(a)(1 - \chi_C(a))$

12. Two sets A and B are said to be *equivalent* if there is a function $f:A \to B$ that is both one-to-one and onto.

a. When are two finite sets A and B equivalent?

An infinite set is called *countable* if it is equivalent to $\mathbb{N}$.

b. Show that $\mathbb{Z}$ is countable.

★**c.** Show that $\mathbb{N} \times \mathbb{N}$ is countable.

d. It can be shown that any infinite subset of a countable set is countable. Use this result and the result of part c to show that $\mathbb{Q}^+$ (the set of positive rationals) is countable.

An infinite set is *uncountable* if it is not countable. For example, it is known that $\mathbb{R}$ is uncountable.

e. Show that "equivalent" is an equivalence relation on the set of all subsets of $\mathbb{R}$.

f. Show that, for any real numbers a and b with $a < b$, the open interval (a, b) is equivalent to $(0, 1)$. (Hint: consider $f:(0, 1) \to (a, b)$ defined by $f(x) = (b - a)x + a$.)

g. Show that $(-1, 1)$ is equivalent to $\mathbb{R}$. (Hint: consider $f:(-1, 1) \to \mathbb{R}$ defined by $f(x) = x/(1 - x^2)$.) It follows from this result and parts e and f that any open interval (a, b) is uncountable.

13. Give an example of a function $f:\mathbb{Z} \to \mathbb{Z}$ such that $f \circ f \neq I_{\mathbb{Z}}$, but $f \circ f \circ f = I_{\mathbb{Z}}$. Now generalize this example. Let $f^1 = f$ and, for $n \geq 2$, define $f^n = f \circ f^{n-1}$. For each $n \geq 2$, give an example of a function $f:\mathbb{Z} \to \mathbb{Z}$ such that $f^n = I_{\mathbb{Z}}$, but $f^m \neq I_{\mathbb{Z}}$ for $1 \leq m < n$.

14. Given $f:A \to B$, prove the following.

a. f is onto $\leftrightarrow \forall D \subseteq B(f(f^{-1}(D)) = D)$.

b. f is one-to-one $\leftrightarrow \forall C \subseteq A(f^{-1}(f(C)) = C)$.

15. Let A be a finite set with $n(A) = m$. Find the cardinality of each of the following sets.

a. the set of relations on A

b. the set of functions on A

c. the set of permutations of A

16. Consider a function $f:A \to B$, where f is onto. Define a relation $\sim$ on A by $x \sim y \leftrightarrow f(x) = f(y)$.

a. Show that $\sim$ is an equivalence relation on A.

Recall that $[a]$ denotes the equivalence class containing the element a. Let $\mathcal{P}$ be the partition of A induced by $\sim$. Define $q:A \to \mathcal{P}$ by $q(a) = [a]$, and define $\bar{f}:\mathcal{P} \to B$ by $\bar{f}([a]) = f(a)$.

b. Show that the function $\bar{f}$ is well-defined, that is, if $[a_1] = [a_2]$, then $\bar{f}([a_1]) = \bar{f}([a_2])$.

c. Show that $\bar{f}$ is one-to-one and onto.

d. Show that $f = \bar{f} \circ q$.

Thus every onto mapping (f) may be "factored" as the composition of a "quotient mapping" (q) and a one-to-one, onto mapping $(\bar{f})$.

(This result is the set-theoretic analogue of the "fundamental morphism theorem" of abstract algebra.)

17. (A problem for those who have had calculus.) Let $\mathcal{D}$ denote the set of differentiable functions defined on $\mathbb{R}$, let $\mathcal{C}$ denote the set of continuous functions defined on $\mathbb{R}$, and let $\mathcal{F}$ be the set of all functions defined on $\mathbb{R}$. Consider the function $\Delta:\mathcal{D} \to \mathcal{F}$ that maps each function $f \in \mathcal{D}$ to its derivative f' in $\mathcal{F}$, that is, $\Delta(f) = f'$.

a. Is the function Δ one-to-one?

b. Let $f \in \mathcal{D}$. What is $\Delta^{-1}(\{f'\})$?

c. Show that $\mathcal{C} \subseteq \text{im } \Delta$. (In fact, it can be shown that $\mathcal{C} \subset \text{im } \Delta$. For example, consider f defined by $f(x) = x^2 \sin(1/x)$, $x \neq 0$; and $f(0) = 0$. It can be shown that f is differentiable but that f' is not continuous at zero.)

18. Suppose that $f:\mathbb{N} \to \mathbb{Z}$ and $g:\mathbb{N} \to \mathbb{Z}$ are functions and, for all $n \geq 2$,

$$f(n) - f(n-1) = g(n) - g(n-1)$$

Prove that $f(n) = g(n) + c$ for some $c \in \mathbb{Z}$. (Hint: consider $h(n) = f(n) - g(n)$.)

19. Each of the following parts recursively defines a function $f:\mathbb{N} \to \mathbb{Z}$ (where $a, b, c \in \mathbb{Z}$). Guess the formula for f and then use induction to prove the formula correct.

a. $f(1) = a + b, f(n) = f(n-1) + a$

b. $f(1) = a + b, f(n) = f(n-1) + 2an - a$

c. $f(1) = a + b + c, f(2) = 4a + 2b + c, f(n) = f(n-1) + (2a - b)n - a + b$

d. $f(1) = a, f(n) = af(n-1)$

e. $f(1) = a + b, f(n) = af(n-1) + b(1-a)$

f. $f(1) = a, f(n) = f(n-1)$

Six Algebraic Structures

6.1 BINARY OPERATIONS

Consider the operation of addition on the set $\mathbb{R}$. Associated with any two real numbers a and b there is a uniquely determined third real number $a + b$. This association can be formally defined as a function $f:\mathbb{R} \times \mathbb{R} \to \mathbb{R}$, where $f((a,b)) = a + b$ for all $(a,b) \in \mathbb{R} \times \mathbb{R}$. As another example, we can define a function $g:\mathbb{R} \times \mathbb{R} \to \mathbb{R}$ by $g((a,b)) = ab$. Each of these functions is an example of a "binary operation" defined on the set of real numbers.

DEFINITION 6.1

Given any nonempty set A, a function $f:A \times A \to A$ is called a *binary operation* on A.

As a general rule we shall denote an abstract binary operation on a set by the symbol $*$. Instead of using the functional notation $*((a,b))$, we shall use the more standard notation $a * b$, which more closely parallels the way in which well-known binary operations (like $+$ and $\cdot$ on $\mathbb{R}$) are displayed.

Example 6.1 (a) Let A be a nonempty set and let $\mathcal{F}(A)$ denote the set of all functions from A to A. If f and g are two functions in $\mathcal{F}(A)$, then $f \circ g \in \mathcal{F}(A)$. Thus composition $\circ$ is a binary operation on $\mathcal{F}(A)$. More specifically, let $\mathcal{S}(A)$ denote the set of permutations of A. By Corollary 5.5, if $f, g \in \mathcal{S}(A)$, then $f \circ g \in \mathcal{S}(A)$, so composition is also a binary operation on $\mathcal{S}(A)$.
(b) Let C denote the set of all valid character strings in some programming language. Many languages (such as FORTRAN77 and some versions of Pascal) define the binary operation of "concatenation" on C. For instance, the concatenation of the strings 'PASC' and 'AL' is the string 'PASCAL'. ■

Consider the set $\mathbb{Z}_m = \{[0], [1], \ldots, [m-1]\}$ of residue classes modulo m. Is it perhaps possible to define, in a natural way, one or more binary operations on $\mathbb{Z}_m$? Indeed, it is possible and well motivated by the natural binary operations of $+$ and $\cdot$ on $\mathbb{Z}$. For $[a], [b] \in \mathbb{Z}_m$, define $\oplus$ and $\odot$ as follows:

1. $[a] \oplus [b] = [a + b]$
2. $[a] \odot [b] = [ab]$

Bear in mind that if $\oplus$ and $\odot$ are to be binary operations on $\mathbb{Z}_m$, then both $[a] \oplus [b]$ and $[a] \odot [b]$ must be uniquely determined by $[a]$ and $[b]$; that is, $\oplus$ and $\odot$ must be well-defined functions from $\mathbb{Z}_m \times \mathbb{Z}_m$ to $\mathbb{Z}_m$. Perhaps the reader can see a bit of a problem with this determination. The problem can be illustrated by the following example. In $\mathbb{Z}_7$, we have $[2] = [-5]$ and $[4] = [18]$. In order for $\oplus$ and $\odot$ to be binary operations, it is necessary that both $[2] \oplus [4] = [-5] \oplus [18]$ and $[2] \odot [4] = [-5] \odot [18]$. Both of these are easily seen to be true. In general, we must show that if $[a] = [a_1]$ and $[b] = [b_1]$ in $\mathbb{Z}_m$, then $[a] \oplus [b] = [a_1] \oplus [b_1]$ and $[a] \odot [b] = [a_1] \odot [b_1]$. If these can be shown to hold, then it follows that $\oplus$ and $\odot$ are binary operations on $\mathbb{Z}_m$.

THEOREM 6.1 For each $m \in \mathbb{N}$, $\oplus$ and $\odot$ are binary operations on $\mathbb{Z}_m$.

Proof We present only the proof that $\odot$ is a binary operation on $\mathbb{Z}_m$. Suppose then, that $[a] = [a_1]$ and $[b] = [b_1]$ in $\mathbb{Z}_m$. We then have that $a \equiv a_1 (\text{mod } m)$ and $b \equiv b_1 (\text{mod } m)$. Hence $a = a_1 + mk$ and $b = b_1 + mt$ for some $k, t \in \mathbb{Z}$, and we obtain $ab = a_1 b_1 + m(a_1 t + b_1 k + mkt)$. Thus $ab \equiv a_1 b_1 (\text{mod } m)$, and so $[ab] = [a_1 b_1]$. Therefore

$$[a] \odot [b] = [a_1] \odot [b_1]$$

so that $\odot$ is a binary operation on $\mathbb{Z}_m$. ∎

In general, if we are asked to prove that $*$ is a binary operation on a set A, then it should be emphasized that there are two properties to be verified:

1. If $a, b \in A$, then $a * b \in A$.
2. If $a, b \in A$, then $a * b$ is uniquely determined.

If property 1 is satisfied, then we say that $*$ is *defined on* A; if property 2 holds, then we say that $*$ is *well-defined*. It is especially important to show property 2 when the elements of A can be represented by different names, such as with $\mathbb{Z}_m$.

Example 6.2 Determine which of the following are binary operations:

(a) $a * b = a/(b^2 + 1)$ on $\mathbb{Z}$

(b) $a * b = \sqrt{(a + b)^2}$ on $\mathbb{Q}$
(c) $[a] * [b] = [a \text{ DIV } b]$ on $\mathbb{Z}_4$

Solution (a) This is not a binary operation on $\mathbb{Z}$; for example, if $a = b = 1$, then $a * b = \frac{1}{2} \notin \mathbb{Z}$.
(b) This is a binary operation on $\mathbb{Q}$. Note that if $a, b \in \mathbb{Q}$, then $a * b = \sqrt{(a + b)^2} = |a + b| \in \mathbb{Q}$.
(c) Here $*$ fails to be well-defined. For instance, $[2] * [2] = [2 \text{ DIV } 2] = [1]$, while $[6] * [2] = [6 \text{ DIV } 2] = [3]$. Hence we have $[2] = [6]$ but $[1] \neq [3]$. ■

Given a binary operation $*$ on a finite set A, it is often helpful to construct the *operation table* for $*$, which is similar to the addition and multiplication tables for integers (with which the reader is no doubt familiar). Such a table has a row and a column for each element of A; the element $a_i * a_j$ is placed in the row corresponding to a_i and the column corresponding to a_j. The operation tables for $\mathbb{Z}_6$ under $\oplus$ and for $\mathbb{Z}_5^\#$ under $\odot$ are shown in Figure 6.1, where $\mathbb{Z}_5^\# = \mathbb{Z}_5 - \{[0]\}$.

$\oplus$	[0]	[1]	[2]	[3]	[4]	[5]
[0]	[0]	[1]	[2]	[3]	[4]	[5]
[1]	[1]	[2]	[3]	[4]	[5]	[0]
[2]	[2]	[3]	[4]	[5]	[0]	[1]
[3]	[3]	[4]	[5]	[0]	[1]	[2]
[4]	[4]	[5]	[0]	[1]	[2]	[3]
[5]	[5]	[0]	[1]	[2]	[3]	[4]

(a) $\mathbb{Z}_6$ under $\oplus$

$\odot$	[1]	[2]	[3]	[4]
[1]	[1]	[2]	[3]	[4]
[2]	[2]	[4]	[1]	[3]
[3]	[3]	[1]	[4]	[2]
[4]	[4]	[3]	[2]	[1]

(b) $\mathbb{Z}_5^\#$ under $\odot$

FIGURE 6.1 Two operation tables

The binary operations $\oplus$ and $\odot$ defined on $\mathbb{Z}_m$ satisfy some of the same properties that hold for $+$ and $\cdot$ on $\mathbb{Z}$. In general, given any binary operation $*$ on a set A, we are interested in what properties are satisfied by $*$. Two of the most common properties investigated are contained in the following definition.

DEFINITION 6.2

Let $*$ be a binary operation on a set A. We say that $*$ is *associative* if $(a * b) * c = a * (b * c)$ for all $a, b, c \in A$. The operation $*$ is called *commutative* if $a * b = b * a$ for all $a, b \in A$.

For example, it is clear that $+$ and $\cdot$ are associative and commutative binary operations on $\mathbb{R}$, and the reader will be asked to show in Exercise 14 that $\oplus$ and $\odot$ are associative and commutative operations on $\mathbb{Z}_m$. However, subtraction is a binary operation on $\mathbb{R}$ that is neither associative nor commutative. That subtraction is not associative can be seen, for example, from the inequality $(2 - 3) - 4 \neq 2 - (3 - 4)$.

In Example 6.1 we saw that composition is a binary operation on the set $\mathcal{F}(A)$ of all functions from a set A to itself. That composition is an associative binary operation follows from Theorem 5.7, which we restate here.

THEOREM 5.7 Let A, B, C, and D be nonempty sets. Given $f:A \to B$, $g:B \to C$, and $h:C \to D$, then $(h \circ g) \circ f = h \circ (g \circ f)$. ■

If we simply apply Theorem 5.7 with $B = C = D = A$, then we obtain the following theorem as a corollary.

THEOREM 6.2 Let A be any nonempty set. Then composition is an associative binary operation on the set $\mathcal{F}(A)$ of all functions from A to A.

Proof Given $f, g, h \in \mathcal{F}(A)$, we have $f:A \to A$, $g:A \to A$, and $h:A \to A$. Applying Theorem 5.7 with $B = C = D = A$, we obtain $(h \circ g) \circ f = h \circ (g \circ f)$. ■

It can easily be seen that function composition is not, in general, a commutative operation. For example, define $f: \mathbb{R} \to \mathbb{R}$ by $f(x) = x^2$ and $g: \mathbb{R} \to \mathbb{R}$ by $g(x) = x + 1$. Then $(f \circ g)(x) = (x + 1)^2$, while $(g \circ f)(x) = x^2 + 1$. So $f \circ g \neq g \circ f$, and hence $\circ$ is not commutative on $\mathcal{F}(\mathbb{R})$.

Given a set A, it is sometimes possible to define several different binary operations on A, although in a given discussion, we may be interested in studying properties of one specific binary operation, say $*$, on A. To avoid any confusion in such cases, we shall refer to A together with $*$ by the notation $(A, *)$. For instance, we could consider any one of $(\mathbb{R}, +)$, or $(\mathbb{R}, \cdot)$, or $(\mathbb{R}, -)$. We could also study the properties satisfied by a pair of binary operations defined on a set, such as $+$ and $\cdot$ on $\mathbb{R}$. In this case, we will use the notation $(\mathbb{R}, +, \cdot)$. Of course, this notation can be extended to three or more binary operations on a set. Systems like $(\mathbb{R}, +)$ and $(\mathbb{R}, +, \cdot)$ are examples of *algebraic structures*.

In addition to associativity and commutativity, there are some other interesting and natural properties that an operation might satisfy. For example, in $(\mathbb{R}, \cdot)$ there is a very special real number, namely 1, that satisfies the condition $1 \cdot a = a \cdot 1 = a$ for all $a \in \mathbb{R}$. Also, in $(\mathbb{R}, \cdot)$, for each $a \in \mathbb{R} - \{0\}$, there is a real number, namely $1/a$, such that $a \cdot (1/a) = 1$. These examples serve to motivate the following definition.

DEFINITION 6.3

Let $*$ be a binary operation on the set A.

1. An element $e \in A$ is called an *identity element* for $(A, *)$ if $a * e = e * a = a$ for all $a \in A$.
2. Suppose $(A, *)$ has an identity element e. If $a \in A$, then an element $a' \in A$ is called an *inverse* of a in $(A, *)$ if $a * a' = a' * a = e$.

Example 6.3 (a) In $(\mathbb{Z}, +)$, zero is the identity element, and the inverse of the integer n is $-n$, since $n + (-n) = (-n) + n = 0$.
(b) In $(\mathcal{S}(A), \circ)$, the function I_A is the identity element (Theorem 5.4) and the inverse of the permutation f is the permutation f^{-1}.
(c) The identity element for concatenation of character strings is the empty string. No nonempty string has an inverse.
(d) Consider $(\mathcal{P}(A), \cap)$. Since for any $B \subseteq A$ it is true that $A \cap B = B \cap A = B$, the set A is the identity element. However, no proper subset of A has an inverse. ■

Given an algebraic structure $(A, *)$, it may or may not be the case that $(A, *)$ has an identity element. Also, given that $(A, *)$ has an identity element, there is the question of the existence of inverses for the elements of A. For example, if E is the set of even integers, then $(E, \cdot)$ has no identity element. On the other hand, $(\mathbb{Z}, \cdot)$ has identity element 1, but inverses exist only for 1 and -1. Don't be fooled by this last statement; it states that if $a \in \mathbb{Z}$ and $a \neq \pm 1$, then there is no $b \in \mathbb{Z}$ such that $ab = 1$.

Suppose that $(A, *)$ is an algebraic structure with identity element e. Is e uniquely determined or can there possibly be more than one identity element for $(A, *)$? If $a \in A$ has an inverse a', is a' uniquely determined? These questions are treated in the following theorem.

THEOREM 6.3 If $(A, *)$ is an algebraic structure with identity element e, then e is the only identity element of $(A, *)$. If $a \in A$ has an inverse a' in $(A, *)$, and $*$ is associative, then a' is uniquely determined.

Proof We prove the uniqueness of the identity element, leaving the uniqueness of inverses for the exercises. Suppose that both e and $\bar{e}$ are identity elements for $(A, *)$. Since e is an identity element, we have $e * \bar{e} = \bar{e}$. Also, since $\bar{e}$ is an identity element, $e * \bar{e} = e$. Thus $e = \bar{e}$, and e is uniquely determined. ■

In what follows, the term *identity* is used interchangeably with the term *identity element*.

Exercises 6.1

1. Which of the following functions $*$ define binary operations on the set of integers? For those that do, indicate whether $*$ is associative and commutative. Also discuss the existence of an identity and inverses.
   a. $m * n = mn + 1$ **b.** $m * n = m + n - 1$
   c. $m * n = n$ **d.** $m * n = 2m + n$
   e. $m * n = (m + n)/2$ **f.** $m * n = 2^{mn}$
2. Show that $\oplus$ is a well-defined operation on $\mathbb{Z}_m$. In other words, show that if $a \equiv a_1 (\text{mod } m)$ and $b \equiv b_1 (\text{mod } m)$, then $a + b \equiv a_1 + b_1 (\text{mod } m)$.
3. Assume $*$ is a binary operation on A. Complete each of the following statements.
   a. $*$ is not associative if______.
   b. $*$ is not commutative if______.
   c. $a \in A$ is not an identity for $*$ if______.
   d. Given that $(A, *)$ has identity e, then $b \in A$ is not the inverse of $a \in A$ if______.
4. Prove: If an element a has an inverse in $(A, *)$ and $*$ is associative, then the inverse is unique. This will complete the proof of Theorem 6.3. (Assume, of course, that $(A, *)$ has an identity.)
5. Give the operation table for each of the following algebraic structures.
   a. $(\mathcal{S}(\{1, 2\}), \circ)$ **b.** $(\mathbb{Z}_5, \oplus)$
   c. $(\mathbb{Z}_4, \odot)$ **d.** $(\mathcal{P}(\{1, 2, 3\}), \cup)$
   e. $(\mathbb{Z}_6, \odot)$ **f.** $(\mathbb{Z}_7^{\#}, \odot)$ $(\mathbb{Z}_7^{\#} = \mathbb{Z}_7 - \{[0]\})$
   g. $(\mathbb{Z}_8^{\#}, \odot)$
6. Let $*$ be a binary operation on A and let B be a nonempty subset of A.
   a. Under what condition is $*$ a binary operation on B?
   Assume that $*$ is also a binary operation on B.
   b. Prove: If $*$ is associative on A, then $*$ is associative on B.
   c. Prove: If $*$ is commutative on A, then $*$ is commutative on B.
   d. Prove or disprove: If there is an identity for $*$ in A, then there is an identity for $*$ in B.
7. This exercise concerns the following partial operation table.

$*$	a	b	c	d
a	a		c	
b				a
c		d	a	
d			b	

   Complete the table in such a way as to make $*$
   a. associative

b. commutative

c. commutative and having an identity

(Note: part a has only one correct answer, but parts b and c have several correct answers.)

8. Consider $(\mathcal{F}(A), \circ)$, where $A = \{1, 2\}$.

a. List the elements of $(\mathcal{F}(A), \circ)$ and construct the operation table.

b. Is $\circ$ commutative in this case?

c. What is the identity and which elements have inverses?

9. a. Determine the smallest subset A of $\mathbb{Z}$ such that $2 \in A$ and $+$ is a binary operation on A.

b. Determine the smallest subset B of $\mathbb{Z}$ such that $2 \in B$ and $+$ is a binary operation on B with an identity and such that each element has an inverse.

c. Determine the smallest subset C of $\mathbb{Q}$ such that $3 \in C$ and division is a binary operation on C.

10. Prove that if A has more than one element, then $\circ$ is not a commutative operation on $\mathcal{F}(A)$.

11. Let T denote the set of all rational numbers except 0 and 1. Define the following six functions on T:

$$f_1(x) = x \qquad f_4(x) = (x - 1)/x$$
$$f_2(x) = 1 - x \qquad f_5(x) = 1/(1 - x)$$
$$f_3(x) = 1/x \qquad f_6(x) = x/(x - 1)$$

Verify that $\circ$ is a binary operation on $F = \{f_1, f_2, \ldots, f_6\}$ and construct the operation table for $(F, \circ)$.

12. Division is a binary operation on the set $\mathbb{R}^+$ of positive real numbers.

a. Is it associative?

b. Is it commutative?

c. Note that $a/1 = a$ for every positive real number a. Does this mean that 1 is an identity for division on $\mathbb{R}^+$?

13. Each of the following parts gives an operation table for a binary operation $*$. Answer the following questions about $*$. (i) Is $*$ associative? (ii) Is $*$ commutative? (iii) Does $*$ have an identity? (iv) If $*$ has an identity, which elements have inverses?

a.

$*$	a	b	c
a	a	b	c
b	b	c	a
c	c	a	b

b.

$*$	a	b	c
a	a	b	a
b	b	a	c
c	a	c	a

c.

$*$	a	b	c
a	b	a	c
b	a	b	c
c	a	c	b

d.

$*$	a	b	c
a	a	a	a
b	b	b	b
c	c	c	c

14. Show that $\oplus$ and $\odot$ are both associative and commutative operations on $\mathbb{Z}_m$.

15. Give an example of a binary operation $*$ on $A = \{a, b, c\}$ that has identity a but in which the inverse of b is not unique.

16. Each of the following parts defines a binary operation $*$ on $\mathbb{Z} \times \mathbb{Z}$. Determine whether (i) $*$ is associative, (ii) $*$ is commutative, and (iii) $*$ has an identity. If $*$ has an identity, then (iv) determine those elements of $\mathbb{Z} \times \mathbb{Z}$ that have inverses.

a. $(a,b) * (c,d) = (a + c, b + d)$
b. $(a,b) * (c,d) = (ac, bd)$
c. $(a,b) * (c,d) = (ad, bc)$
d. $(a,b) * (c,d) = (ad + bc, bd)$

6.2 SEMIGROUPS AND GROUPS

In the last section we saw that it is often possible to define, in a natural way, one or more binary operations on a given set. Of particular interest is the case where these binary operations satisfy a specified set of properties. Such algebraic structures are the main objects of study in the branch of mathematics called modern algebra. Many of the topics covered in this section will seem very abstract to the student at first, and will take some getting used to. Suffice it to say, however, that there are many interesting and significant applications of modern algebra to such diverse areas as quantum mechanics, satellite communication, microwave transmission, electrical switching networks, and crystallography. In the rest of this chapter we shall survey several of the more important algebraic structures. We shall also investigate the particular application of the theory of Boolean algebras to electrical switching networks.

Two algebraic structures for which the theory is highly developed are "semigroups" and "groups." In this section we introduce these and explore some of their elementary properties.

DEFINITION 6.4

Let $*$ be a binary operation on a nonempty set A. If $*$ is associative, then $(A, *)$ is called a *semigroup*.

Example 6.4 (a) It follows from Theorem 6.2 that $(\mathcal{F}(A), \circ)$ (the set of all functions from A to A under the operation of composition) is a semigroup for any nonempty set A.

(b) The set of character strings (in a given programming language) under the operation of concatenation is a semigroup.
(c) Let A be a set, and consider $\mathcal{P}(A)$ under the operation of intersection. We know that for any $B, C, D \in \mathcal{P}(A)$, it is true that $(B \cap C) \cap D = B \cap (C \cap D)$; in other words, the operation of intersection is associative. Therefore, $(\mathcal{P}(A), \cap)$ is a semigroup. ■

It is easily checked that $(\mathbb{R}, +)$ is a semigroup with identity element zero, and that each real number a has inverse element $-a$. If $\mathbb{R}^{\#} = \mathbb{R} - \{0\}$, then $(\mathbb{R}^{\#}, \cdot)$ is a semigroup with identity element 1, and each $a \in \mathbb{R}^{\#}$ has inverse $1/a \in \mathbb{R}^{\#}$. In general, semigroups with identity in which each element has an inverse belong to a very special class of algebraic structures.

DEFINITION 6.5

An algebraic structure $(G, *)$ is called a *group* provided the following conditions are satisfied:

1. The operation $*$ is associative.
2. $(G, *)$ has an identity element.
3. Each $g \in G$ has an inverse in $(G, *)$.

Example 6.5 (a) Consider the semigroup $(\mathbb{Z}_6, \oplus)$, whose operation table is shown in Figure 6.1(a). For any $[a] \in \mathbb{Z}_6$, we have that $[a] \oplus [0] = [0] \oplus [a] = [0 + a] = [a]$. Thus $[0]$ is the identity in $(\mathbb{Z}_6, \oplus)$. Furthermore, $[a] \oplus [-a] = [-a] \oplus [a] = [0]$, so the inverse of $[a]$ is $[-a]$. Therefore $(\mathbb{Z}_6, \oplus)$ is a group. (Note that $[-a] = [6 - a]$ in $\mathbb{Z}_6$; thus, for example, the inverse of $[2]$ is $[-2] = [6 - 2] = [4]$.)
(b) Consider the semigroup $(\mathbb{Z}_5^{\#}, \odot)$, whose operation table is shown in Figure 6.1(b). It can easily be seen from the operation table that $[1]$ is the identity. We can also see that the inverses of $[1]$, $[2]$, $[3]$, and $[4]$ are $[1]$, $[3]$, $[2]$, and $[4]$, respectively. Thus $(\mathbb{Z}_5^{\#}, \odot)$ is a group.
(c) Let A be a nonempty set and recall that $\mathcal{S}(A)$ denotes the set of all permutations of A. Then $(\mathcal{S}(A), \circ)$ is a semigroup. Furthermore, we have already shown (Example 6.3) that I_A is the identity and that the inverse of the permutation f is the permutation f^{-1}. Thus $(\mathcal{S}(A), \circ)$ is a group. In particular, if $X = \{1, 2, \ldots, m\}$, then $(\mathcal{S}(X), \circ)$ is called the *symmetric group of degree m* and is denoted by S_m. ■

Let m be a positive integer and let p be a prime. The reader will be asked to show in the exercises that both $(\mathbb{Z}_m, \oplus)$ and $(\mathbb{Z}_p^{\#}, \odot)$ are groups.

If $(G, *)$ is a group, then the results of the previous section imply directly that G has a unique identity element and that each $a \in G$ has a unique inverse element. We shall denote the inverse of a by a^{-1}. With regard to inverses in a group $(G, *)$, what can be said about $(a^{-1})^{-1}$, or about $(a * b)^{-1}$? These questions are answered by the following result.

THEOREM 6.4 Let $(G, *)$ be a group and let a and b be any elements of G. Then the following are true:

1. $(a^{-1})^{-1} = a$
2. $(a * b)^{-1} = b^{-1} * a^{-1}$

Proof First let a be any element of G and consider $(a^{-1})^{-1}$. Note that this notation means "the inverse of a^{-1}." But it is also true that $a * a^{-1} = a^{-1} * a = e$, where e is the identity of $(G, *)$. This can be interpreted as saying that a is also an inverse element for a^{-1}. Since the inverse of an element is unique, we conclude that $a = (a^{-1})^{-1}$.

Next let a and b be any elements of G and consider $(a * b)^{-1}$, which denotes the inverse of the element $a * b$. As above, to show that $(a * b)^{-1} = b^{-1} * a^{-1}$, it suffices to show that $b^{-1} * a^{-1}$ is the inverse of $a * b$:

$$\begin{aligned}(a * b) * (b^{-1} * a^{-1}) &= [(a * b) * b^{-1}] * a^{-1} \\ &= [a * (b * b^{-1})] * a^{-1} \\ &= (a * e) * a^{-1} \\ &= a * a^{-1} \\ &= e\end{aligned}$$

Similarly, it can be shown that $(b^{-1} * a^{-1}) * (a * b) = e$, so it follows that $(a * b)^{-1} = b^{-1} * a^{-1}$. ■

In the context of Theorem 6.4, it is very tempting to write $(a * b)^{-1} = a^{-1} * b^{-1}$ for all $a, b \in G$. However, this is true if and only if $*$ is commutative (see Exercise 14).

In general, if $(G, *)$ is a group in which $*$ is commutative, then $(G, *)$ is called an *abelian group,* after the famous Norwegian mathematician Niels Abel (1802–1829). A group that is not abelian is called *nonabelian.*

Example 6.6 Recall that S_3 denotes the group of permutations of $\{1, 2, 3\}$ under the operation of composition. The six elements of this group are

$$\begin{array}{llll} \varepsilon: & 1 \to 1 & \beta: \ 1 \to 3 & \delta: \ 1 \to 2 \\ & 2 \to 2 & \quad\ \ 2 \to 2 & \quad\ \ 2 \to 3 \\ & 3 \to 3 & \quad\ \ 3 \to 1 & \quad\ \ 3 \to 1 \\ \alpha: & 1 \to 1 & \gamma: \ 1 \to 2 & \rho: \ 1 \to 3 \\ & 2 \to 3 & \quad\ \ 2 \to 1 & \quad\ \ 2 \to 1 \\ & 3 \to 2 & \quad\ \ 3 \to 3 & \quad\ \ 3 \to 2 \end{array}$$

(Such notation will be used to describe permutations of a small, finite set. The notation indicates, for example, that $\beta(1) = 3$, $\beta(2) = 2$, and $\beta(3) = 1$.) Show that S_3 is a nonabelian group.

Solution To show that S_3 is nonabelian, it suffices to exhibit two elements of S_3 that do not commute. Consider α and β, and note that

$$(\alpha \circ \beta)(1) = \alpha(\beta(1)) = \alpha(3) = 2$$

whereas

$$(\beta \circ \alpha)(1) = \beta(\alpha(1)) = \beta(1) = 3$$

This shows that $\alpha \circ \beta \neq \beta \circ \alpha$. In fact, it can be readily verified that $\alpha \circ \beta = \delta$ and $\beta \circ \alpha = \rho$. Exercise 9 will ask for a complete operation table for the group S_3. ■

If $(G, *)$ is a group with identity e, and $a \in G$, then we define inductively the *powers* of a as follows:

$$a^0 = e$$
$$a^n = a * a^{n-1}$$
$$a^{-n} = (a^{-1})^n$$

where n is a positive integer. Thus a^m has meaning for any integer m; we also refer to m as an *exponent*. The rules for working with exponents in a group $(G, *)$ are identical to the familiar rules for exponents of real numbers.

THEOREM 6.5 Let $(G, *)$ be a group and let $a \in G$. For any $m, n \in \mathbb{Z}$, the following properties hold:

1. $a^{-n} = (a^{-1})^n = (a^n)^{-1}$
2. $a^m * a^n = a^{m+n} = a^n * a^m$
3. $(a^m)^n = a^{mn}$

Proof We shall prove 1 and leave the remaining parts as exercises.

To prove that $a^{-n} = (a^{-1})^n$ for all $n \in \mathbb{Z}$, it suffices, in view of the definition, to prove it for $n \leq 0$. If $n = 0$, then observe that $a^{-0} = a^0 = e = (a^{-1})^0$. If $n < 0$, let $n = -k$, where $k \in \mathbb{N}$. Then we have that

$$\begin{aligned} a^{-n} &= a^k \\ &= ((a^{-1})^{-1})^k \qquad \text{(by Theorem 6.4)} \\ &= (a^{-1})^{-k} \qquad \text{(by definition)} \\ &= (a^{-1})^n \end{aligned}$$

To prove that $(a^{-1})^n = (a^n)^{-1}$ for all integers n, we first proceed by induction to show the result for $n \geq 0$. Let S be the set of nonnegative integers for which the result holds. To see that $0 \in S$, simply note that

$$(a^0)^{-1} = e^{-1} = e = (a^{-1})^0$$

Assume $k \in S$ for some arbitrary $k \geq 0$; the induction hypothesis is that $(a^k)^{-1} = (a^{-1})^k$. To see that $k + 1 \in S$, consider the following string of equalities:

$$\begin{aligned}(a^{k+1})^{-1} &= (a * a^k)^{-1} \\ &= (a^k)^{-1} * a^{-1} \qquad \text{(by Theorem 6.4)} \\ &= (a^{-1})^k * a^{-1} \qquad \text{(by IHOP)} \\ &= (a^{-1})^{k+1}\end{aligned}$$

Thus $k + 1 \in S$, and it follows that S is the set of all nonnegative integers.

The result $(a^{-1})^n = (a^n)^{-1}$ is next established for $n < 0$; as before, let $n = -k$, with $k \in \mathbb{N}$. Then

$$\begin{aligned}(a^n)^{-1} &= (a^{-k})^{-1} \\ &= ((a^{-1})^k)^{-1} \qquad \text{(by definition)} \\ &= ((a^k)^{-1})^{-1} \qquad \text{(by the result for } k \geq 0) \\ &= a^k \qquad \text{(by Theorem 6.4)} \\ &= a^{-n} \\ &= (a^{-1})^n \qquad \text{(by the first part of the proof)}\end{aligned}$$

Hence we have shown that $(a^n)^{-1} = (a^{-1})^n$ for all $n \in \mathbb{Z}$. ■

Example 6.7 Find each of the following:

(a) 2^3 in $(\mathbb{Z}, +)$ (b) $[2]^3$ in $(\mathbb{Z}_5, \oplus)$ (c) $[2]^3$ in $(\mathbb{Z}_5^{\#}, \odot)$

(d) 2^{-3} in $(\mathbb{Z}, +)$ (e) $[2]^{-3}$ in $(\mathbb{Z}_5, \oplus)$ (f) $[2]^{-3}$ in $(\mathbb{Z}_5^{\#}, \odot)$

Solution Here it is important to bear in mind that the meaning of an expression such as $[2]^3$ is based on the definition of the operation in the group under consideration.

(a) $2^3 = 2 + 2 + 2 = 6$

(b) $[2]^3 = [2] \oplus [2] \oplus [2] = [6] = [1]$

(c) $[2]^3 = [2] \odot [2] \odot [2] = [8] = [3]$

(d) $2^{-3} = (2^3)^{-1} = 6^{-1} = -6$

(e) $[2]^{-3} = ([2]^3)^{-1} = [1]^{-1} = [4]$

(f) $[2]^{-3} = ([2]^3)^{-1} = [3]^{-1} = [2]$ ■

If $(G, *)$ is a group, then it is certainly possible that $a^m = e$ for some $a \in G$ and some positive integer m, where e is the identity of G. Then, by the principle of well-ordering, there must be a smallest positive integer n such that $a^n = e$.

DEFINITION 6.6

Let $(G, *)$ be a group with identity e and let $a \in G$. If there exists a positive integer m such that $a^m = e$, then the smallest positive integer n for which $a^n = e$ is called the *order of the element* a in $(G, *)$. We write $|a| = n$. If there is no positive integer m such that $a^m = e$, then we say that a has *infinite order* and write $|a| = \infty$. If $n(G) = n$, then n is the *order of the group* G and we write $|G| = n$ (or $|G| = \infty$ if G has infinite order).

Example 6.8 Find the order of each of the following:

(a) $[2]$ in $(\mathbb{Z}_6, \oplus)$
(b) $[2]$ in $(\mathbb{Z}_5^\#, \odot)$
(c) 2 in $(\mathbb{Z}, +)$
(d) δ in S_3

Solution To find the order of a in G we must either show $|a| = \infty$ or else find the smallest positive integer n such that $a^n = e$, where e is the identity of G.

(a) In $(\mathbb{Z}_6, \oplus)$, we have $[2]^2 = [4]$ and $[2]^3 = [0]$, so $|[2]| = 3$.
(b) In $(\mathbb{Z}_5^\#, \odot)$, we have $[2]^2 = [4]$, $[2]^3 = [3]$, and $[2]^4 = [1]$, so $|[2]| = 4$.
(c) In $(\mathbb{Z}, +)$, $2^n = 2 + 2 + \cdots + 2 = 2n$ for each $n \in \mathbb{N}$, and thus $|2| = \infty$.
(d) In S_3 it can be observed that $\delta \circ \delta = \rho$ and $(\delta \circ \delta) \circ \delta = \rho \circ \delta = \varepsilon$; hence $|\delta| = 3$. ■

THEOREM 6.6 Let $(G, *)$ be a group with identity e and let $x \in G$. Then the following hold:

1. $|x| = 1$ if and only if $x = e$.
2. $|x| = |x^{-1}|$
3. If $|x| = n$ and $x^m = e$, then $n \mid m$.

Proof We leave the proof to Exercise 29. ■

Example 6.9 Consider the group $(\mathbb{Z}, +)$, and let $m \in \mathbb{Z}$. Observe that

$$1^m = m$$

(the operation is addition) and so each integer can be expressed as a power of the element 1 in the group $(\mathbb{Z}, +)$. As another example of this phenomenon, in $(\mathbb{Z}_5^\#, \odot)$ observe that $[2]^1 = [2]$, $[2]^2 = [4]$, $[2]^3 = [3]$, and $[2]^4 = [1]$, so that each element of $(\mathbb{Z}_5^\#, \odot)$ can be expressed as a power of $[2]$. This leads us to the next definition. ■

DEFINITION 6.7

A group $(G, *)$ is called *cyclic* if there is an element $a \in G$ such that

$$G = \{a^n \mid n \in \mathbb{Z}\}$$

The element a is called a *generator* for $(G, *)$; we write $(G, *) = \langle a \rangle$.

It follows from Example 6.9 that both $(\mathbb{Z}, +)$ and $(\mathbb{Z}_5^{\#}, \odot)$ are cyclic groups; $(\mathbb{Z}, +) = \langle 1 \rangle$ and $(\mathbb{Z}_5^{\#}, \odot) = \langle [2] \rangle$.

THEOREM 6.7 Let $(G, *)$ be a group with identity e, and assume that $(G, *)$ is cyclic with $(G, *) = \langle a \rangle$.

1. If $|a| = \infty$, then $a^r = a^t$ implies $r = t$, and hence
$$G = \{. . . , a^{-2}, a^{-1}, e, a, a^2, . . .\}$$
2. If $|a| = n$, then $G = \{e, a, a^2, . . . , a^{n-1}\}$, and the elements $e, a, a^2, . . . , a^{n-1}$ are distinct.

Proof We prove part 2 only, leaving part 1 to Exercise 30.

Assume $|a| = n$, and let m be an arbitrary integer. By the division algorithm, $m = nq + r$, where $0 \leq r < n$, and so

$$a^m = a^{nq+r} = a^{nq} * a^r = (a^n)^q * a^r = e^q * a^r = e * a^r = a^r$$

Thus a^m is one of $e, a, a^2, . . . , a^{n-1}$. To see that $e, a, a^2, . . . , a^{n-1}$ are distinct, suppose that $a^s = a^t$ for some s and t, where $0 \leq t \leq s \leq n - 1$. Then $a^s * a^{-t} = a^t * a^{-t} = e$, which says that $a^{s-t} = e$. Since the order of a is n, it follows from Theorem 6.6, part 3, that $n \mid (s - t)$. But since $0 \leq s - t < n$, we conclude that $s - t = 0$, or that $s = t$. This shows that $e, a, a^2, . . . , a^{n-1}$ are distinct. ■

COROLLARY 6.7 If $(G, *)$ is a cyclic group with $(G, *) = \langle a \rangle$, then $|G| = |a|$. ■

Several other interesting properties of groups will be left to the exercises. Suffice it to say at this point that group theory is a highly developed subject, rich in applications; our treatment is quite elementary. Further study of the subject is better left to a first course in modern algebra.

Exercises 6.2

1. Assume that $(\{x,y,z\}, *)$ is a group with identity x. Give its operation table.

2. a. Prove: If a and b are elements of a group $(G, *)$ of orders m and

n, respectively, and $a * b = c = b * a$, then the order of c divides mn.

b. Give an example to show that the result of part a may not hold if $a * b \neq b * a$.

3. Find the order of each element in the following groups.

a. The group $(\mathbb{Z}_6, \oplus)$

b. The group S_3

c. The *Klein four-group,* whose operation table is shown in Figure 6.2(a)

d. The *quaternion group,* Q_8, whose operation table is shown in Figure 6.2(b)

$*$	a	b	c	d
a	a	b	c	d
b	b	a	d	c
c	c	d	a	b
d	d	c	b	a

(a) The Klein four-group

$\cdot$	1	-1	i	$-i$	j	$-j$	k	$-k$
1	1	-1	i	$-i$	j	$-j$	k	$-k$
-1	-1	1	$-i$	i	$-j$	j	$-k$	k
i	i	$-i$	-1	1	k	$-k$	j	$-j$
$-i$	$-i$	i	1	-1	$-k$	k	$-j$	j
j	j	$-j$	$-k$	k	-1	1	i	$-i$
$-j$	$-j$	j	k	$-k$	1	-1	$-i$	i
k	k	$-k$	$-j$	j	$-i$	i	-1	1
$-k$	$-k$	k	j	$-j$	i	$-i$	1	-1

(b) The quaternion group

FIGURE 6.2

4. Let $(G, *)$ be a group and let $x, y, z \in G$. Prove the *cancellation laws:*

a. If $x * y = x * z$, then $y = z$.

b. If $y * x = z * x$, then $y = z$.

5. In each of the following parts, there is only one way to complete the operation table so as to get a group. Do so.

a.

$*$	a	b	c
a			c
b			
c			

b.

$*$	a	b	c	d
a				
b	b	c		
c				
d				

c.

$*$	a	b	c	d
a				
b	b	a		
c			a	
d				

d.

$*$	a	b	c	d	e
a					
b	b	c			
c		d			
d				b	
e					

6. Verify that for the operation table of a finite group $(G, *)$, every element of G occurs exactly once in each row and in each column. (Hint: use the cancellation laws.)

7. Verify that each of the following algebraic structures is a semigroup. Discuss the existence of an identity element and inverses. Also note whether the operation is commutative. Which are groups?
   a. $(\mathbb{Z}, +)$
   b. $(\mathbb{Q}^+, \cdot)$, where $\mathbb{Q}^+ = \{r \in \mathbb{Q} \mid r > 0\}$
   c. $(E, +)$, where E is the set of even integers
   d. $(E, \cdot)$
   e. $(D, \cdot)$, where D is the set of odd integers
   f. $(M_k, +)$, where M_k, for $k \in \mathbb{N}$, is the set of integer multiples of k
   g. $(\mathcal{P}(A), \cup)$

8. Verify that for any $m \in \mathbb{N}$, the algebraic structure $(\mathbb{Z}_m, \oplus)$ is a cyclic group.

9. Give the operation table for each of the following groups.
   a. S_3 **b.** $(\mathbb{Z}_5, \oplus)$ **c.** $(\mathbb{Z}_7^\#, \odot)$

10. **a.** Verify that $(\mathbb{Z}_p^\#, \odot)$ is a group for any prime p.
   b. Show that, if $m \in \mathbb{N}$ and m is composite, then $\odot$ is not a binary operation on $\mathbb{Z}_m^\#$. (Hint: if $m = ab$, $1 < a < b < m$, what is $[a] \odot [b]$?)

11. **a.** For which of the operations $*$ found in Exercise 1 of Section 6.1 is $(\mathbb{Z}, *)$ a group?
   b. Are any of the groups abelian? cyclic?

12. Verify that the algebraic structure $(F, \circ)$ of Exercise 11 of Section 6.1 is a nonabelian group.

13. Prove: If x and y are elements of a group $(G, *)$ and $x * y = y$, then x is the identity of G.

14. Let $(G, *)$ be a group. Prove that $(G, *)$ is abelian if and only if

$$(a * b)^{-1} = a^{-1} * b^{-1}$$

for all $a, b \in G$.

15. Let $(G, *)$ be a group and let $x, y \in G$. Prove that $x^{-1} = y^{-1}$ if and only if $x = y$.

16. For which of the operations $*$ given in Exercise 16 of Section 6.1 is $(\mathbb{Z} \times \mathbb{Z}, *)$ a group?

17. Let $(G, *)$ be a group. Prove that $(G, *)$ is abelian if and only if

$$(a * b)^2 = a^2 * b^2$$

for all $a, b \in G$.

18. **a.** Show that if a is a nonidentity element of a group $(G, *)$, then $a = a^{-1}$ if and only if $|a| = 2$.
   b. Show that every group that has even order contains an element of order 2. (Hint: pair each nonidentity element with its inverse.)

c. Let $(G, *)$ be a group such that $a^2 = e$ for all $a \in G$, where e is the identity of G. Show that $(G, *)$ is abelian.

19. **a.** Find the order of each element in $(\mathbb{Z}_{12}, \oplus)$. (Note that the order of each element divides 12.)

b. In the group $(\mathbb{Z}_{30}, \oplus)$, find the elements of order 1, 2, 3, 5, 6, 10, 15, and 30.

c. Let $(G, *) = \langle a \rangle$ be a cyclic group of order n. For $0 \leq m \leq n - 1$, what is the order of the element a^m in $(G, *)$?

20. Let $(G, *)$ be a group. If H is a nonempty subset of G, the operation $*$ is a binary operation on H, and $(H, *)$ is a group, then we say that $(H, *)$ is a *subgroup* of $(G, *)$. (For example, if $G = \mathbb{Z}_6$ and $H = \{[0], [2], [4]\}$, then $(H, \oplus)$ is a subgroup of $(G, \oplus)$.) Prove that $(H, *)$ is a subgroup of $(G, *)$ if and only if both of the following hold:

(i) For all $a, b \in H$, the quantity $a * b \in H$.
(ii) For all $a \in H$, the inverse $a^{-1} \in H$.

21. Find four subgroups of each of the following groups.

a. $(\mathbb{Z}_{12}, \oplus)$ **b.** S_3
c. $(\mathbb{Z}_7^{\#}, \odot)$ **d.** the quaternion group Q_8 of Figure 6.2(b)

22. Let $(G, *)$ be a group and let a and b be nonidentity elements of G.

a. Find $|a|$ if $b^2 = e$ and $bab^{-1} = a^2$.
b. Find $|a|$ if $|b| = 3$ and $ab = ba^2$.

23. Let D_4 be the subset of S_4 that contains the following eight permutations:

e:	r_1:	r_2:	r_3:
$1 \to 1$	$1 \to 2$	$1 \to 3$	$1 \to 4$
$2 \to 2$	$2 \to 3$	$2 \to 4$	$2 \to 1$
$3 \to 3$	$3 \to 4$	$3 \to 1$	$3 \to 2$
$4 \to 4$	$4 \to 1$	$4 \to 2$	$4 \to 3$

h:	v:	d_1:	d_2:
$1 \to 4$	$1 \to 2$	$1 \to 1$	$1 \to 3$
$2 \to 3$	$2 \to 1$	$2 \to 4$	$2 \to 2$
$3 \to 2$	$3 \to 4$	$3 \to 3$	$3 \to 1$
$4 \to 1$	$4 \to 3$	$4 \to 2$	$4 \to 4$

a. Show that D_4 (under composition) is a subgroup of S_4 by constructing the operation table for D_4.
b. Find the inverse and order of each element in D_4.
c. Verify that D_4 is not cyclic and find a subgroup of D_4 of order 4 that is not cyclic.

(The group D_4 is called the *group of symmetries of the square*.)

24. Use the result of Exercise 20 to show that if $(G, *)$ is a group and $a \in G$, then

$$H = \{a^m \mid m \in \mathbb{Z}\}$$

is a subgroup of G. We call H the *cyclic subgroup of G generated by a*.

25. Let $(G, *)$ be a cyclic group with $(G, *) = \langle a \rangle$.
   a. Show that $(G, *) = \langle a^{-1} \rangle$.
   b. Prove: If $|a| = n$, $1 \le m < n$, and $\gcd(m, n) = 1$, then $(G, *) = \langle a^m \rangle$. (Hint: if $\gcd(m, n) = 1$, then there exist integers s and t such that $1 = ms + nt$.)

26. Show that every subgroup of a cyclic group is cyclic.

27. Use the result of Exercise 26 to find all subgroups of each of the following groups.
   a. $(\mathbb{Z}, +)$ **b.** $(\mathbb{Z}_{15}, \oplus)$ **c.** $(\mathbb{Z}_{11}^{\#}, \odot)$ **d.** $(\mathbb{Z}_7, \oplus)$

28. Prove parts 2 and 3 of Theorem 6.5.

29. Prove Theorem 6.6.

30. Prove Theorem 6.7 in the case $|a| = \infty$.

6.3 RINGS, INTEGRAL DOMAINS, AND FIELDS

A group $(G, *)$ is an algebraic structure composed of a nonempty set G, together with a single binary operation $*$. In this section we focus our attention on algebraic structures with two binary operations. In particular, we shall study "rings," "integral domains," and "fields." Like groups, these structures have interested mathematicians for some time and continue to be very popular. Applications exist to the solution of polynomial equations and to algebraic coding theory, to mention just a couple.

To begin, consider the structure $(\mathbb{Z}, +, \cdot)$; it is clear that $(\mathbb{Z}, +)$ is an abelian group and that $(\mathbb{Z}, \cdot)$ is a semigroup. Furthermore, we can combine the two operations $+$ and $\cdot$ to obtain properties such as

$$a \cdot (b + c) = (a \cdot b) + (a \cdot c)$$

which holds for all $a, b, c \in \mathbb{Z}$. This is the distributive law for ordinary arithmetic in $\mathbb{Z}$.

As another example, consider $(\mathbb{Z}_m, \oplus, \odot)$, where m is any positive integer. It has already been noted that $(\mathbb{Z}_m, \oplus)$ is a group; in fact, it is an abelian group since $[a] \oplus [b] = [a + b] = [b + a] = [b] \oplus [a]$ for all $[a]$, $[b] \in \mathbb{Z}_m$. We have also seen that $\odot$ is associative, and hence $(\mathbb{Z}_m, \odot)$ is a semigroup. Moreover, we can check that

$$[a] \odot ([b] \oplus [c]) = ([a] \odot [b]) \oplus ([a] \odot [c])$$

holds for all $[a], [b], [c] \in \mathbb{Z}_m$.

Examples such as $(\mathbb{Z}, +, \cdot)$ and $(\mathbb{Z}_m, \oplus, \odot)$ serve to motivate the following definition.

DEFINITION 6.8

Let $+'$ and $\cdot'$ denote binary operations on a nonempty set S. The algebraic structure $(S, +', \cdot')$ is called a *ring* if the following conditions hold:

1. $(S, +')$ is an abelian group.
2. $(S, \cdot')$ is a semigroup.
3. $a \cdot' (b +' c) = (a \cdot' b) +' (a \cdot' c)$ and $(b +' c) \cdot' a = (b \cdot' a) +' (c \cdot' a)$ hold for all $a, b, c \in S$.

The group $(S, +')$ is called the *additive group* of the ring. The identity element of $(S, +')$ is denoted by z and is called the *zero element* of the ring. We refer to $+'$ as the *ring addition* and to $\cdot'$ as the *ring multiplication*. Consequently, in the group $(S, +')$ we shall denote the inverse of an element $a \in S$ by $-a$ and refer to $-a$ as the *additive inverse* of a. The properties in condition 3 of the definition are called the *distributive laws*.

As the reader has probably noted, it is somewhat cumbersome to have to write $+'$ and $\cdot'$ to denote the operations in a general ring. We therefore agree simply to use $+$ and $\cdot$ to denote the operations, with the understanding that, in general, they are not the ordinary operations of addition and multiplication of real numbers. Also, if $(S, +, \cdot)$ is a ring and $a, b \in S$, then we shall agree to rewrite $a + (-b)$ as $a - b$, and we refer to the operation "$-$" as "subtraction."

Example 6.10 Let $\mathbb{Z}[\sqrt{2}]$ denote the set of all real numbers of the form $a + b\sqrt{2}$, where a and b are integers. Show that $(\mathbb{Z}[\sqrt{2}], +, \cdot)$ is a ring.

Solution First, to show that $(\mathbb{Z}[\sqrt{2}], +)$ is an abelian group, it suffices to show that it is a subgroup of $(\mathbb{R}, +)$. To do this we shall apply the result of Exercise 20 of Section 6.2. Let $a + b\sqrt{2}$ and $c + d\sqrt{2}$ belong to $\mathbb{Z}[\sqrt{2}]$. Then

$$(a + b\sqrt{2}) + (c + d\sqrt{2}) = (a + c) + (b + d)\sqrt{2} \in \mathbb{Z}[\sqrt{2}]$$

and

$$-(a + b\sqrt{2}) = (-a) + (-b)\sqrt{2} \in \mathbb{Z}[\sqrt{2}]$$

Hence $(\mathbb{Z}[\sqrt{2}], +)$ is a subgroup of $(\mathbb{R}, +)$. Next we must show that $(\mathbb{Z}[\sqrt{2}], \cdot)$ is a semigroup. Here it suffices to show that $\mathbb{Z}[\sqrt{2}]$ is closed under multiplication. Associativity will then follow since $(\mathbb{R}, \cdot)$ is a semigroup. With $a + b\sqrt{2}$ and $c + d\sqrt{2}$ members of $\mathbb{Z}[\sqrt{2}]$,

$$(a + b\sqrt{2})(c + d\sqrt{2}) = (ac + 2bd) + (ad + bc)\sqrt{2} \in \mathbb{Z}[\sqrt{2}]$$

Finally, we must verify that the distributive laws hold, but again these are inherited from $(\mathbb{R}, +, \cdot)$. Therefore $(\mathbb{Z}[\sqrt{2}], +, \cdot)$ is a ring. ■

Example 6.11 Recall that $\mathscr{F}(\mathbb{R})$ denotes the set of all functions from $\mathbb{R}$ to $\mathbb{R}$. We may define the operations $+$ and $\cdot$ on $\mathscr{F}(\mathbb{R})$ as follows: for $f, g \in \mathscr{F}(\mathbb{R})$,

$$\text{(i)} \quad (f + g)(x) = f(x) + g(x)$$
$$\text{(ii)} \quad (f \cdot g)(x) = f(x) \cdot g(x)$$

(In words, (i) says that the image of x under the function $f + g$ is the sum of the image of x under f and the image of x under g.) Show that $(\mathscr{F}(\mathbb{R}), +, \cdot)$ is a ring.

Solution We shall only outline the verification, leaving the details to Exercise 2. To show that $(\mathscr{F}(\mathbb{R}), +)$ is an abelian group, it must be shown that $+$ is both associative and commutative. The identity for $+$ is the "zero function," $z: \mathbb{R} \to \mathbb{R}$ defined by $z(x) = 0$. The additive inverse of a function f is the function $-f: \mathbb{R} \to \mathbb{R}$ defined by $(-f)(x) = -f(x)$. Next it must be shown that $\cdot$ is associative; thus $(\mathscr{F}(\mathbb{R}), \cdot)$ is a semigroup. Finally, for the distributive laws, let $f, g, h \in \mathscr{F}(\mathbb{R})$; then for any $x \in \mathbb{R}$,

$$\begin{aligned}
(f \cdot (g + h))(x) &= f(x) \cdot (g + h)(x) \\
&= f(x) \cdot [g(x) + h(x)] \\
&= [f(x) \cdot g(x)] + [f(x) \cdot h(x)] \\
&= (f \cdot g)(x) + (f \cdot h)(x) \\
&= [(f \cdot g) + (f \cdot h)](x)
\end{aligned}$$

Since this is true for all $x \in \mathbb{R}$, we have $f \cdot (g + h) = (f \cdot g) + (f \cdot h)$.

Verifying the other distributive law is similar. Therefore, $(\mathscr{F}(\mathbb{R}), +, \cdot)$ is a ring. ■

Each of the rings mentioned so far is an example of a ring $(S, +, \cdot)$ in which the multiplication $\cdot$ is commutative; that is,

$$x \cdot y = y \cdot x$$

for all $x, y \in S$. Such rings are called *commutative rings*. Also note that each of the rings mentioned possesses an identity element with respect to the ring multiplication. In general, a ring $(S, +, \cdot)$ is called a *ring with identity* if it has an identity element with respect to the ring multiplication. In keeping with common practice, we denote this element by e and call it the *multiplicative identity* of $(S, +, \cdot)$. Thus for all $a \in S$,

$$e \cdot a = a \cdot e = a$$

Example 6.12 The ring of integers $(\mathbb{Z}, +, \cdot)$ is a commutative ring with multiplicative identity 1. Verify that each of the rings $(\mathbb{Z}_m, \oplus, \odot)$ and $(\mathscr{F}(\mathbb{R}), +, \cdot)$ is a commutative ring with identity.

Solution For $[a], [b] \in \mathbb{Z}_m$, note that $[a] \odot [b] = [a \cdot b] = [b \cdot a] = [b] \odot [a]$, and $[1] \odot [a] = [a]$. Thus $(\mathbb{Z}_m, \oplus, \odot)$ is a commutative ring with identity $[1]$.

Let $f, g \in \mathcal{F}(\mathbb{R})$, and let $e:\mathbb{R} \to \mathbb{R}$ be defined by $e(x) = 1$. Then $(f \cdot g)(x) = f(x) \cdot g(x) = g(x) \cdot f(x) = (g \cdot f)(x)$, and $(e \cdot f)(x) = e(x) \cdot f(x) = 1 \cdot f(x) = f(x)$. This shows that $(\mathcal{F}(\mathbb{R}), +, \cdot)$ is a commutative ring with identity e. ■

Example 6.13 As an example of a noncommutative ring without identity, consider the set $A = \{z, a, b, c\}$, with $+$ and $\cdot$ defined by the operation tables in Figure 6.3. It is easy to check that $(A, +)$ is an abelian group; in fact, there is a great resemblance between this group and the Klein four-group of Figure 6.2(a). It can also be checked that $\cdot$ is associative but not commutative, and that there is no identity with respect to $\cdot$. ■

+	z	a	b	c
z	z	a	b	c
a	a	z	c	b
b	b	c	z	a
c	c	b	a	z

·	z	a	b	c
z	z	z	z	z
a	z	a	b	c
b	z	z	z	z
c	z	a	b	c

FIGURE 6.3 Operation tables for Example 6.13

We now proceed to some of the basic properties of rings.

THEOREM 6.8 Let $(S, +, \cdot)$ be a ring and let $a, b, c \in S$. Then the following conditions hold.

1. If $c + c = c$, then $c = z$.
2. $a \cdot z = z \cdot a = z$
3. $(-a) \cdot b = a \cdot (-b) = -(a \cdot b)$
4. $(-a) \cdot (-b) = a \cdot b$
5. $a \cdot (b - c) = (a \cdot b) - (a \cdot c)$
6. $(b - c) \cdot a = (b \cdot a) - (c \cdot a)$

Proof We prove parts 2 and 3 and leave the other parts for the exercises. To establish part 2, note that

$$a \cdot z = a \cdot (z + z) = (a \cdot z) + (a \cdot z)$$

From $a \cdot z = (a \cdot z) + (a \cdot z)$ and part 1, it follows that $a \cdot z = z$.
To show that $(-a) \cdot b = -(a \cdot b)$, first see that

$$((-a) \cdot b) + (a \cdot b) = (-a + a) \cdot b = z \cdot b = z$$

so $(-a) \cdot b$ is the additive inverse of $a \cdot b$. Since additive inverses are unique, it must be that $(-a) \cdot b = -(a \cdot b)$. Similarly, $a \cdot (-b) = -(a \cdot b)$. ■

Recall that in a group $(G, *)$, the expression a^n, $n \geq 0$, was defined by

$$a^0 = e \tag{1}$$

$$a^n = a * a^{n-1}, \quad n > 0 \tag{2}$$

and the following properties of exponents were given:

$$a^{-n} = (a^n)^{-1} = (a^{-1})^n \tag{3}$$

$$a^m * a^n = a^{m+n} \tag{4}$$

$$(a^m)^n = a^{mn} \tag{5}$$

for all $m, n \in \mathbb{Z}$. If $(S, +, \cdot)$ is a ring, then $(S, +)$ is an abelian group. Since the additive notation is being used in this group, we wish to restate properties (1) through (5) with this notation. For example, in place of $a^2 = a * a$, we write $2a = a + a$, and instead of $a^3 = a * a * a$, we write $3a = a + a + a$. In general, equations (1) and (2) become

$$0a = z \tag{1'}$$

$$na = a + (n - 1)a \tag{2'}$$

and we refer to na as a *multiple* of a. Properties (3) through (5) become

$$(-n)a = -(na) = n(-a) \tag{3'}$$

$$ma + na = (m + n)a \tag{4'}$$

$$n(ma) = (mn)a \tag{5'}$$

Similar properties involving ring multiplication are contained in the following theorem.

THEOREM 6.9 If $(S, +, \cdot)$ is a ring and $a, b \in S$, then for any integers m and n the following hold.

1. $m(a \cdot b) = (ma) \cdot b = a \cdot (mb)$
2. $(mn)(a \cdot b) = (ma) \cdot (nb)$

Proof We shall prove part 1 and leave part 2 as an exercise.

We first use induction to show that $m(a \cdot b) = (ma) \cdot b$ holds for all integers $m \geq 0$. If $m = 0$, then $0(a \cdot b) = z = z \cdot b = (0a) \cdot b$, and if $m = 1$, then $1(a \cdot b) = a \cdot b = (1a) \cdot b$, so the result holds in these two cases. Let k be an arbitrary integer, $k \geq 1$, and assume that $k(a \cdot b) = (ka) \cdot b$. Then

$$\begin{aligned}
(k + 1)(a \cdot b) &= k(a \cdot b) + 1(a \cdot b) && \text{(by definition)}\\
&= [(ka) \cdot b] + [(1a) \cdot b] && \text{(by IHOP)}\\
&= [(ka) + (1a)] \cdot b && \text{(by the distributive law)}\\
&= [(k + 1)a] \cdot b && \text{(by definition)}
\end{aligned}$$

This shows that the result holds for $k + 1$ and hence, by induction, $m(a \cdot b) = (ma) \cdot b$ holds for all $m \geq 0$.

To show that $m(a \cdot b) = (ma) \cdot b$ for $m < 0$, let $m = -n$, with $n \in \mathbb{N}$. Then

$$\begin{aligned} m(a \cdot b) &= (-n)(a \cdot b) \\ &= -(n(a \cdot b)) && \text{(by property (3'))} \\ &= -((na) \cdot b) && \text{(from the proof for } m \geq 0\text{)} \\ &= (-(na)) \cdot b && \text{(by Theorem 6.8, part 3)} \\ &= ((-n)a) \cdot b && \text{(by property (3'))} \\ &= (ma) \cdot b \end{aligned}$$

In a similar fashion, it can be shown that $m(a \cdot b) = a \cdot (mb)$ holds for any integer m. ■

Before proceeding further, there is one very simple ring that should be mentioned: that is the ring consisting of a single element. This element is necessarily the zero element, z, so that $z + z = z$ and $z \cdot z = z$ are the addition and multiplication in the ring. Thus the ring has an identity element, namely z. We shall call this ring the *trivial ring*.

THEOREM 6.10 If $(S, +, \cdot)$ is a ring with identity and is not the trivial ring, then $e \neq z$.

Proof Proceed by contradiction and suppose that $e = z$. Since $S \neq \{z\}$, there is some element $a \in S$ such that $a \neq z$. But then

$$a = a \cdot e = a \cdot z = z$$

which is a contradiction. Hence $e \neq z$. ■

Suppose now that $(S, +, \cdot)$ is a ring with identity and is not the trivial ring. What can be said about the multiplicative structure $(S, \cdot)$ of the ring? For example, must each element of S possess a multiplicative inverse? The answer to this question is clearly no if we consider the element z. Since $a \cdot z = z$ for all $a \in S$, it seems more reasonable to consider $(S^{\#}, \cdot)$, where $S^{\#} = S - \{z\}$. Again the answer is no, in general, for consider the ring $(\mathbb{Z}_{12}, \oplus, \odot)$. The element $[2]$ satisfies the condition $[6] \odot [2] = [0]$, the zero element of the ring. If there were an element $[a] \in \mathbb{Z}_{12}$ for which $[2] \odot [a] = [1]$, then we would obtain

$$\begin{aligned} [6] &= [6] \odot [1] \\ &= [6] \odot ([2] \odot [a]) \\ &= ([6] \odot [2]) \odot [a] \\ &= [0] \odot [a] \\ &= [0] \end{aligned}$$

a contradiction. Thus the element $[2]$ has no multiplicative inverse.

DEFINITION 6.9

Let $(S, +, \cdot)$ be a ring. An element $a \in S$, where $a \neq z$, is called a *zero divisor* of the ring if there is an element $b \in S$, where $b \neq z$, such that either $a \cdot b = z$ or $b \cdot a = z$.

From the preceding discussion we see that [2] and [6] are zero divisors in the ring $(\mathbb{Z}_{12}, \oplus, \odot)$.

Example 6.14 Consider again the ring $(\mathcal{F}(\mathbb{R}), +, \cdot)$. As mentioned, the zero of this ring is the zero function z, defined by $z(x) = 0$ for all x. To exhibit zero divisors in this ring, we must find functions $f, g \in \mathcal{F}(\mathbb{R})$ such that $f \neq z$, $g \neq z$, and $f \cdot g = z$, that is, $f(x) \cdot g(x) = 0$ for all x. Such examples are numerous and are easy to construct. For instance, define f and g by

$$f(x) = \begin{cases} 0 & \text{if } x < 0 \\ 1 & \text{if } x \geq 0 \end{cases} \qquad \text{and} \qquad g(x) = \begin{cases} -1 & \text{if } x < 0 \\ 0 & \text{if } x \geq 0 \end{cases}$$ ■

Suppose next that $(S, +, \cdot)$ is a ring with identity and no zero divisors. Will it then be the case that each element $a \in S^{\#}$ has a multiplicative inverse? The answer is no; in fact, we already have seen an example of a ring with identity and no zero divisors in which infinitely many nonzero elements do not have multiplicative inverses. The ring is $(\mathbb{Z}, +, \cdot)$. Notice that each element different from ± 1 has no multiplicative inverse. Yet rings like $(\mathbb{Z}, +, \cdot)$ do possess some added structure in that they have no zero divisors.

DEFINITION 6.10

Let $(S, +, \cdot)$ be a commutative ring with identity $e \neq z$. If $(S, +, \cdot)$ has no zero divisors, then $(S, +, \cdot)$ is called an *integral domain*.

Example 6.15 Show that each of the following rings is an integral domain.
(a) $(\mathbb{Z}[\sqrt{2}], +, \cdot)$ (b) $(\mathbb{Z}_p, \oplus, \odot)$, where p is prime

Solution First note that each of these rings is commutative and has an identity ($1 = 1 + 0\sqrt{2}$ is the identity of $\mathbb{Z}[\sqrt{2}]$). To show that a commutative ring with identity, $(S, +, \cdot)$, has no zero divisors, it suffices to prove

$$xy = z \rightarrow (x = z \vee y = z)$$

for all $x, y \in S$.

(a) Consider first the ring $(\mathbb{Z}[\sqrt{2}], +, \cdot)$, and let $x = a + b\sqrt{2}$ and $y = c + d\sqrt{2}$ be elements of this ring. Suppose that $xy = 0$. Then

$$(ac + 2bd) + (ad + bc)\sqrt{2} = 0$$

which means that $ac + 2bd = 0 = ad + bc$. From this it can be shown that either both a and b are 0 or both c and d are 0, from which it follows that either x or y equals 0. However, there is an easier way to obtain this result. Keep in mind that $\mathbb{Z}[\sqrt{2}] \subseteq \mathbb{R}$, so for the same x and y, we have that x and y are real numbers and $xy = 0$. From this we immediately obtain that $x = 0$ or $y = 0$, which shows that $(\mathbb{Z}[\sqrt{2}], +, \cdot)$ has no zero divisors.
(b) Next, for $\mathbb{Z}_p$, where p is prime, suppose that $[x] \odot [y] = [0]$. Then $[xy] = [0]$, which implies that $p \mid xy$. Hence, by Euclid's lemma, either $p \mid x$ or $p \mid y$, so that either $[x] = [0]$ or $[y] = 0$. Therefore $(\mathbb{Z}_p, \oplus, \odot)$ is an integral domain when p is prime. ■

The definition of an integral domain is often given in terms of the "cancellation law," a condition that turns out to be equivalent to having no zero divisors.

DEFINITION 6.11

Let $(S, +, \cdot)$ be a commutative ring. We say that $(S, +, \cdot)$ satisfies the *cancellation law* if, for all $a, b, c \in S$, the condition $a \cdot b = a \cdot c$ and $a \neq z$ implies that $b = c$.

THEOREM 6.11 Let $(S, +, \cdot)$ be a commutative ring. Then $(S, +, \cdot)$ has no zero divisors if and only if $(S, +, \cdot)$ satisfies the cancellation law.

Proof Assume first that $(S, +, \cdot)$ has no zero divisors. Let $a, b, c \in S$, $a \neq z$, and suppose that $a \cdot b = a \cdot c$. Then $(a \cdot b) - (a \cdot c) = z$, so by the distributive law, $a \cdot (b - c) = z$. Since $a \neq z$ and S has no zero divisors, it follows that $b - c = z$. Hence $b = c$. Thus $(S, +, \cdot)$ satisfies the cancellation law.

For the converse, assume that $(S, +, \cdot)$ satisfies the cancellation law, and suppose that $a \cdot b = z$ for some $a, b \in S$. If $a = z$ there is nothing to prove, so assume $a \neq z$. By Theorem 6.8, part 2, $a \cdot z = z$, so we have that $a \cdot b = a \cdot z$. Since S satisfies the cancellation law and $a \neq z$, we may conclude that $b = z$. Hence $(S, +, \cdot)$ has no zero divisors. ■

COROLLARY 6.11 Let $(S, +, \cdot)$ be a commutative ring with identity. If $(S, +, \cdot)$ satisfies the cancellation law, then $(S, +, \cdot)$ is an integral domain. ■

If $(S, +, \cdot)$ is a commutative ring with identity $e \neq z$, then the multiplicative structure $(S^{\#}, \cdot)$ is a commutative semigroup with identity. In general, $(S^\#, \cdot)$ need not be a group, as in the case of $(\mathbb{Z}^\#, \cdot)$. In fact, $(S^\#, \cdot)$ is an abelian group if and only if each element has a multiplicative inverse. The resulting ring $(S, +, \cdot)$ is then called a "field."

DEFINITION 6.12

Let $(F, +, \cdot)$ be a commutative ring with identity. If $(F^\#, \cdot)$ is an abelian group, then $(F, +, \cdot)$ is called a *field*.

As already noted, we shall denote the multiplicative identity of a field $(F, +, \cdot)$ by e. In addition, if $a \in F^\#$, then the multiplicative inverse of a will be denoted by the familiar notation a^{-1}.

Example 6.16 The multiplicative inverse of the real number a, where $a \neq 0$, is $1/a$. It follows that each of the rings $(\mathbb{R}, +, \cdot)$ and $(\mathbb{Q}, +, \cdot)$ are fields, called the *field of real numbers* and the *field of rational numbers*, respectively. ■

Example 6.17 It can be shown that $(\mathbb{Z}[\sqrt{2}], +, \cdot)$ is not a field (see Exercise 19). However, consider the ring $(\mathbb{Q}[\sqrt{2}], +, \cdot)$, where

$$\mathbb{Q}[\sqrt{2}] = \{a + b\sqrt{2} \mid a, b \in \mathbb{Q}\}$$

In a similar manner as for $(\mathbb{Z}[\sqrt{2}], +, \cdot)$, we can verify that $(\mathbb{Q}[\sqrt{2}], +, \cdot)$ is a commutative ring with identity. Show that $(\mathbb{Q}[\sqrt{2}], +, \cdot)$ is a field.

Solution We must show that each element $a + b\sqrt{2}$, where $a, b \in \mathbb{Q}$ and not both a and b are zero, has a multiplicative inverse. That is, we must find an element $c + d\sqrt{2}$ such that

$$(a + b\sqrt{2})(c + d\sqrt{2}) = 1 + 0\sqrt{2}$$

From this equation we obtain $ac + 2bd = 1$ and $ad + bc = 0$. These equations must be solved for c and d. We leave it to the reader to check that the solution is $c = a/(a^2 - 2b^2)$ and $d = -b/(a^2 - 2b^2)$. Therefore, $(\mathbb{Q}[\sqrt{2}], +, \cdot)$ is a field. ■

THEOREM 6.12 If $(F, +, \cdot)$ is a field, then it is an integral domain.

Proof In order to prove that $(F, +, \cdot)$ is an integral domain, it suffices by Corollary 6.11 to show that $(F, +, \cdot)$ satisfies the cancellation law. So suppose that $a, b, c \in F$, with $a \neq z$, and $a \cdot b = a \cdot c$. Since $a \neq z$, a^{-1} exists. Thus

$$a^{-1} \cdot (a \cdot b) = a^{-1} \cdot (a \cdot c)$$
$$(a^{-1} \cdot a) \cdot b = (a^{-1} \cdot a) \cdot c$$
$$e \cdot b = e \cdot c$$
$$b = c$$

Hence, $(F, +, \cdot)$ satisfies the cancellation law. ■

It has been pointed out that $(\mathbb{Z}, +, \cdot)$ is an integral domain that is not a field, and thus the collection of all fields is a proper subset of the set of all integral domains. Indeed, using the notations

$$\mathcal{R} = \text{set of all rings}$$
$$\mathcal{C} = \text{set of all commutative rings}$$
$$\mathcal{D} = \text{set of all integral domains}$$
$$\mathcal{F} = \text{set of all fields}$$

the following inclusion relations hold:

$$\mathcal{F} \subset \mathcal{D} \subset \mathcal{C} \subset \mathcal{R}$$

Under certain restrictions it happens that every integral domain is a field.

THEOREM 6.13 Every finite integral domain is a field.

Proof Let $(D, +, \cdot)$ be a finite integral domain, say $D = \{a_1, a_2, \ldots, a_m\}$. To show that $(D, +, \cdot)$ is a field, we need show only that each nonzero element of D has a multiplicative inverse. Let $a \in D$, with $a \neq z$, and consider the elements $a \cdot a_1, a \cdot a_2, \ldots, a \cdot a_m$. These elements are distinct, for if $a \cdot a_i = a \cdot a_j$ for some i and j, then we can apply the cancellation law (since $a \neq z$) to conclude that $a_i = a_j$ (or $i = j$). Since $a \cdot a_1, a \cdot a_2, \ldots, a \cdot a_m$ are m distinct elements of D and $n(D) = m$, we may conclude that

$$D = \{a \cdot a_1, a \cdot a_2, \ldots, a \cdot a_m\}$$

Thus the identity element e must be one of the elements $a \cdot a_1, a \cdot a_2, \ldots, a \cdot a_m$, say $a \cdot a_i = e$. It follows that $a^{-1} = a_i$. Therefore, $(D, +, \cdot)$ is a field. ■

COROLLARY 6.13 The ring $(\mathbb{Z}_m, \oplus, \odot)$ is a field if and only if m is prime. ■

Exercises 6.3

1. Each of the following parts gives a set S and two binary operations # and $*$ on S. Determine whether $(S, \#, *)$ is a ring.
   a. $S = \mathbb{R}$, $x \# y = x + y - 1$, $x * y = xy$
   b. $S = \mathbb{Z}$, $x \# y = x + y - 1$, $x * y = x + y - xy$
   c. $S = \mathbb{R}^+$, $x \# y = xy$, $x * y = x + y$
   d. $S = \mathbb{R}^+$, $x \# y = xy$, $x * y = x^y$
2. Complete the solution of Example 6.11, showing that $(\mathcal{F}(\mathbb{R}), +, \cdot)$ is a ring.

a. Show that $+$ is associative and commutative.
b. Show that $z: \mathbb{R} \to \mathbb{R}$ defined by $z(x) = 0$ is the additive identity.
c. Show that the additive inverse of the function f is the function $-f: \mathbb{R} \to \mathbb{R}$ defined by $(-f)(x) = -f(x)$.
d. Show that $\cdot$ is associative.

3. Consider the ring $(\mathbb{Z}, \#, *)$ found in Exercise 1b.
a. What is the zero element?
b. What is the multiplicative identity?
c. Is this ring commutative?
d. Is this ring an integral domain?
e. Is this ring a field?

4. Consider the set $\mathbb{R} \times \mathbb{R}$ of ordered pairs of real numbers. Define the binary operations $+$ and $\cdot$ on $\mathbb{R} \times \mathbb{R}$ by

$$(a,b) + (c,d) = (a + c,\ b + d)$$
$$(a,b)\cdot(c,d) = (ac - bd,\ ad + bc)$$

Note that $+$ and $\cdot$ are also binary operations on $\mathbb{Z} \times \mathbb{Z}$.
a. Show that $(\mathbb{Z} \times \mathbb{Z}, +, \cdot)$ is a commutative ring with identity.
b. Is $(\mathbb{Z} \times \mathbb{Z}, +, \cdot)$ an integral domain? A field?
c. Show that $(\mathbb{R} \times \mathbb{R}, +, \cdot)$ is a field.

5. Consider the ring $(\mathbb{Q}, \#, *)$, where $\#$ and $*$ are defined by

$$x \# y = x + y - 1$$
$$x * y = x + y - xy$$

Is this ring a field?

6. For $k \in \mathbb{N}$, let M_k denote the set of integer multiples of k. Show that $(M_k, +, \cdot)$ (where $+$ and $\cdot$ denote ordinary addition and multiplication) is a commutative ring but has no identity for $k \geq 2$.

7. Let A be a nonempty set and consider $\mathcal{P}(A)$ under the operations of $*$ and $\cap$, where $B * C = (B - C) \cup (C - B)$ is the symmetric difference of B and C. Show that $(\mathcal{P}(A), *, \cap)$ is a commutative ring with identity.

8. Prove Theorem 6.8, parts 1, 4, and 5.

9. Is $(\mathcal{F}(\mathbb{R}), +, \circ)$ a ring?

10. Let $(S, +, \cdot)$ be a commutative ring with identity.
a. Prove: If $a \in S$ has a multiplicative inverse, then a is not a zero divisor of S.
b. What about the converse of the statement in part a?
c. Prove: If S is finite and $a \in S$ is not a zero divisor of S, then a has a multiplicative inverse.

11. Consider the ring $(\mathbb{Z}_m, \oplus, \odot)$ and let $[a]$ be an element of $\mathbb{Z}_m$ such that $[a] \neq [0]$.
a. Show that if $\gcd(a, m) > 1$, then $[a]$ is a zero divisor.

b. Show that $[a]$ has a multiplicative inverse if and only if $\gcd(a, m) = 1$.

c. Show that $(\mathbb{Z}_m, \oplus, \odot)$ is not an integral domain if m is composite.

12. Prove Theorem 6.9, part 2.

13. Let E be the set of even integers. Define an operation $*$ on E by

$$x * y = xy/2$$

Show that $(E, +, *)$ is an integral domain.

14. Define operations $+$ and $\cdot$ on $\mathbb{R} \times \mathbb{R}$ as follows:

$$(a,b) + (c,d) = (a + c, b + d)$$
$$(a,b) \cdot (c,d) = (ac, bd)$$

a. Show that $(\mathbb{R} \times \mathbb{R}, +, \cdot)$ is a commutative ring with identity.

b. Is $(\mathbb{R} \times \mathbb{R}, +, \cdot)$ an integral domain? (Compare with Exercise 4.)

15. We have noted $(\mathbb{Z}_4, \oplus, \odot)$ is not a field. Show that the operation tables below define a field $(F, +, \cdot)$ of order 4.

$+$	z	e	a	b
z	z	e	a	b
e	e	z	b	a
a	a	b	z	e
b	b	a	e	z

$\cdot$	z	e	a	b
z	z	z	z	z
e	z	e	a	b
a	z	a	b	e
b	z	b	e	a

(It can be proven that there exists a finite field of order m if and only if m is a power of a prime.)

16. Let $(S, +, \cdot)$ be a ring with identity e. Show that $(-e) \cdot a = -a$ for all $a \in S$.

17. Let $(F, +, \cdot)$ be a field, and let $a, b \in F$, where $a \neq z$. Show that the equation

$$(a \cdot x) + b = z$$

has a (unique) solution $x \in F$.

18. (For those students who know about matrices.) Let $\mathcal{M}_n$ denote the set of all n by n matrices. Show that $\mathcal{M}_n$, under the operations of matrix addition and multiplication, is a noncommutative ring with identity.

19. Consider the ring $(\mathbb{Z}[\sqrt{2}], +, \cdot)$. For $r = a + b\sqrt{2} \in \mathbb{Z}[\sqrt{2}]$, define $N(r)$ by

$$N(r) = a^2 - 2b^2$$

a. Let $r, s \in \mathbb{Z}[\sqrt{2}]$. Show that $N(rs) = N(r)N(s)$.

b. Show that $r \in \mathbb{Z}[\sqrt{2}]$ has a multiplicative inverse if and only if $|N(r)| = 1$.

c. Show that the ring $(\mathbb{Z}[\sqrt{2}], +, \cdot)$ is not a field.

6.4 BOOLEAN ALGEBRAS

Recall that a lattice is a poset $(A, \lesssim)$ in which every two elements possess both a least upper bound (lub) and a greatest lower bound (glb). For each $a, b \in A$, both lub($\{a, b\}$) and glb($\{a, b\}$) are uniquely determined elements of A; notationally, lub($\{a, b\}$) $= a \vee b$ and glb($\{a, b\}$) $= a \wedge b$. The quantity $a \vee b$ is called the *join* of a and b, and $a \wedge b$ is called the *meet* of a and b. It is clear that $\vee$ and $\wedge$ are binary operations on A; we can then consider the algebraic structure $(A, \vee, \wedge)$.

THEOREM 6.14 If $(A, \lesssim)$ is a lattice, then the following properties hold for all $a, b, c \in A$:

1. *Commutative laws:*

$$a \vee b = b \vee a \quad \text{and} \quad a \wedge b = b \wedge a$$

2. *Associative laws:*

$$a \vee (b \vee c) = (a \vee b) \vee c \quad \text{and} \quad a \wedge (b \wedge c) = (a \wedge b) \wedge c$$

3. *Idempotent laws:*

$$a \vee a = a \quad \text{and} \quad a \wedge a = a$$

4. *Absorption laws:*

$$(a \vee b) \wedge a = a \quad \text{and} \quad (a \wedge b) \vee a = a$$

Proof We shall demonstrate the proof technique involved by considering the property $(a \vee b) \wedge a = a$. Let $d = (a \vee b) \wedge a$; then clearly $d \lesssim a \vee b$ and $d \lesssim a$. It suffices, by antisymmetry, to show that $a \lesssim d$. Since $a \lesssim a \vee b$ and $a \lesssim a$, we may conclude that $a \lesssim (a \vee b) \wedge a$. Hence $a \lesssim d$. Thus $a = d$.

The proofs of the remaining properties are handled in a similar manner and are left to the exercises. ■

Beyond those listed in Theorem 6.14, there are some other important properties that certain lattices may satisfy. Recall that a lattice $(L, \lesssim)$ is called a *distributive lattice* (see Chapter 4, Problem 27) provided

$$a \wedge (b \vee c) = (a \wedge b) \vee (a \wedge c)$$

and

$$a \vee (b \wedge c) = (a \vee b) \wedge (a \vee c)$$

A most prominent example of a distributive lattice is provided by the lattice $(\mathcal{P}(U), \subseteq)$ of all subsets of a set U. In this case $A \vee B = A \cup B$ and $A \wedge B = A \cap B$, and we proved in Chapter 2 (Theorem 2.6) that

$$A \cap (B \cup C) = (A \cap B) \cup (A \cap C)$$

and

$$A \cup (B \cap C) = (A \cup B) \cap (A \cup C)$$

Let us examine $(\mathcal{P}(U), \subseteq)$ further to see if it satisfies any other interesting and notable properties. Right off we know that ϕ and U are elements of $\mathcal{P}(U)$ such that

$$A \cup \phi = A \qquad \text{and} \qquad A \cap U = A$$

for all $A \in \mathcal{P}(U)$. Thus it follows that ϕ is an identity element with respect to $\cup$ and U is an identity element with respect to $\cap$.

DEFINITION 6.13

Let $(L, \preceq)$ be a lattice. An element $1 \in L$ is called a *unity* for L if $a \wedge 1 = a$ for all $a \in L$. An element $0 \in L$ is called a *zero element* for L if $a \vee 0 = a$ for all $a \in L$. If $(L, \preceq)$ has both a unity element and a zero element, we shall simply write that $(L, \preceq)$ is a *lattice with* 0 *and* 1.

In view of the above definition, we can say that $(\mathcal{P}(U), \subseteq)$ is a distributive lattice with unity element U and zero element ϕ.

In addition to the above properties, the lattice $(\mathcal{P}(U), \subseteq)$ has the property that for each $A \in \mathcal{P}(U)$ there is an element $A' \in \mathcal{P}(U)$ such that

$$A \cup A' = U \qquad \text{and} \qquad A \cap A' = \phi$$

Recall that A' is called the complement of A in U.

DEFINITION 6.14

Let $(L, \preceq)$ be a lattice with 0 and 1. An element $a' \in L$ is called the *complement* of a in L provided

$$a \vee a' = 1 \qquad \text{and} \qquad a \wedge a' = 0$$

If each element of L has a complement in L, then we call $(L, \preceq)$ a *complemented lattice*.

Example 6.18 Let L be the set of positive divisors of 30, and for each $a, b \in L$, define $a \preceq b$ provided $a \mid b$. Show that $(L, \preceq)$ is a lattice with zero and unity that is both distributive and complemented.

Solution Note that $L = \{1, 2, 3, 5, 6, 10, 15, 30\}$. It is easily shown that $(L, \preceq)$ is a poset. In addition, for $a, b \in L$, the quantity $a \wedge b$ is that element $d \in L$ satisfying

1. $d \mid a$ and $d \mid b$.
2. If $c \mid a$ and $c \mid b$ for some $c \in L$, then $c \mid d$.

It follows that $a \wedge b = \gcd(a, b)$. Similarly, it can be shown that $a \vee b = \text{lcm}(a, b)$. So $(L, \preceq)$ is a lattice.

For each $a \in L$, it is clear that $\text{lcm}(a, 1) = a$. Hence $a \vee 1 = a$ for all $a \in L$, which shows that 1 is a zero element for $(L, \preceq)$. Also, $\gcd(a, 30) = a$ for each $a \in L$, so 30 is a unity for $(L, \preceq)$.

To see that $(L, \preceq)$ is complemented, it should be noted that 30 has no repeated prime factors; thus, if $d \mid 30$, then $\gcd(d, 30/d) = 1$. Hence we define $a' = 30/a$ for each $a \in L$ and note that both $a \wedge a' = 1$ (the zero element) and $a \vee a' = 30$ (the unity).

For the distributive laws, let $a, b, c \in L$, and write

$$a = 2^{\alpha_1} \cdot 3^{\alpha_2} \cdot 5^{\alpha_3}$$
$$b = 2^{\beta_1} \cdot 3^{\beta_2} \cdot 5^{\beta_3}$$
$$c = 2^{\gamma_1} \cdot 3^{\gamma_2} \cdot 5^{\gamma_3}$$

Since $30 = 2 \cdot 3 \cdot 5$, each of the exponents above is either 0 or 1. To show the distributive law $a \wedge (b \vee c) = (a \wedge b) \vee (a \wedge c)$, we must verify that

$$\gcd(a, \text{lcm}(b, c)) = \text{lcm}(\gcd(a, b), \gcd(a, c))$$

Suppose that $\gcd(a, \text{lcm}(b, c)) = 2^{\delta_1} \cdot 3^{\delta_2} \cdot 5^{\delta_3}$ and $\text{lcm}(\gcd(a, b), \gcd(a, c)) = 2^{\varepsilon_1} \cdot 3^{\varepsilon_2} \cdot 5^{\varepsilon_3}$. We must show that $\delta_1 = \varepsilon_1$, $\delta_2 = \varepsilon_2$, and $\delta_3 = \varepsilon_3$. Note that

$$\delta_1 = \min(\alpha_1, \max(\beta_1, \gamma_1))$$

and

$$\varepsilon_1 = \max(\min(\alpha_1, \beta_1), \min(\alpha_1, \gamma_1))$$

Thus it is not hard to verify that $\delta_1 = \varepsilon_1$. Similarly it follows that $\delta_2 = \varepsilon_2$ and $\delta_3 = \varepsilon_3$. Therefore, $a \wedge (b \vee c) = (a \wedge b) \vee (a \wedge c)$. We leave verification of the other distributive law to the reader. ■

It has already been established that $(\mathcal{P}(U), \subseteq)$ is a distributive, complemented lattice with zero and unity. It turns out that this special lattice is the classical example of a very important algebraic structure.

DEFINITION 6.15

A lattice with zero and unity that is both distributive and complemented is called a *Boolean algebra*.

In addition to $(\mathcal{P}(U), \subseteq)$ being a Boolean algebra, we see that the lattice $(L, \leq)$ of Example 6.18 is also. In fact, if $n = p_1 p_2 \cdots p_r$, where $p_1, p_2, \ldots, p_r$ are distinct primes, and L is the set of positive divisors of n, then $(L, |)$ is a Boolean algebra (see Exercise 6).

Example 6.19 Let L be the set of positive divisors of 72. Show that $(L, |)$ is not a Boolean algebra.

Solution It suffices to show that one of the properties for a Boolean algebra does not hold. Just as in Example 6.18, it can be shown that $(L, |)$ is a lattice with unity 72 and zero 1. However, $(L, |)$ is not complemented; for instance, there is no $a' \in L$ for which both $24 \wedge a' = 1$ and $24 \vee a' = 72$. ∎

Example 6.20 Determine whether each of the lattices whose Hasse diagrams are shown in Figure 6.4 is a Boolean algebra.

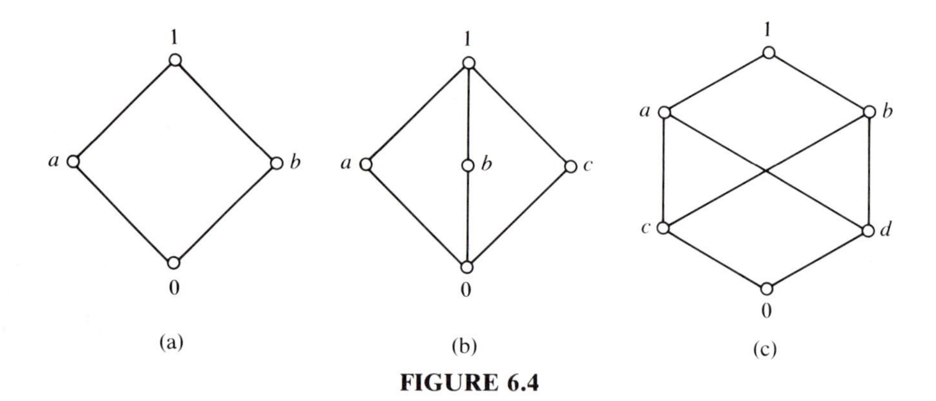

FIGURE 6.4

Solution The lattice of Figure 6.4(a) is a Boolean algebra. The unity and zero are shown as 1 and 0, and $b = a'$. Verifying the distributive laws takes a bit of effort, but it can be done.

The lattice in Figure 6.4(b) fails to satisfy the distributive laws; for instance, $a \vee (b \wedge c) = a \vee 0 = a$, but $(a \vee b) \wedge (a \vee c) = 1 \wedge 1 = 1$. Thus it is not a Boolean algebra.

An example of a lattice that is distributive but not complemented is shown in Figure 6.4(c). Note that the element a does not have a complement; for example, $b \neq a'$ since $a \wedge b = c \neq 0$, and $d \neq a'$ since $a \wedge d = d \neq 0$. ∎

It should be emphasized at this point that our definition of a Boolean algebra originates with a lattice $(L, \leq)$. The binary operations $\vee$ and $\wedge$

are defined by $a \vee b = \text{lub}(\{a, b\})$ and $a \wedge b = \text{glb}(\{a, b\})$. If $(B, \lesssim)$ is a Boolean algebra, then the following properties are satisfied by $\vee$ and $\wedge$:

1. *Commutative laws:* For $a, b \in B$,

$$a \vee b = b \vee a \qquad \text{and} \qquad a \wedge b = b \wedge a$$

2. *Associative laws:* For $a, b, c \in B$,

$$(a \vee b) \vee c = a \vee (b \vee c) \qquad \text{and} \qquad (a \wedge b) \wedge c = a \wedge (b \wedge c)$$

3. *Distributive laws:* For $a, b, c \in B$,

$$a \vee (b \wedge c) = (a \vee b) \wedge (a \vee c)$$

and

$$a \wedge (b \vee c) = (a \wedge b) \vee (a \wedge c)$$

4. *Existence of zero and unity:* There exist elements 0 and 1 in B such that, for every $a \in B$,

$$a \vee 0 = a \qquad \text{and} \qquad a \wedge 1 = a$$

5. *Complements:* For each $a \in B$ there is an element $a' \in B$ such that

$$a \wedge a' = 0 \qquad \text{and} \qquad a \vee a' = 1$$

A student who consults the literature will find that frequently a Boolean algebra is defined to be an algebraic structure $(B, \vee, \wedge)$, where $\vee$ and $\wedge$ are binary operations on B that satisfy precisely properties 1 through 5. In fact, if $(A, \vee, \wedge)$ is such an algebraic structure, then $(A, \lesssim)$ will be a Boolean algebra (in the sense of Definition 6.15) if we define

$$\mathrm{a} \lesssim b \leftrightarrow a \vee b = b$$

for any $a, b \in A$.

Exercises 6.4

1. Each of the lattices in Figure 6.5 (page 228) has a 0 and 1 as shown. First, determine if the lattice is complemented. If so, find a complement for each element. Next, if the lattice is complemented, determine whether it is distributive. Is the lattice a Boolean algebra?

2. Show that each element in a Boolean algebra has a unique complement.

3. Draw the Hasse diagram and construct the operation tables for $\vee$ and $\wedge$ for the Boolean algebra of positive divisors of each of the following numbers.

a. 5 **b.** 10 **c.** 30

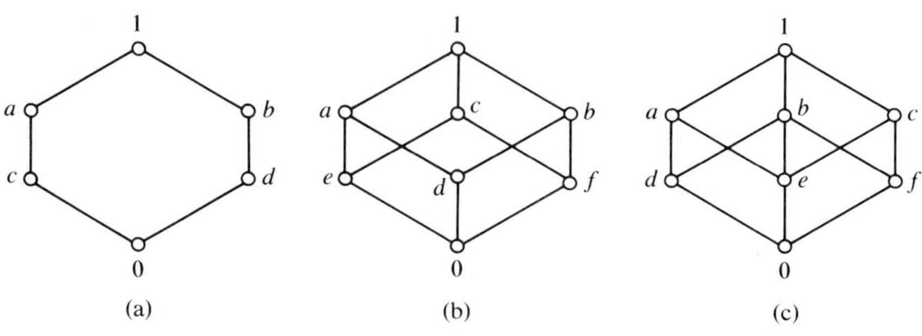

FIGURE 6.5

4. Show that any finite Boolean algebra has an even number of elements.

5. Draw the Hasse diagram and construct the operation tables for $\vee$ and $\wedge$ for the Boolean algebra of positive divisors of n, where
  a. n is prime;
  b. n is a product of two distinct primes: $n = pq$;
  c. n is a product of three distinct primes: $n = pqr$.

6. **a.** Show that the lattice of positive divisors of $n \in \mathbb{N}$ is complemented if and only if n is a product of distinct primes. (Hint: if p^2 divides n, where p is prime, show that p has no complement.)
  b. Use part a to show that the lattice of positive divisors of $n \in \mathbb{N}$ is a Boolean algebra if and only if n is a product of distinct primes.

7. If n is a positive integer and n is the product of m distinct primes, $n = p_1 p_2 \cdots p_m$, and B is the Boolean algebra of positive divisors of n, how many elements does B have?

8. **a.** Show that $(\{0, 1\}, \leq)$ is a Boolean algebra, where $0 \leq 1$. Complete each of the following: $0 \wedge 1 =$ ____, $0 \vee 1 =$ ____, $0' =$ ____, $1' =$ ____.
  b. Consider the set $B = \{(x_1, x_2, \ldots, x_n) \mid x_i = 0 \text{ or } 1, \quad 1 \leq i \leq n\}$, with $\wedge$ and $\vee$ defined as follows:

$$(x_1, x_2, \ldots, x_n) \wedge (y_1, y_2, \ldots, y_n) = (z_1, z_2, \ldots, z_n)$$

  where $z_i = x_i y_i$, $1 \leq i \leq n$, and

$$(x_1, x_2, \ldots, x_n) \vee (y_1, y_2, \ldots, y_n) = (w_1, w_2, \ldots, w_n),$$

  where $w_i = (x_i + y_i)$ MOD 2. Show that $\wedge$ and $\vee$ satisfy properties 1 through 5 on page 227; thus it follows that $(B, \vee, \wedge)$ is a Boolean algebra.
  c. What partial ordering is induced on B if $\leq$ is defined as at the bottom of page 227?

9. Consider the Boolean algebra of positive divisors of $n = pqr$, where p, q, and r are distinct primes. Simplify each of the following expressions.

a. $p \wedge q'$ **b.** $p \vee q'$ **c.** $(p \wedge q) \vee r$
d. $(p \vee q) \wedge r$ **e.** $p' \wedge q'$ **f.** $p' \vee q'$
g. $(p' \wedge q') \vee q$ **h.** $(p \vee q) \wedge (q \vee r)$

10. For elements a and b in a Boolean algebra B, show that the following conditions are equivalent:

(i) $a \lesssim b$ (ii) $a \wedge b' = 0$ (iii) $a' \vee b = 1$ (iv) $b' \lesssim a'$

11. Show that every Boolean algebra B satisfies DeMorgan's laws for all $a, b \in B$:

(i) $(a \vee b)' = a' \wedge b'$ (ii) $(a \wedge b)' = a' \vee b'$

12. Prove: For all elements a, b, and c of a Boolean algebra B,

$$(a \lesssim b \wedge c) \leftrightarrow (a \lesssim b \text{ and } a \lesssim c)$$

13. Prove: If B is a Boolean algebra with $a, b, c \in B$, such that $a \vee c = b \vee c$ and $a \wedge c = b \wedge c$, then $a = b$.

14. Complete the proof of Theorem 6.14 by showing that a lattice satisfies the commutative laws, the associative laws, the idempotent laws, and the absorption law $(a \wedge b) \vee a = a$.

★15. Does there exist a nondistributive lattice of order 5 that has a 0 and 1, other than the one of Figure 6.4(b)?

16. We defined a Boolean algebra as a special kind of lattice. As mentioned on page 227, a different approach is to define a Boolean algebra as an algebraic structure $(B, \vee, \wedge)$, in which $\vee$ and $\wedge$ satisfy properties 1 through 5 on page 227. To show that these approaches are equivalent, define a relation $\lesssim$ on B by $a \lesssim b \leftrightarrow a \vee b = b$. Verify that $\lesssim$ is a partial-order relation on B and that $(B, \lesssim)$ is a lattice.

6.5 SWITCHING CIRCUITS

Boolean algebra can be used in a very neat and clear way to analyze special types of electrical circuits. To explain the nature of such circuits, we must first agree on terminology. To begin with, a *switch* is a device having two possible states, *open* and *closed*. Current will flow through the point at which a switch is located if and only if the switch is closed. Figure 6.6 shows a diagram of a circuit containing a single switch a. The points s and t are called *terminals* of the circuit, and it is clear that current will flow from s to t if and only if switch a is closed.

In general, we will consider circuits that contain only switches and two

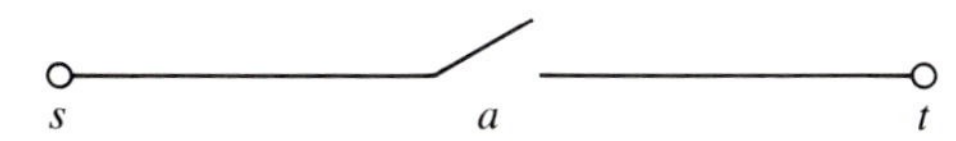

FIGURE 6.6 A circuit with one switch

terminals, s and t (and wire, of course); such circuits are called *switching circuits*. A switching circuit is said to be *closed* if current can pass from s to t, and it is *open* otherwise. Given a switching circuit, we are interested in those combinations of switching states for which the circuit is closed.

For example, consider the circuit of Figure 6.7. It seems evident that this circuit is closed if and only if either a or b is closed. This combination of switches a and b is denoted (strangely enough!) by $a \vee b$. Moreover, switches a and b are said to be in *parallel*.

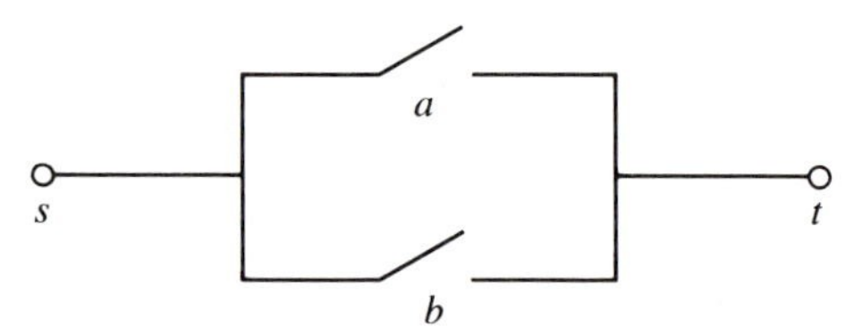

FIGURE 6.7 A parallel circuit

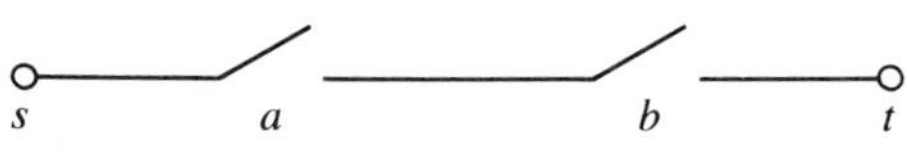

FIGURE 6.8 A series circuit

A contrasting situation is provided by the circuit of Figure 6.8. In this case it is easy to see that the circuit is closed if and only if both a and b are closed. Here a and b are said to be in *series*, and the circuit is denoted by $a \wedge b$.

Starting with a set S of switches, we can consider the set C of all switching circuits that can be constructed using combinations of series and parallel circuits. We call such circuits *series-parallel circuits*. (Keep in mind that $S \subseteq C$.) In effect, we have now defined two binary operations $\vee$ and $\wedge$ on C and, corresponding to any meaningful expression involving $\vee$, $\wedge$, and the elements of S, there is a uniquely determined series-parallel circuit. Conversely, each series-parallel circuit uniquely determines such an expression. Consequently, we shall draw no distinction between a given series-parallel circuit and its associated expression. Thus, for example, we have the circuit $[a \wedge (b \vee c)] \vee a$, which is displayed in Figure 6.9.

Notice in the circuit of Figure 6.9 that there are two occurrences of the switch a; this simply means that the associated switches are either both open, or both closed, simultaneously.

For each switch $a \in S$, we shall assume there is a switch $a' \in S$ that is open when a is closed and closed when a is open. The switch a' is called the *complement* of a.

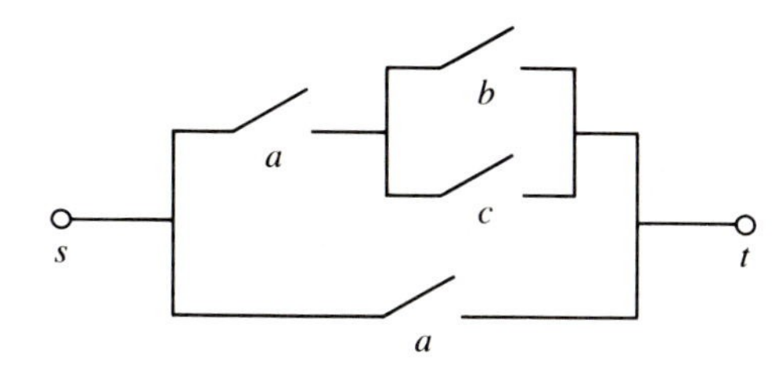

FIGURE 6.9 The circuit $[a \wedge (b \vee c)] \vee a$

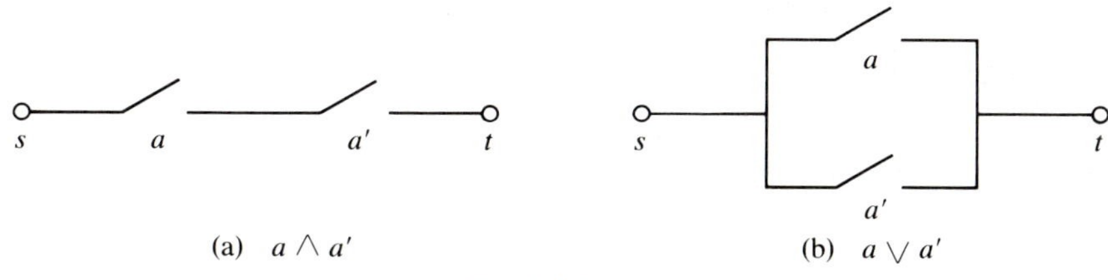

(a) $a \wedge a'$ (b) $a \vee a'$

FIGURE 6.10

Consider the circuits $a \wedge a'$ and $a \vee a'$, displayed in Figure 6.10. The circuit in (a) is always open, whereas the circuit in (b) is always closed. Should we adopt special notations for these circuits? Indeed, we shall denote a circuit that is always open by 0, and one that is always closed by 1. To this end we write

$$a \wedge a' = 0 \qquad \text{and} \qquad a \vee a' = 1$$

In general, two circuits C_1 and C_2 are called *equal,* denoted $C_1 = C_2$, provided the switch positions that close C_1 are the same as those that close C_2. With this definition of equality, we can legitimately consider the mathematical system $(C, \vee, \wedge)$. It will turn out that $(C, \vee, \wedge)$ is a Boolean algebra. The zero and unity elements are the circuits 0 and 1, respectively, just defined. To see that the distributive law

$$a \vee (b \wedge c) = (a \vee b) \wedge (a \vee c)$$

holds, for instance, consider the circuits in Figure 6.11.

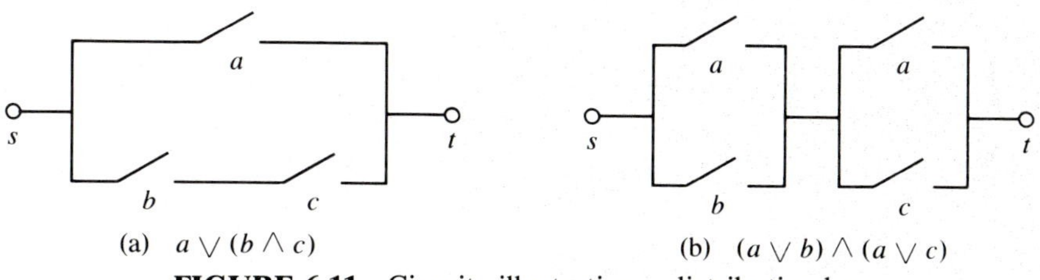

(a) $a \vee (b \wedge c)$ (b) $(a \vee b) \wedge (a \vee c)$

FIGURE 6.11 Circuits illustrating a distributive law

A four-minute perusal should be sufficient to determine that the two circuits are closed if and only if either a is closed or both b and c are closed.

With the knowledge that $(C, \vee, \wedge)$ is a Boolean algebra, we can now apply the properties of a Boolean algebra to a given circuit, possibly simplifying it and saving considerably on switches and wire.

Example 6.21 Simplify the circuit of Figure 6.12.

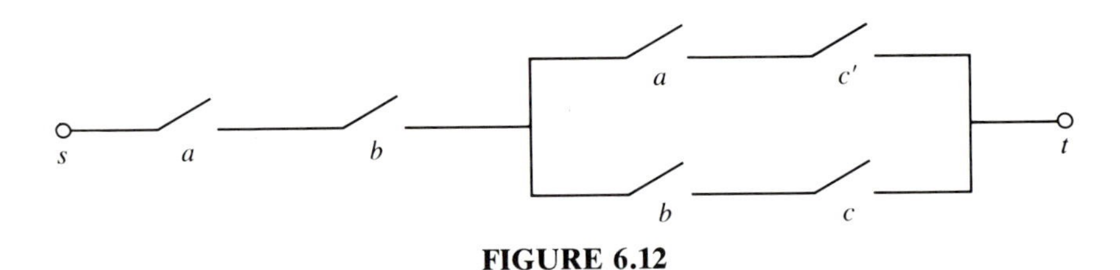

FIGURE 6.12

Solution This circuit is given by the expression

$$a \wedge b \wedge [(a \wedge c') \vee (b \wedge c)]$$

Using the properties of a Boolean algebra, we obtain

$$\begin{aligned} a \wedge b \wedge [(a \wedge c') \vee (b \wedge c)] & \\ &= [(a \wedge b) \wedge (a \wedge c')] \vee [(a \wedge b) \wedge (b \wedge c)] \\ &= [(a \wedge a) \wedge (b \wedge c')] \vee [a \wedge (b \wedge b) \wedge c] \\ &= [a \wedge (b \wedge c')] \vee [(a \wedge b) \wedge c] \\ &= [(a \wedge b) \wedge c')] \vee [(a \wedge b) \wedge c] \\ &= (a \wedge b) \wedge (c' \vee c) \\ &= (a \wedge b) \wedge 1 \\ &= a \wedge b \end{aligned}$$

Thus we may replace the given circuit by the circuit $a \wedge b$. ■

Consider the drawing of the circuit $(a \wedge b) \vee (a' \wedge b')$ in Figure 6.13(a). Such a drawing is used to study the properties of the given circuit and is not intended to show how such a circuit is actually implemented. In practice we would want the switches a and a', for example, to be physically connected in some way, so that switch a is open if and only if switch a' is closed. A drawing that shows how a given circuit would actually be wired is called a *wiring diagram*. A wiring diagram for the circuit $(a \wedge b) \vee (a' \wedge b')$ is given in Figure 6.13(b).

Example 6.22 Often a circuit with certain specialized properties must be designed. For example, suppose we have a garage light that we wish to be able to control

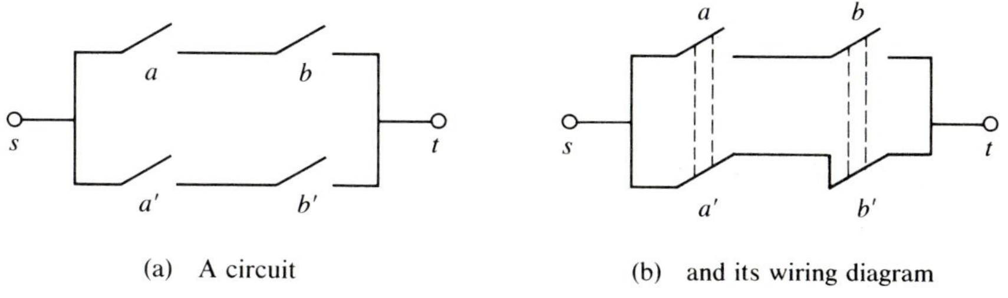

(a) A circuit (b) and its wiring diagram

FIGURE 6.13

from either the house or the garage. In other words, this light will have two switches (one in the house and one in the garage), and it must be possible to turn the light on or off from either switch.

To aid us in the design of this circuit, let us construct a table that shows the possible positions of the two switches.

House	*Garage*	*Light*
open	open	?
open	closed	?
closed	open	?
closed	closed	?

Now, regardless of whether the house switch is open or closed, we must be able to control the light from the garage. Thus we must begin to complete the table something like this:

House	*Garage*	*Light*
open	open	off
open	closed	on
closed	open	?
closed	closed	?

In the last two rows of the table, the light must be "on" in one and "off" in the other. Can we complete the table as follows?

House	*Garage*	*Light*
open	open	off
open	closed	on
closed	open	off
closed	closed	on

In this case, notice that the light is being completely controlled by the switch in the garage, so this won't do. Thus we are led to the following solution.

House	*Garage*	*Light*
open	open	off
open	closed	on
closed	open	on
closed	closed	off

Note that this circuit has the property that the light is on if and only if exactly one of the switches is closed. Recalling the truth table for the "exclusive or" from Chapter 1, we see that a circuit with the desired properties is

$$(H \wedge G') \vee (H' \wedge G)$$ ■

Boolean algebras receive considerable interest currently due to their usefulness in the design of computers and also because of applications to telephone switching networks.

We would be remiss if we did not mention the rather obvious relationship between Boolean algebras and propositional calculus. It is easy to check that the properties satisfied by propositions under the operations of disjunction ($\vee$) and conjunction ($\wedge$) are precisely the axioms for a Boolean algebra. Here the unity is the tautology T, the zero is the contradiction F, and the complement of a proposition p is the proposition $\sim p$. Indeed, it is common to allude to the Boolean algebra of propositions.

Exercises 6.5

1. Simplify each of the circuits in Figure 6.14.

2. Draw and simplify the circuit

$$(a \vee b) \wedge (a \vee c) \wedge (b \vee c)$$

3. a. Draw the circuit represented by

$$a' \vee (b \wedge c) \vee (b \wedge c')$$

b. Use the properties of a Boolean algebra to simplify the circuit in part a.

4. a. Draw the circuit represented by

$$(a \vee b \vee c) \wedge (a' \vee b \vee c) \wedge (a' \vee b' \vee c)$$

b. Use the properties of a Boolean algebra to simplify the circuit in part a and then draw the simplified circuit.

5. Draw the circuit represented by each of the following.

a. $(a \wedge (b \vee c' \vee d)) \vee c$

b. $(a \wedge b' \wedge c') \vee (a' \wedge b) \vee (a' \wedge c)$

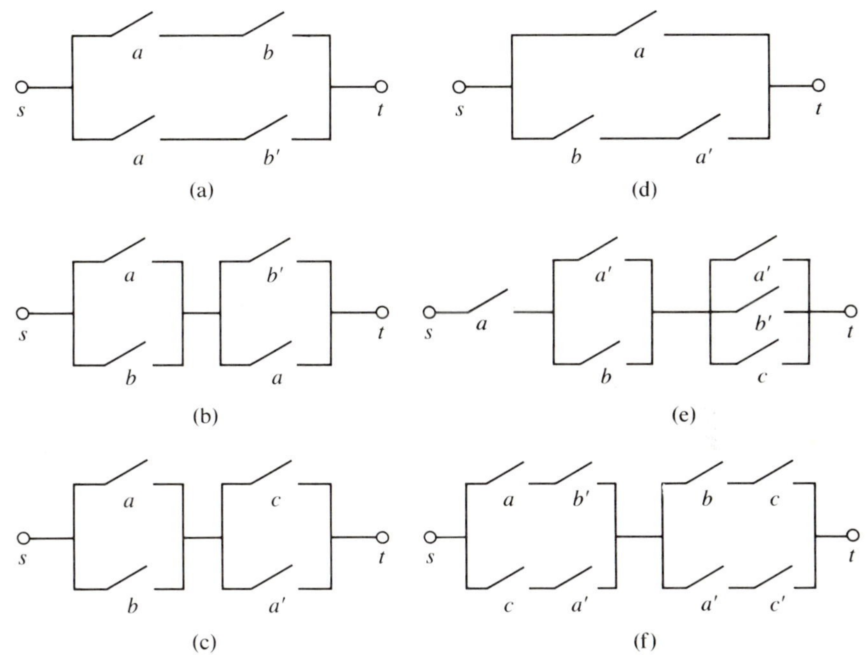

FIGURE 6.14

6. Draw a circuit that is closed if and only if

  a. exactly one of a and b is closed;

  b. at least one of a and b is closed;

  c. at least two of a, b, and c are closed;

  d. exactly two of a, b, and c are closed;

  e. at most two of a, b, and c are closed.

7. Farmer Joe has a barn light that can be switched on or off from either the barn, the garage, or the house. Design such a circuit.

8. Suppose in the previous exercise that the switch in the barn is a "master switch": when it is off the light is off, and when it is on the light may be controlled from either the house or the garage. Design such a circuit.

9. **a.** Design a circuit that allows four people to vote yes or no, and that allows current to pass only if a majority vote yes.

  b. Add a fifth person to the above situation who will act (only) as a tiebreaker.

6.6 ISOMORPHISMS

Consider the square S whose vertices A, B, C, and D are situated at the respective points $(1,1)$, $(-1,1)$, $(-1,-1)$, and $(1,-1)$ of a coordinate system, as shown in Figure 6.15. Let π_0, π_1, π_2, and π_3 denote the following rotations of the square S about (0,0):

π_0: counterclockwise rotation through 0°
π_1: counterclockwise rotation through 90°
π_2: counterclockwise rotation through 180°
π_3: counterclockwise rotation through 270°

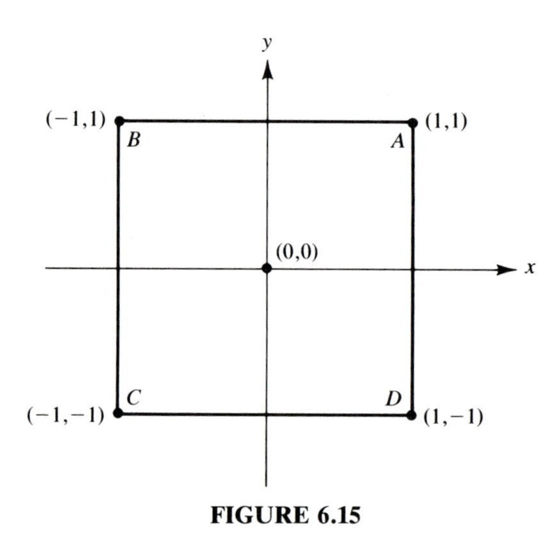

FIGURE 6.15

Then $\Gamma = \{\pi_0, \pi_1, \pi_2, \pi_3\}$ is a set of functions on S and, in fact, is a group under the operation of composition. Its operation table is easily constructed, as seen in Figure 6.16.

$\circ$	π_0	π_1	π_2	π_3
π_0	π_0	π_1	π_2	π_3
π_1	π_1	π_2	π_3	π_0
π_2	π_2	π_3	π_0	π_1
π_3	π_3	π_0	π_1	π_2

FIGURE 6.16 An operation table for $(\Gamma, \circ)$

Next, consider the group $(\mathbb{Z}_4, \oplus)$; its operation table is in Figure 6.17.

$\oplus$	[0]	[1]	[2]	[3]
[0]	[0]	[1]	[2]	[3]
[1]	[1]	[2]	[3]	[0]
[2]	[2]	[3]	[0]	[1]
[3]	[3]	[0]	[1]	[2]

FIGURE 6.17 An operation table for $(\mathbb{Z}_4, \oplus)$

The two groups $(\Gamma, \circ)$ and $(\mathbb{Z}_4, \oplus)$ are different, yet the reader will no doubt notice a similarity between the two operation tables. How can we make this resemblance more precise? It is tempting to say that π_0 corresponds to [0], π_1 to [1], π_2 to [2], and π_3 to [3]. Moreover, as is evident from the tables, such a correspondence is preserved by the operations $\circ$ and $\oplus$; that is, if $\pi_i \circ \pi_j = \pi_k$ in Γ, then $[i] \oplus [j] = [k]$ in $\mathbb{Z}_4$. But this correspondence is just a function, so what we are saying is that we have a function $f:\Gamma \to \mathbb{Z}_4$, defined by $f(\pi_i) = [i]$ for $i = 0,1,2,3$. Moreover, the function f satisfies the interesting property

$$f(\pi_i \circ \pi_j) = f(\pi_i) \oplus f(\pi_j)$$

(At first glance this property may seem confusing, but keep in mind that $f(\pi_i)$ and $f(\pi_j)$ are elements of $\mathbb{Z}_4$, so the expression $f(\pi_i) \oplus f(\pi_j)$ makes sense.)

Therefore, even though the two groups $(\Gamma, \circ)$ and $(\mathbb{Z}_4, \oplus)$ are not equal, they are the same in the sense we have just demonstrated. We shall say that $(\Gamma, \circ)$ is "isomorphic to" $(\mathbb{Z}_4, \oplus)$.

DEFINITION 6.16

Given groups $(G, *)$ and $(H, \#)$, a function $f:G \to H$ is called an *isomorphism* if the following conditions hold:

1. f is one-to-one and onto.
2. $f(a * b) = f(a) \# f(b)$, for all $a, b \in G$.

We say that $(G, *)$ is *isomorphic to* $(H, \#)$ and write $(G, *) \cong (H, \#)$.

Example 6.23 Consider the function $f:(\mathbb{R}, +) \to (\mathbb{R}^+, \cdot)$ defined by $f(x) = 2^x$. (Recall that $\mathbb{R}^+$ denotes the set of positive reals.) Show that f is an isomorphism.

Solution The function f is one-to-one, since $f(x_1) = f(x_2)$ implies that $2^{x_1} = 2^{x_2}$, which implies that $x_1 = x_2$. Also f is onto, since for each $y \in \mathbb{R}^+$ there is an $x \in \mathbb{R}$ such that $2^x = y$, namely, $x = \log_2 y$. Finally, let $a, b \in \mathbb{R}$. Then $f(a + b) = 2^{a+b} = 2^a \cdot 2^b = f(a) \cdot f(b)$. Therefore f is an isomorphism, which shows that the groups $(\mathbb{R}, +)$ and $(\mathbb{R}^+, \cdot)$ are isomorphic. ∎

Example 6.24 Determine which of the following functions are isomorphisms.

(a) $f:(\mathbb{Z}, +) \to (\mathbb{Z}, +)$; $f(n) = n + 1$
(b) $f:(\mathbb{Z}, +) \to (\mathbb{Z}_7, \oplus)$; $f(n) = [n]$
(c) $f:(\mathbb{Z}_6, \oplus) \to (\mathbb{Z}_6, \oplus)$; $f([a]) = [5a]$
(d) $f:(\mathbb{Z}, +) \to (\mathbb{Q}, +)$; $f(n) = n$

Solution (a) This function f is clearly one-to-one and onto. But for $m, n \in \mathbb{Z}$, it happens that $f(m + n) = (m + n) + 1$, whereas $f(m) + f(n) = (m + 1) + (n + 1) = m + n + 2$. Thus $f(m + n) \neq f(m) + f(n)$, and so f is not an isomorphism. (Clearly, however, every group is isomorphic to itself, the identity function being an isomorphism. This points out the distinction between determining whether two given groups are isomorphic, and the problem of verifying whether a given function is an isomorphism.)

(b) This function f is not one-to-one; for example, $f(0) = f(7)$. Also, since $\mathbb{Z}_7$ has only seven elements, any function $f:\mathbb{Z} \to \mathbb{Z}_7$ will not be one-to-one. So the groups $(\mathbb{Z}, +)$ and $(\mathbb{Z}_7, \oplus)$ are not isomorphic.

(c) We first show that f is one-to-one. Suppose that $f([a]) = f([b])$. Then we have

$$\begin{aligned}[5a] = [5b] &\to 6 \mid (5a - 5b)\\ &\to 6 \mid 5(a - b)\\ &\to 6 \mid (a - b)\end{aligned}$$

so $[a] = [b]$. Thus f is one-to-one, and since $\mathbb{Z}_6$ is finite, f is also onto by Theorem 5.1. In addition, $f([a] \oplus [b]) = f([a + b]) = [5(a + b)] = [5a + 5b] = [5a] \oplus [5b] = f([a]) \oplus f([b])$. Hence f is an isomorphism.

(d) Here $\operatorname{im} f = \mathbb{Z}$, so f is clearly not onto. (In the Chapter Problems we will see that, in fact, the groups $(\mathbb{Z}, +)$ and $(\mathbb{Q}, +)$ are not isomorphic.) ■

We have seen that the groups $(\Gamma, \circ)$ and $(\mathbb{Z}_4, \oplus)$ are isomorphic. It is sometimes said that these groups "are the same, up to isomorphism." A central problem in group theory is to determine, for a given positive integer n, all the nonisomorphic groups of order n. For example, the reader will be asked to show in Exercise 13 that there are exactly two nonisomorphic groups of order 4: the group $(\mathbb{Z}_4, \oplus)$ and the Klein four-group of Figure 6.2(a).

We can compare two algebraic structures of other types, semigroups or rings, for example, and in each case we can define the concept of isomorphism in an appropriate manner. In each setting, the notion of isomorphism is used to determine when two structures are essentially the same.

Consider again the function f of Example 6.24(b), where $f:\mathbb{Z} \to \mathbb{Z}_7$ is defined by $f(n) = [n]$. As was mentioned, f is not one-to-one; however

$$\begin{aligned}f(m + n) &= [m + n]\\ &= [m] \oplus [n]\\ &= f(m) \oplus f(n)\end{aligned}$$

So f does satisfy property 2 of the definition of an isomorphism for groups. Thus, even though $(\mathbb{Z}, +)$ is not isomorphic to $(\mathbb{Z}_7, \oplus)$, the two groups are related in this sense. We call such a function a "homomorphism."

DEFINITION 6.17

Given groups $(G, *)$ and $(H, \#)$, a function $h:G \to H$ is called a *homomorphism* if $g(a * b) = g(a) \# g(b)$ for all $a, b \in G$.

Example 6.25 In this example we wish to exhibit a homomorphism from the group $(\mathbb{Z}_{12}, \oplus)$ to the group $(\mathbb{Z}_4, \oplus)$. To avoid confusion, let us use the notation $[x]_m$ to denote the element $[x]$ in the group $\mathbb{Z}_m$. Now, define $h:(\mathbb{Z}_{12}, \oplus) \to (\mathbb{Z}_4, \oplus)$ by $h([x]_{12}) = [x]_4$. Then

$$\begin{aligned}
h([0]_{12}) &= h([4]_{12}) = h([8]_{12}) = [0]_4 \\
h([1]_{12}) &= h([5]_{12}) = h([9]_{12}) = [1]_4 \\
h([2]_{12}) &= h([6]_{12}) = h([10]_{12}) = [2]_4 \\
h([3]_{12}) &= h([7]_{12}) = h([11]_{12}) = [3]_4
\end{aligned}$$

The fact that h is a homomorphism can be seen from the following:

$$\begin{aligned}
h([a]_{12} \oplus [b]_{12}) &= h([a + b]_{12}) \\
&= [a + b]_4 \\
&= [a]_4 \oplus [b]_4 \\
&= h([a]_{12}) \oplus h([b]_{12})
\end{aligned}$$

■

We shall not attempt to develop further the theory of isomorphisms or homomorphisms, leaving this for a course in modern algebra. It will suffice for our purposes to address some of the elementary properties in the exercises and problems.

Before leaving this topic, however, perhaps the reader is curious as to the definition of isomorphism for Boolean algebras. This turns out to be a case where the notion of isomorphism greatly clarifies matters.

DEFINITION 6.18

Let $(B_1, \vee, \wedge)$ and $(B_2, \vee', \wedge')$ be Boolean algebras. A function $\theta:B_1 \to B_2$ is called an *isomorphism* provided the following conditions hold:

1. θ is one-to-one and onto.
2. $\theta(a \vee b) = \theta(a) \vee' \theta(b)$
3. $\theta(a \wedge b) = \theta(a) \wedge' \theta(b)$

for all $a, b \in B_1$. We write $(B_1, \vee, \wedge) \cong (B_2, \vee', \wedge')$.

Example 6.26 Let $(L, \vee, \wedge)$ be the Boolean algebra of positive divisors of 30, and let $(\mathcal{P}(A), \cup, \cap)$ be the Boolean algebra of all subsets of $A = \{a, b, c\}$. Prove that $(L, \vee, \wedge) \cong (\mathcal{P}(A), \cup, \cap)$.

Solution We must define a mapping $\theta: L \to \mathcal{P}(A)$ that satisfies the conditions of Definition 6.18. Begin by defining $\theta(2) = \{a\}$, $\theta(3) = \{b\}$, and $\theta(5) = \{c\}$. Conditions 2 and 3 of the definition imply that

$$\text{(a)} \quad \theta(x \vee y) = \theta(x) \cup \theta(y)$$
$$\text{(b)} \quad \theta(x \wedge y) = \theta(x) \cap \theta(y)$$

for all $x, y \in L$. These conditions require that

$$\begin{aligned}
\theta(6) &= \theta(2 \vee 3) = \theta(2) \cup \theta(3) = \{a\} \cup \{b\} = \{a, b\} \\
\theta(10) &= \theta(2 \vee 5) = \{a, c\} \\
\theta(15) &= \theta(3 \vee 5) = \{b, c\} \\
\theta(30) &= \theta(2 \vee 15) = \{a\} \cup \{b, c\} = \{a, b, c\} \\
\theta(1) &= \theta(2 \wedge 3) = \theta(2) \cap \theta(3) = \{a\} \cap \{b\} = \phi
\end{aligned}$$

The function θ so defined is clearly one-to-one and onto. We must still prove that the function θ meets conditions (a) and (b) above. To show this, it suffices to consider expressions of the following types:

$$\begin{array}{ll}
1 \vee x, & 1 \wedge x \\
x \vee x, & x \wedge x \\
x \vee 30, & x \wedge 30 \\
p \vee q, & p \wedge q \\
p \vee qr, & p \wedge qr \\
pq \vee qr, & pq \wedge qr
\end{array}$$

where p, q, and r are distinct prime factors of 30, and x is any element of L. We have already established that $\theta(p \vee q) = \theta(p) \cup \theta(q)$. Consider $x \wedge 30$ and $p \vee qr$. For $x \wedge 30$ we see that

$$\theta(x \wedge 30) = \theta(x) = \theta(x) \cap A = \theta(x) \cap \theta(30)$$

and for $p \vee qr$ we have that

$$\begin{aligned}
\theta(p \vee qr) = \theta(pqr) = \theta(30) = A &= \{a\} \cup \{b\} \cup \{c\} \\
&= \theta(p) \cup (\theta(q) \cup \theta(r)) \\
&= \theta(p) \cup \theta(q \vee r) \\
&= \theta(p) \cup \theta(qr)
\end{aligned}$$

The remaining cases follow in a similar fashion and are left for the reader. Thus $(L, \vee, \wedge) \cong (\mathcal{P}(A), \cup, \cap)$. ■

A Boolean algebra $(B, \vee, \wedge)$ is referred to as a finite Boolean algebra if the set B is finite. Example 6.26 shows that the finite Boolean algebras

$(L, \vee, \wedge)$ and $(\mathcal{P}(A), \cup, \cap)$ are isomorphic. Is every finite Boolean algebra isomorphic to the Boolean algebra of subsets of some finite set A? The answer is yes, and this fact is known as the *fundamental isomorphism theorem for Boolean algebras*.

THEOREM 6.15 If $(B, \vee, \wedge)$ is a finite Boolean algebra, then $(B, \vee, \wedge)$ is isomorphic to the Boolean algebra $(\mathcal{P}(A), \cup, \cap)$ of subsets of some finite set A. ■

In other words, Theorem 6.15 states that if $(B, \vee, \wedge)$ is a finite Boolean algebra, then, up to isomorphism, $(B, \vee, \wedge)$ is nothing more than the Boolean algebra of all subsets of a set with m elements for some nonnegative integer m. Recall that if A is a set with m elements, then the power set $\mathcal{P}(A)$ has 2^m elements. So we have the following consequence of Theorem 6.15.

COROLLARY 6.15 If $(B, \vee, \wedge)$ is a finite Boolean algebra, then $n(B) = 2^m$ for some nonnegative integer m. ■

Exercises 6.6

1. Let $(G, *)$ and $(H, \#)$ be groups with identities e_G and e_H, respectively, and let $f: G \to H$ be a homomorphism. If $n \in \mathbb{N}$ and $a \in G$, prove each of the following properties.
   a. $f(e_G) = e_H$
   b. $f(a^{-1}) = [f(a)]^{-1}$
   c. $f(a^n) = [f(a)]^n$
   d. $f(a^{-n}) = [f(a)]^{-n}$
2. Show that any function $f: (\mathbb{Z}_6, \oplus) \to (\mathbb{Z}_4, \oplus)$ with $f([1]) = [1]$ cannot be a homomorphism.
3. Let (G, $*$) be a group.
   a. Let $f: (\mathbb{Z}_m, \oplus) \to (G, *)$ be a homomorphism. Show that f is completely determined by $f([1])$.
   b. Let $g: (\mathbb{Z}, +) \to (G, *)$ be a homomorphism. Show that g is completely determined by $g(1)$.
4. Verify that $f: (\mathbb{Z}_4, \oplus) \to (\mathbb{Z}_5^\#, \odot)$ determined by $f([1]) = [2]$ is an isomorphism.
5. Find the homomorphism h for each of the following parts.
   a. from $(\mathbb{Z}_{12}, \oplus)$ to $(\mathbb{Z}_8, \oplus)$ with $h([1]) = [2]$
   b. from $(\mathbb{Z}_{12}, \oplus)$ to $(\mathbb{Z}_8, \oplus)$ with $h([1]) = [4]$
   c. from $(\mathbb{Z}, +)$ to $(\mathbb{Z}, +)$ with $h(1) = 2$
   d. from $(\mathbb{Z}, +)$ to $(\mathbb{Z}_8, \oplus)$ with $h(1) = [2]$
6. Let H be the set of rational numbers of the form 2^m, $m \in \mathbb{Z}$; that is,

$$H = \{. . . , \tfrac{1}{4}, \tfrac{1}{2}, 1, 2, 4, . . .\}$$

a. Show that $(H, \cdot)$ is a subgroup of $(\mathbb{Q}^+, \cdot)$.
b. Show that $(H, \cdot)$ is isomorphic to $(\mathbb{Z}, +)$.

7. Let δ and γ denote the following elements of the group S_3:

$$\delta: \begin{array}{l} 1 \to 2 \\ 2 \to 3 \\ 3 \to 1 \end{array} \qquad \gamma: \begin{array}{l} 1 \to 2 \\ 2 \to 1 \\ 3 \to 3 \end{array}$$

a. Find the homomorphism $f: \mathbb{Z}_6 \to S_3$ such that $f([1]) = \delta$.
b. Is there a homomorphism $g: S_3 \to \mathbb{Z}_6$ with $g(\delta) = [2]$ and $g(\gamma) = [0]$?
c. Is there a homomorphism $h: S_3 \to \mathbb{Z}_6$ with $h(\delta) = [0]$ and $h(\gamma) = [3]$?

8. Let M_k denote the set of integer multiples of $k \in \mathbb{N}$. Show that the groups $(M_k, +)$ and $(\mathbb{Z}, +)$ are isomorphic.

9. Show that the groups $(\mathbb{Z}_6, \oplus)$ and $(\mathbb{Z}_7^\#, \odot)$ are isomorphic.

10. Consider the group D_4 that was introduced in Exercise 23 of Section 6.2.
a. Show that $H = \{e, r_2, h, v\}$ is a subgroup of D_4.
b. Show that H is isomorphic to the Klein four-group (see Exercise 3 of Section 6.2).

11. Let $(G, *)$ and $(H, \#)$ be groups, and let $f: G \to H$ be a homomorphism. Show that if $x \in G$ and the order of x in G is finite, then the order of $f(x)$ in H is finite and divides the order of x in G.

12. Let $(G, *)$ and $(H, \#)$ be groups, and let $f: G \to H$ be an isomorphism. Prove: If $x \in G$ and the order of x in G is finite, then the order of $f(x)$ in H equals the order of x in G. What does this say about the number of elements of a given order in both G and H?

13. a. Let $(G, *)$ be a group of order 3, where $G = \{e, a, b\}$. Show that $(G, *)$ is isomorphic to $(\mathbb{Z}_3, \oplus)$.
b. Show that $(\mathbb{Z}_4, \oplus)$ and the Klein four-group are nonisomorphic.
c. Let $(G, *)$ be a group of order 4, where $G = \{e, a, b, c\}$. Show that $(G, *)$ is isomorphic either to $(\mathbb{Z}_4, \oplus)$ or to the Klein four-group.
d. Show that $(\mathbb{Z}_6, \oplus)$ and S_3 are nonisomorphic. (In fact, these are the only groups of order 6, up to isomorphism.)
e. Show that Q_8 and D_4 are nonisomorphic.

14. Let $(G, *)$ and $(H, \#)$ be groups, and let $f: G \to H$ be a homomorphism.
a. Show that $(\text{im } f, \#)$ is a subgroup of $(H, \#)$.
b. Show that $(\text{im } f, \#)$ is abelian if $(G, *)$ is abelian.
c. If $(G, *)$ is abelian, is $(H, \#)$ necessarily abelian?

15. Let $g: (\mathbb{Z}_m, \oplus) \to (\mathbb{Z}_n, \oplus)$ be a function such that $g([1]) = [a]$, where $0 \leq a \leq n - 1$. If g is a homomorphism, what can be said about a? (Hint: use the result of Exercise 11.)

16. Use the result of the previous exercise to determine all homomorphisms for each of the following pairs of groups.
a. from $(\mathbb{Z}_6, \oplus)$ to $(\mathbb{Z}_4, \oplus)$
b. from $(\mathbb{Z}_6, \oplus)$ to $(\mathbb{Z}_9, \oplus)$
c. from $(\mathbb{Z}_6, \oplus)$ to $(\mathbb{Z}_7, \oplus)$
d. from $(\mathbb{Z}_c, \oplus)$ to $(\mathbb{Z}_d, \oplus)$, where $\gcd(c, d) = 1$

17. Find all isomorphisms for each of the following pairs of groups.
a. from $(\mathbb{Z}_6, \oplus)$ to $(\mathbb{Z}_7^\#, \odot)$
b. from $(\mathbb{Z}_8, \oplus)$ to $(\mathbb{Z}_8, \oplus)$
c. from $(\mathbb{Z}, +)$ to $(E, +)$, where E is the set of even integers
d. from $(\mathbb{Z}, +)$ to $(\mathbb{Z}, +)$

18. Complete the verification in Example 6.26 that $(L, \vee, \wedge)$ and $(\mathcal{P}(A), \cup, \cap)$ are isomorphic Boolean algebras.

19. Let $(G, *)$ be a cyclic group with generator a. Prove the following.
a. If $(G, *)$ has order n, then $(G, *)$ is isomorphic to $(\mathbb{Z}_n, \oplus)$.
b. If $(G, *)$ is infinite, then $(G, *)$ is isomorphic to $(\mathbb{Z}, +)$.

20. The purpose of this exercise is to prove Theorem 6.15. Let $(B, \lesssim)$ be a finite Boolean algebra. An element $a \in B$ is called an *atom* if a is an immediate successor of the zero element, that is, $0 < a$ and there is no element b such that $0 < b < a$.
a. Let $C = \{c_1, c_2, \ldots, c_r\}$ and $D = \{d_1, d_2, \ldots, d_t\}$ be sets of atoms of B. Show that $c_1 \vee c_2 \vee \cdots \vee c_r = d_1 \vee d_2 \vee \cdots \vee d_t$ if and only if $C = D$. (Hint: If $C \neq D$, then assume without loss of generality that $c_1 \notin D$. Now consider $c_1 \wedge (c_1 \vee c_2 \vee \cdots \vee c_r)$ and $c_1 \wedge (d_1 \vee d_2 \vee \cdots \vee d_t)$.)

Let A be the set of atoms of B. For $b \in B$, let A_b be the subset of A defined by $A_b = \{a \in A \mid a \lesssim b\}$.
b. Show that A_b is nonempty if and only if $b \neq 0$.
c. For $b, c \in B$, show that $b \lesssim c$ if and only if $A_b \subseteq A_c$.
d. For $b \in B$, $b \neq 0$, show that b is equal to the join of the atoms in A_b. (Hint: Let c be the join of the atoms in A_b. To show $c \lesssim b$, show $b \wedge c = c$. To show $b \lesssim c$, use part c.)

It follows from parts a and d that the function $\theta: B \to \mathcal{P}(A)$ defined by $\theta(b) = A_b$ is one-to-one and onto. This shows that if $n(A) = m \geq 1$, then $n(B) = 2^m$.
e. Show that, in fact, θ is an isomorphism.

21. Let $(G, *)$ and $(H, \#)$ be groups, with e_H the identity of H. Define $k: G \to H$ by $k(x) = e_H$ for all $x \in G$. Show that k is a homomorphism.

CHAPTER PROBLEMS

1. For each of the following functions $*$, (i) verify that $*$ is a binary operation on $\mathbb{Q}$, (ii) determine whether $*$ is associative, (iii) determine whether $*$ is commutative, (iv) determine whether $*$ has an identity, and (v) if $*$ has an identity, determine which rational numbers have inverses.

a. $x * y = (x + y)/2$ **b.** $x * y = x + y + 2$
c. $x * y = 3xy$ **d.** $x * y = xy/2$
e. $x * y = x(y + 1)$ **f.** $x * y = x + y + xy$

2. The operations we have studied in this chapter have been *binary operations,* so called because such an operation requires *two operands*. In general, operations on a set may have 1, 2, or more operands. The case of operations requiring a *single operand* is particularly important; such operations are called *unary operations*. For example, consider the operation of "minus" on the set of reals; given the number r as the operand, this operation returns as a result the number $-r$.

a. Give an example of a unary operation on the set $\mathbb{R}^{\#}$ of nonzero reals.
b. Give an example of a unary operation on the power set $\mathcal{P}(A)$ of a given set A.
c. Give as many examples as you can of unary operations in Pascal.
d. What is another name for a unary operation on a set A?

3. In the usual notation for binary operations, the operator symbol is written between the operands, so it is called *infix* notation. An alternative notation is known as *postfix* (or *reverse Polish*) notation; here each operator is written immediately after its rightmost operand. For example, in postfix notation $x + y$ is written $x\ y\ +$. Postfix notation has several advantages over infix notation. For one, it is suitable for both unary and binary operations, whereas true infix notation assumes binary operations. Perhaps more importantly, postfix notation is completely unambiguous. The problem of ambiguity for infix notation is illustrated by the expression $x + y * z$. Does it mean $(x + y) * z$ or $x + (y * z)$? Note that with infix notation this problem must be resolved by using parentheses, or by agreeing on certain precedence rules for the operations. With postfix notation, however, $(x + y) * z$ is written $x\ y + z\ *$, whereas $x + (y * z)$ is written $x\ y\ z\ *\ +$, so there is no problem! Find the value of each of the following postfix expressions. (Here $-$ denotes unary minus and $*$ denotes multiplication.)

a. $3\ 5 + 7\ *$ **b.** $3 - 8\ +$ **c.** $5\ 7 * 4 - 6 * +$

Write each of the following expressions using postfix notation.

d. $-3 * (2 + 6)$ **e.** $-((3 * 8) + (2 * 7))$

4. Another common notation for expressions is known as *prefix* (or *Polish*) notation, in which the operator symbol is written immediately before its leftmost operand; for example, $x + y$ is written $+\ x\ y$. A special version of prefix notation is called *Cambridge Polish* notation, and it is used in the programming language LISP. It encloses each (sub)expression within parentheses, so, for instance, $x + y$ is written $(+\ x\ y)$ and $-x * y$ is written $(*\ (-\ x)\ y)$. (Here again $*$ denotes multiplication and $-$ denotes unary minus.) Write each of the following expressions using Cambridge Polish notation.

a. $(3 + 5) * 7$ **b.** $3 + (5 * 7)$

c. $-(4 * 9)$ **d.** $(5 * 7) + (-4 * 6)$

Find the value of each of the following expressions.

e. $(*\ (-\ 3)\ (+\ 2\ 6))$ **f.** $(-\ (+\ (*\ 3\ 8)\ (*\ 2\ 7)))$

5. Let S be a finite set having n elements. Determine the number of
  a. binary operations on S;
  b. commutative binary operations on S;
  c. binary operations on S having an identity.

6. For each of the following subsets of $\mathcal{F}(\mathbb{R})$, determine whether composition is a binary operation on the subset. If so, discuss the properties of the resulting semigroup.
  a. the set $\mathcal{C}$ of continuous functions
  b. the set $\mathcal{D}$ of differentiable functions
  c. the set $\mathcal{P}$ of polynomial functions
  d. the set $\mathcal{P}_n$ of polynomials of degree less than or equal to n

7. Let $(G, *)$ be a group and let H be a subset of G. Prove that $(H, *)$ is a subgroup of $(G, *)$ if and only if (i) H is nonempty, and (ii) for all $a, b \in H$, the quantity $a * b^{-1} \in H$. (See Exercise 20 of Section 6.2.)

8. Let $(G, *)$ be a group and let $C = \{c \in G \mid a * c = c * a \text{ for all } a \in G\}$. The set C is called the *center* of G.
  a. Use the result of Problem 7 to show that $(C, *)$ is a subgroup of $(G, *)$.
  b. What is C if G is abelian?
  c. Find the center of S_3.
  d. Find the center of D_4.

9. Let $\mathcal{F}(\mathbb{R})^*$ denote the set of all functions $f: \mathbb{R} \to \mathbb{R}$ such that $f(a) \neq 0$ for all $a \in \mathbb{R}$. Show that $(\mathcal{F}(\mathbb{R})^*, \cdot)$ is a group, where $(f \cdot g)(x) = f(x)g(x)$.

10. Let $(G, *)$ be a group and let $(H, *)$ be a subgroup. Define a relation $\sim$ on G by $a \sim b \leftrightarrow a * b^{-1} \in H$.
  a. Show that $\sim$ is an equivalence relation on G.
  b. For $x \in G$, let $[x]$ denote the equivalence class containing x. Show that $y \in [x] \leftrightarrow y = h * x$ for some $h \in H$. It follows that $[x] = Hx = \{h * x \mid h \in H\}$.
  c. If G is finite, use part b to show that for $x \in G$, $n([x]) = |H|$.
  d. It follows that if G is finite, then $|H|$ divides $|G|$. Why? (This fact is known as *Lagrange's theorem*.)
  e. As a corollary to Lagrange's theorem, we obtain that if G is finite and $a \in G$, then the order of a divides $|G|$. Why?

11. Consider the following two elements of S_4:

$$\alpha\colon \begin{array}{l} 1 \to 2 \\ 2 \to 3 \\ 3 \to 1 \\ 4 \to 4 \end{array} \qquad \beta\colon \begin{array}{l} 1 \to 1 \\ 2 \to 3 \\ 3 \to 4 \\ 4 \to 2 \end{array}$$

a. Find the subgroup of S_4 of minimum order that contains both α and β. This subgroup is denoted A_4.

b. Give an example of a subgroup of A_4 isomorphic to each of the following groups.
(i) $(\mathbb{Z}_2, \oplus)$ (ii) $(\mathbb{Z}_3, \oplus)$ (iii) the Klein four-group

c. Show that A_4 contains no subgroup of order 6. (This provides a counterexample to the statement: If n divides the order of a finite group $(G, *)$, then $(G, *)$ has a subgroup of order n.)

12. Let $(G, *)$ be a semigroup. Suppose that for any two fixed elements $a, b \in G$, both of the equations $a * x = b$ and $y * a = b$ have solutions x and y in G. Prove that $(G, *)$ is a group.

13. Let $(G, *)$ be a group and let $(H, *)$ and $(K, *)$ be subgroups.

a. Show that $(H \cap K, *)$ is a subgroup of $(G, *)$.

b. Give an example to show that $H \cup K$ is not necessarily a subgroup of G.

14. Let $\mathcal{S}_G$ denote the collection of all subgroups of a group $(G, *)$.

a. Show that $\mathcal{S}_G$ is partially ordered by inclusion.
In fact, it can be shown that the poset $(\mathcal{S}_G, \subseteq)$ is a lattice, called the lattice of subgroups of the group $(G, *)$. Draw the Hasse diagram for the lattice of subgroups of each of the following groups.

b. S_3 **c.** $(\mathbb{Z}_{12}, \oplus)$ **d.** D_4

15. In each of the following parts, determine whether H is a subgroup of G.

a. $G = (\mathbb{R}^{\#}, \cdot)$, $H = \mathbb{Q}^{\#}$
b. $G = (\mathbb{Q}^{+}, \cdot)$, $H = \mathbb{N}$
c. $G = (\mathbb{Z}, +)$, $H = E$
d. $G = (\mathbb{Z}, +)$, $H = \{-1, 0, 1\}$
e. $G = (\mathbb{Q}^{\#}, \cdot)$, $H = \{-1, 1\}$

16. Let a and b be elements of an abelian group $(G, *)$, of orders m and n, respectively. Prove: If $\gcd(m, n) = 1$, then $a * b$ has order mn.

17. Give an example of elements $f, g \in \mathcal{S}(\mathbb{N})$ (the symmetric group on the set of positive integers) such that each of f and g has finite order in $(\mathcal{S}(\mathbb{N}), \circ)$, but $f \circ g$ has infinite order.

18. Let $(G, *)$ be an abelian group and let $H = \{x \in G \mid x^n = e\}$, where n is a fixed positive integer and e is the identity of G.

a. Show that $(H, *)$ is a subgroup of $(G, *)$.

b. Give an example to show that the result of part a may not hold if G is nonabelian.

19. Let A be a nonempty subset of $\{1, 2, \ldots, n\}$. Let H_A be that subset of S_n consisting of those permutations f such that $f(a) = a$ for all $a \in A$. Show that H_A is a subgroup of S_n.

20. If $(G, *)$ is a group and $(H, *)$ is a subgroup of G such that $\{e\} \subset H \subset G$, then $(H, *)$ is called a *proper subgroup* of G.

a. Prove that if $(G, *)$ has no proper subgroups, then $(G, *)$ is cyclic.

b. Show that $(\mathbb{Z}_m, \oplus)$ has no proper subgroups if and only if m is prime.

21. An element s in a ring $(S, +, \cdot)$ is called *idempotent* if $s \cdot s = s$. Find the idempotent elements in each of the following rings.

a. $(\mathbb{Z}, +, \cdot)$ **b.** $(\mathbb{Z}[\sqrt{2}], +, \cdot)$

c. $(\mathscr{F}(\mathbb{R}), +, \cdot)$ **d.** $(\mathbb{Z}_6, \oplus, \odot)$

22. Let $(S, +, \cdot)$ be a ring and let T be a nonempty subset of S. We say that T is a *subring* of S if $(T, +, \cdot)$ is a ring. Show that T is a subring of S if and only if both of the following hold:

(i) $(T, +)$ is a subgroup of $(S, +)$;
(ii) If $t_1, t_2 \in T$, then $t_1 \cdot t_2 \in T$.

23. Let $(S, +, \cdot)$ be a commutative ring with identity. An element $a \in S$ is called *invertible* if a has a multiplicative inverse in S.

a. Show that the invertible elements of S form a group under the ring multiplication.

The set of invertible elements in the ring $(\mathbb{Z}_n, \oplus, \odot)$ is denoted U_n.

b. Find U_8. To what other group is $(U_8, \odot)$ isomorphic?

c. Find U_{10}. To what other group is $(U_{10}, \odot)$ isomorphic?

24. Consider the ring $(\mathscr{F}(\mathbb{R}), +, \cdot)$. Determine whether each of the sets $\mathscr{C}$, $\mathscr{D}$, $\mathscr{P}$, and $\mathscr{P}_n$, which were defined in Problem 6, is a subring of the ring $(\mathscr{F}(\mathbb{R}), +, \cdot)$.

25. Let $(F, +, \cdot)$ be a field. For $a, b \in F$, where $b \neq z$, define the operation "/" (called "division") on F by $a/b = a \cdot b^{-1}$. Show the following.

a. $(a/b) \cdot (c/d) = (a \cdot c)/(b \cdot d)$

b. $(a/b) + (c/d) = [(a \cdot d) + (b \cdot c)]/(b \cdot d)$

26. Let $(S, +, \cdot)$ be a ring with identity; S is said to be *Boolean* if every element of S is idempotent (see Problem 21).

a. Prove that if S is Boolean, then S is commutative and $2s = z$ for all $s \in S$.

b. Show that the ring $(\mathscr{P}(A), *, \cap)$, where $*$ denotes symmetric difference, is Boolean.

27. Let $(S, +, \cdot)$ be an integral domain. If there exists an $n \in \mathbb{N}$ such that $ne = z$, then the smallest such n is called the *characteristic* of S. If no such n exists, then S is said to have *characteristic zero*.

a. Show that if S has characteristic n, then $ns = z$ for all $s \in S$.

b. Show that if S has characteristic $n > 0$, then n is prime.

c. For each prime p, give an example of an integral domain of characteristic p. Give an example of an integral domain of characteristic zero.

28. Let $(G, *)$ be an abelian group and define $f: G \to G$ by $f(x) = x^2$. Show that f is a homomorphism.

29. Let $(B, \vee, \wedge)$ be a Boolean algebra with $a, b, c \in B$.
  a. Prove: If $a \vee c = b \vee c$ and $a \vee c' = b \vee c'$, then $a = b$.
  b. Give an example to show that the condition $a \vee c = b \vee c$ alone is not sufficient for $a = b$.

30. If $\theta: (A, \vee, \wedge) \to (B, \vee', \wedge')$ is a Boolean algebra isomorphism, prove the following:
  a. θ maps the unity (zero) of A to the unity (zero) of B.
  b. $\theta(a') = \theta(a)'$ for each $a \in A$.

31. Draw the circuit represented in each of the following parts.
  a. $(a \wedge b \wedge c) \vee d$
  b. $(a \vee b) \wedge (c \vee d)$
  c. $a \vee (b \wedge c) \vee d$
  d. $((a \vee b) \wedge c) \vee d$
  e. $(a \wedge (b \vee (c \wedge d))) \vee e$

32. Draw and simplify the circuit represented by
  a. $(a \wedge b) \vee (a' \wedge b \wedge c')$
  b. $(a' \vee (b \wedge c')) \wedge (a \vee c)$

33. Let $(G, *)$ be a group. An isomorphism $f: G \to G$ is called an *automorphism* of G. Let a be a fixed element of G.
  a. Show that $f_a: G \to G$ defined by $f_a(x) = a * x$ is a permutation of G. Is f_a an automorphism?
  b. Define $g_a: G \to G$ by $g_a(x) = a * x * a^{-1}$. Show that g_a is an automorphism of G. What is g_a if G is abelian?
  c. Define $h: G \to G$ by $h(x) = x^{-1}$. Show that h is an automorphism of G if and only if G is abelian.

34. Prove that the group $(\mathbb{Q}, +)$ is not cyclic.

35. Let a, b, c, and d be real numbers with $ad - bc \neq 0$. If $c = 0$, define $f: \mathbb{R} \to \mathbb{R}$ by $f(x) = (ax + b)/d$. If $c \neq 0$, define $f: \mathbb{R} \to \mathbb{R}$ as follows:

$$f(x) = \begin{cases} \dfrac{ax + b}{cx + d} & \text{if } x \neq -d/c \\ a/c & \text{if } x = -d/c \end{cases}$$

  a. Given such an f, show that f is one-to-one and onto, and find f^{-1}.
  b. Show that the set of all such functions forms a group under composition (and hence is a subgroup of $(\mathcal{S}(\mathbb{R}), \circ)$).

36. Define the binary operation $*$ on $\mathbb{R}$ by $a * b = \max\{a, b\}$. Show that $(\mathbb{R}, *)$ is a semigroup. Is $*$ commutative? Is there an identity?

37. Let $(L, \preceq)$ be a totally ordered set. Show that $(L, \preceq)$ is a distributive lattice.

38. Show that any finite lattice contains a zero (unity). (Hint: see Chapter Problem 23 of Chapter 4.)

Seven Graph Theory

7.1 INTRODUCTION

It is the object of this chapter to introduce the student to some of the basic ideas and results of graph theory, including several applications of the subject.

Even though graph theory is a very old subject, dating as far back as 1736, it is currently one of the more popular and fertile branches of mathematics. Perhaps a reason for its conspicuously slow start is that for over one hundred years mathematicians and other scholars found no relevant mathematical problems to which the subject could be applied. Indeed, it seems to be the case that it was used mostly to aid in the solution of several well-known puzzles and games.

It was around the middle of the nineteenth century that a very famous problem, called the *four-color problem,* surfaced in a letter from Augustus DeMorgan to Sir William Rowan Hamilton. The letter is dated October 23, 1852, and part of its content appears in the book *Graph Theory 1736–1936,* by Norman L. Biggs, E. Keith Lloyd, and Robin J. Wilson. It is sufficiently interesting to bear repeating here.

> A student of mine asked me today to give him a reason for a fact which I did not know was a fact—and do not yet. He says that if a figure be anyhow divided and the compartments differently coloured so that figures with any portion of common boundary line are differently coloured—four colours may be wanted, but not more—the following is the case in which four are wanted [see Figure 7.1(a)]. Query cannot a necessity for five or more be invented. As far as I see at this moment if four ultimate compartments have each boundary line in common with one of the others, three of them inclose the fourth and prevent any fifth from connexion with it. If this be true, four colours will colour any possible map without any necessity for colour meeting colour except at a point.
>
> Now it does seem that drawing three compartments with common boundary ABC two and two you cannot make a fourth take boundary from all, except by enclosing one [see Figures 7.1(b) and (c)]. But it is tricky work, and I am not

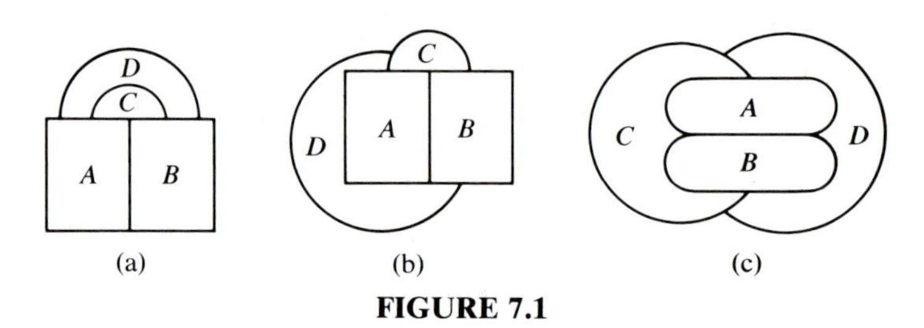

FIGURE 7.1

> sure of the convolutions—what do you say? And has it, if true, been noticed? My pupil says he guessed it in colouring a map of England. The more I think of it, the more evident it seems. If you retort with some very simple case which makes me out a stupid animal, I think I must do as the Sphynx did. . . .

In ordinary language, the four-color problem can be stated as follows:

> A map consisting of n countries with well-defined boundaries is given. Two countries are said to be adjacent if their common boundary contains a nontrivial segment (neither the empty set nor isolated points). Can this map be colored with four (or fewer) different colors so that no two adjacent countries are of the same color?

For over one hundred years, many fruitless attempts were made to solve the problem, and several fallacious solutions were submitted to professional journals. In some cases such alleged solutions were thought to be valid and were published. To make a long story short, in 1977 the four-color problem finally became the *four-color theorem* (in other words, the answer to the above question is yes) with the publication of the celebrated paper, "Every Planar Map is Four-Colorable," by Kenneth Appel, Wolfgang Haken, and John Koch. The article appeared in the Illinois Journal of Mathematics. Appel and Haken also wrote the article, "The Solution of the Four-Color Map Problem" for *Scientific American* in 1977. Besides the fact that the problem was famous, the solution of the four-color problem was an important instance in which the computer was used to aid in a mathematical proof. In fact, the solution required about 1200 hours of computing time!

Exercises 7.1

1. Draw a map having six countries, such that each country shares a boundary with exactly four others.
2. Consider maps such as the one shown in Figure 7.2(a), which are formed by some finite number of straight-line segments that join points on different sides (or opposite corners) of a given rectangle.
   a. Color the map of Figure 7.2(a) using only two colors.

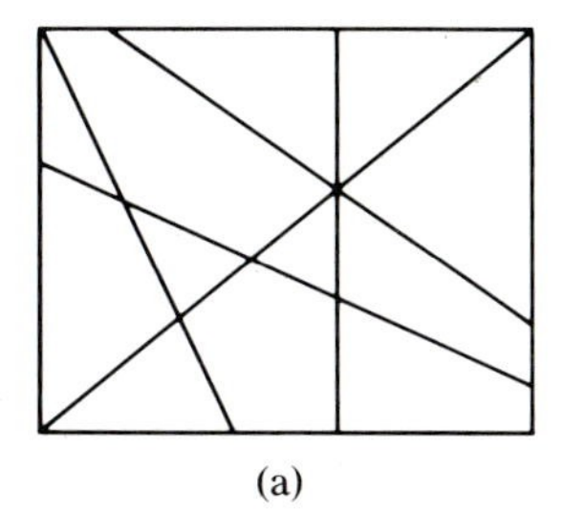
(a)

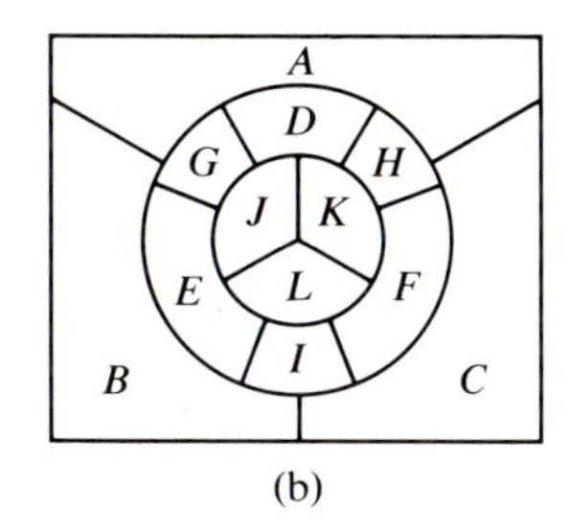

(b)

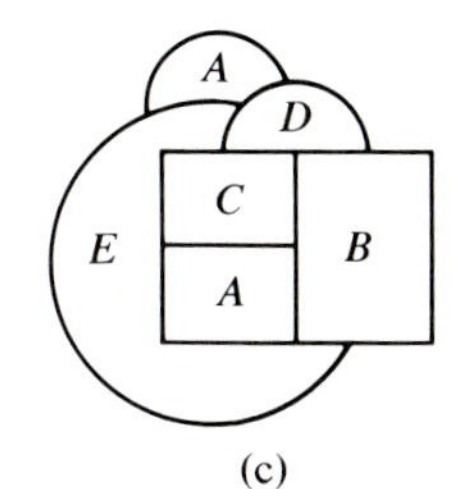

(c)

FIGURE 7.2

b. Use induction (on the number of straight-line segments) to show that any such map may be colored using two colors.

3. Consider the map of Figure 7.2(b); note that each country is adjacent to exactly five others.

a. Color it using four colors.

b. Show that it is not possible to color the map with only three colors.

4. Consider maps such as the one in Figure 7.2(c), which are allowed to contain countries (such as A) whose territory consists of two disconnected regions.

a. Show that such maps must be excluded in order that the four-color theorem hold.

b. Give an example of a map with six countries, two of which are disconnected, requiring six colors.

7.2 PRELIMINARY NOTIONS

The first thing that must be agreed upon is the definition of a graph. Unfortunately, at present there is not universal agreement about this. The fact is that there are essentially two different definitions that pervade the literature. We will present the one we find most useful to work with. Later, we will see how the other definition is a generalization.

DEFINITION 7.1

A *graph* G consists of a finite nonempty set V and a set E of two-element subsets of V. The set V is called the *vertex set* of G, while E is called the *edge set* of G. The graph G is denoted by the ordered pair (V, E).

Given a graph $G = (V, E)$, each $v \in V$ is called a *vertex* of G, while each $e \in E$ is called an *edge* of G. If $e = \{u, v\}$ is an edge, we shall agree to denote it by uv (or vu). In this case u and v are referred to as *adjacent vertices*. In addition, u and v are said to be *incident* with the edge e, and e is *incident* with the vertices u and v (or e *joins* u and v). If e_1 and e_2 are distinct edges of G having a vertex in common, then e_1 and e_2 are called *adjacent edges*. Finally, it will sometimes be convenient to write $V(G)$ and $E(G)$ in place of V and E, respectively; this will especially be the case if a given discussion concerns several graphs.

Example 7.1 Let G be the graph (V, E), where $V = \{u, v, x, y\}$ and $E = \{uv, vx, xy, yu, ux\}$. Then G has four vertices and five edges. The only nonadjacent vertices of G are v and y. The edges uv and vx are adjacent edges, since both are incident with the vertex v. The edges uv and xy are nonadjacent. ■

A graph $G = (V, E)$ can be represented geometrically in a plane if each vertex is represented by a point and each edge by a line segment or simple curve joining its incident vertices.

Example 7.2 Three different ways of representing the graph G of Example 7.1 are shown in Figure 7.3.

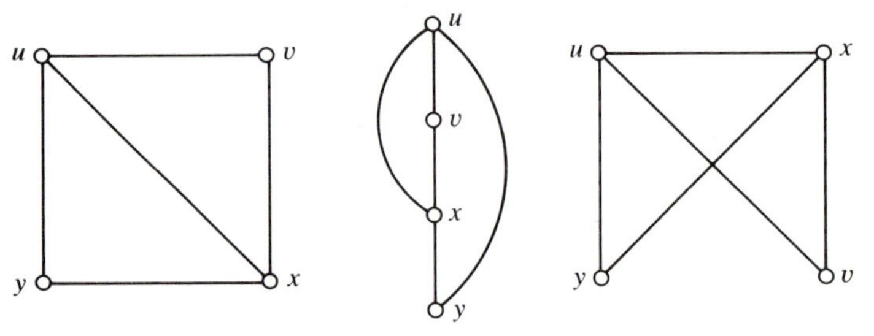

FIGURE 7.3 Three representations of the graph of Example 7.1 ■

In view of Example 7.2, it is clear that there is not a unique way to represent a graph geometrically. The relative positions of the points and curves have no special significance.

In certain applications it is necessary to allow the possibility of any finite number of edges joining two distinct vertices x and y; if two or more edges join x and y, then we refer to the edges as *multiple edges*. In addition, we may wish to allow a vertex to be adjacent to itself; the associated "edge" is called a *loop*. If we allow multiple edges and loops, then the resulting structure is called a *multigraph*. A multigraph with multiple edges joining vertices x and y and loops at the vertices x and u is displayed in Figure 7.4.

Two graphs G_1 and G_2 are called *equal*, denoted $G_1 = G_2$, provided $V(G_1) = V(G_2)$ and $E(G_1) = E(G_2)$. Keeping this in mind, consider the

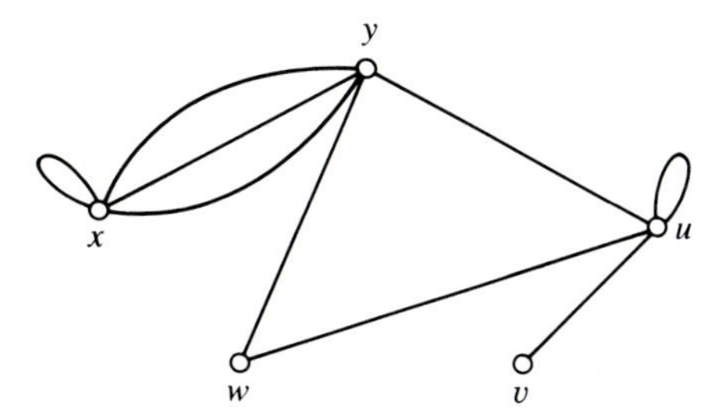

FIGURE 7.4 A multigraph

graphs G_1 and G_2 represented in Figure 7.5. It is clear that $G_1 \neq G_2$, since the two vertex sets are not the same. However, if one represents G_2 as in Figure 7.6, then, except for the labelling of the vertices, the two graphs appear alike.

In fact, we can establish a one-to-one function $\phi: V(G_1) \to V(G_2)$ for which the condition

$$ab \in E(G_1) \leftrightarrow \phi(a)\phi(b) \in E(G_2)$$

is satisfied. In other words, the vertices a and b are adjacent in G_1 if and only if the vertices $\phi(a)$ and $\phi(b)$ are adjacent in G_2. One such function is defined by $\phi(x_1) = u_1$, $\phi(x_2) = u_3$, $\phi(x_3) = u_5$, $\phi(x_4) = u_2$, and $\phi(x_5) = u_4$. Such a mapping ϕ establishes a very strong relationship between the graphs G_1 and G_2. Indeed, even though $G_1 \neq G_2$, the existence of ϕ implies that the graphs are structurally the same.

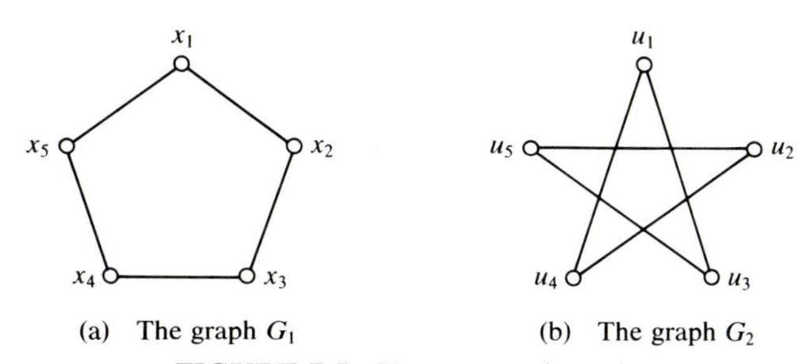

(a) The graph G_1 (b) The graph G_2

FIGURE 7.5 Two unequal graphs

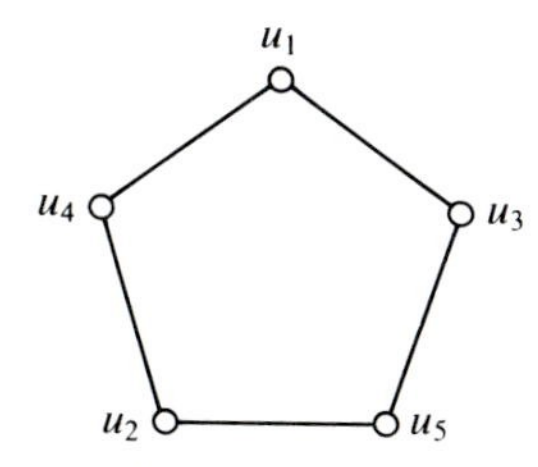

FIGURE 7.6 Another representation of the graph G_2

DEFINITION 7.2

Two graphs G_1 and G_2 are called *isomorphic,* denoted $G_1 \cong G_2$, provided there is a function $\phi: V(G_1) \to V(G_2)$ satisfying the following conditions:

1. ϕ is one-to-one and onto;
2. For all $x, y \in V(G_1)$, $xy \in E(G_1) \leftrightarrow \phi(x)\phi(y) \in E(G_2)$.

The function ϕ is called an *isomorphism.*

In words, condition 2 states that the mapping ϕ *preserves adjacency.*

Example 7.3 Find an isomorphism for the graphs G_1 and G_2 of Figure 7.7.

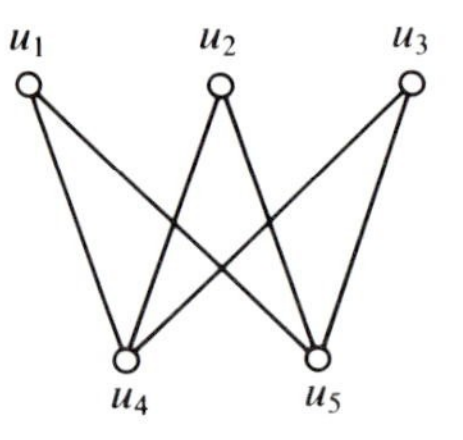

(a) The graph G_1

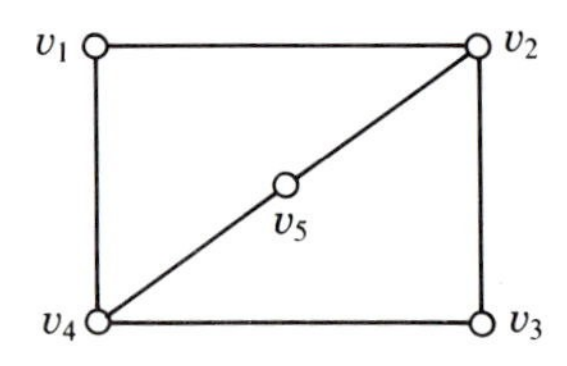

(b) The graph G_2

FIGURE 7.7 Two isomorphic graphs

Solution Let $\phi: V(G_1) \to V(G_2)$ be given by $\phi(u_1) = v_1$, $\phi(u_2) = v_3$, $\phi(u_3) = v_5$, $\phi(u_4) = v_2$, and $\phi(u_5) = v_4$. It can readily be verified that ϕ is one-to-one, onto, and preserves adjacency. The reader may have found a different isomorphism; if $A = \{u_1, u_2, u_3\}$, $B = \{u_4, u_5\}$, $C = \{v_1, v_3, v_5\}$, and $D = \{v_2, v_4\}$, then any one-to-one function f: $V(G_1) \to V(G_2)$ with $f(A) = C$ and $f(B) = D$ is an isomorphism. ■

Recall that the definition of a graph states explicitly that $V(G)$ is a nonempty finite set. If G is a graph with p vertices and q edges, then G is said to have *order* p and *size* q.

Throughout this chapter we shall have occasion to use some very special graphs, which we now define. A graph in which every two distinct vertices are adjacent is called a *complete graph.* Clearly any two complete graphs of the same order are isomorphic, so it makes good sense to refer to *the* complete graph of order p; this graph is denoted K_p.

DEFINITION 7.3

A graph $G = (V, E)$ is called a *bipartite graph* if $V = V_1 \cup V_2$, where V_1 and V_2 are nonempty and

1. $V_1 \cap V_2 = \phi$;
2. Each $e \in E(G)$ joins a vertex of V_1 and a vertex of V_2.

The sets V_1 and V_2 are called the *partite sets* of G. We write $V = (V_1, V_2)$ to denote the associated partition of V.

The graph of Figure 7.8 depicts a bipartite graph with partite sets $V_1 = \{v_1, v_2, v_3\}$ and $V_2 = \{u_1, u_2, u_3, u_4\}$, and edge set $E = \{v_1u_1, v_1u_3, v_2u_2, v_2u_3, v_2u_4, v_3u_4\}$.

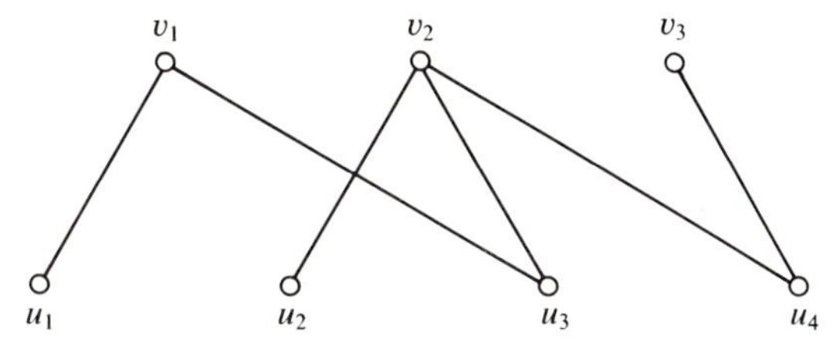

FIGURE 7.8 A bipartite graph

Let G be a bipartite graph with partite sets V_1 and V_2, where $n(V_1) = r$ and $n(V_2) = s$. If each $u \in V_1$ is adjacent to each $v \in V_2$, then G is called a *complete bipartite graph*. Let G' be another complete bipartite graph with $V(G') = (X_1, X_2)$, such that $n(X_1) = r$ and $n(X_2) = s$. Then it is not difficult to see that G and G' are isomorphic, and thus, up to isomorphism, there is just one complete bipartite graph with partite sets of cardinalities r and s. If $r \leq s$, then this graph is denoted $K_{r,s}$. For example, notice that Figure 7.7 shows two different representations of the graph $K_{2,3}$.

Example 7.4 Find all nonisomorphic complete bipartite graphs of order 6.

Solution The three nonisomorphic complete bipartite graphs of order 6 are $K_{1,5}$, $K_{2,4}$, and $K_{3,3}$. ■

Example 7.5 Show that the graph G_1 of Figure 7.5 is not bipartite.

Solution We proceed by contradiction. Suppose G_1 is bipartite with $V(G_1) = (V_1, V_2)$, and assume without loss of generality that $x_1 \in V_1$. Since x_2 is adjacent to x_1, it must be that $x_2 \in V_2$. Similarly we can argue that $x_3 \in V_1$, that $x_4 \in V_2$, and that $x_5 \in V_1$. But now since x_1 is adjacent to x_5, we see that x_1 must belong to V_2. But then $x_1 \in V_1 \cap V_2$, which is a contradiction. Therefore, G_1 is not bipartite. ■

In any given discussion of graphs, we are interested in determining structural properties. Of particular significance in this direction is the number of edges incident with a given vertex u of a given graph G. We call this number the *degree* of u and denote it "deg u." The maximum and minimum vertex degrees in G are denoted by $\Delta(G)$ and $\delta(G)$, respectively. For example, for the graph G of Figure 7.8, $\deg v_1 = 2$, $\deg u_1 = 1 = \delta(G)$, and $\deg v_2 = 3 = \Delta(G)$.

At this point we can state a very basic and useful result concerning the degrees of the vertices in a graph.

THEOREM 7.1 If G is a graph of order p and size q with vertex set $V = \{v_1, v_2, \ldots, v_p\}$, then

$$\sum_{i=1}^{p} \deg v_i = 2q$$

Proof In summing the degrees of the vertices of G, each edge is counted twice, once for each of its incident vertices. Thus the result follows. ■

COROLLARY 7.1 For any graph G, the number of vertices of odd degree is even.

Proof Let $U = \{u_1, u_2, \ldots, u_r\}$ be the set of vertices of odd degree and $W = \{w_1, w_2, \ldots, w_s\}$ be the set of vertices of even degree. Also, let

$$m = \sum_{i=1}^{r} \deg u_i \quad \text{and} \quad n = \sum_{j=1}^{s} \deg w_j$$

If G has size q, then by the theorem $m + n = 2q$. Since $\deg w_j$ is even for $j = 1, 2, \ldots, s$, we see that n is even. Hence it follows that m must be even. Since $\deg u_i$ is odd for $i = 1, 2, \ldots, r$, this implies that r is even. ■

A graph G is called *k-regular* (or *regular of degree k*) provided $\deg x = k$ for all $x \in V(G)$. Thus, for example, the complete graph K_p is $(p - 1)$-regular and the complete bipartite graph $K_{r,r}$ is r-regular. Also notice that G is k-regular if and only if $\delta(G) = \Delta(G) = k$. The graphs G_1, G_2, and G_3 of Figure 7.9 are 1-regular, 2-regular, and 3-regular, respectively.

Much of the value that can be derived from a study of graph theory lies in its application to real-life problems. For instance, graph theory can be nicely used to rephrase or model problems from such areas as transportation, communications, scheduling, and chemistry. In the example that

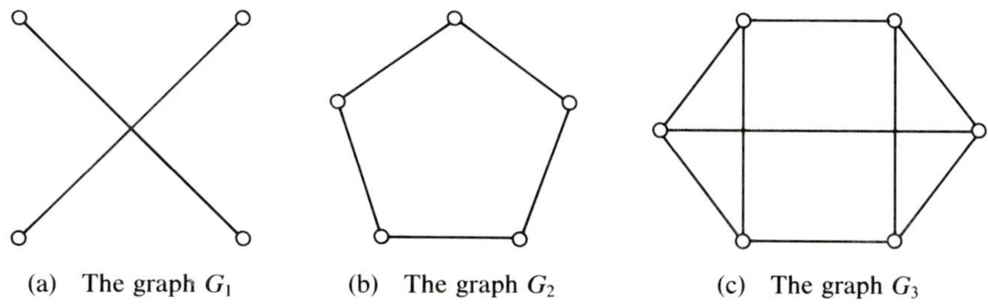

(a) The graph G_1 (b) The graph G_2 (c) The graph G_3

FIGURE 7.9 Regular graphs

follows and in later examples, we shall illustrate how this modeling is done.

Example 7.6 In the country of Freedonia (a tropical island paradise) there are eight cities, A, B, C, D, E, F, G, and H, with the usual property that highways exist between certain pairs of cities. We can depict a road map of Freedonia as a graph as follows:

(i) Represent each city by a vertex.
(ii) If two cities are (directly) linked by a highway, then join the corresponding vertices with an edge.

Assuming that highways directly join cities A and B, A and C, A and D, B and C, B and E, B and G, B and H, C and E, C and F, D and H, E and F, E and G, F and G, F and H, and G and H, this graph is shown in Figure 7.10.

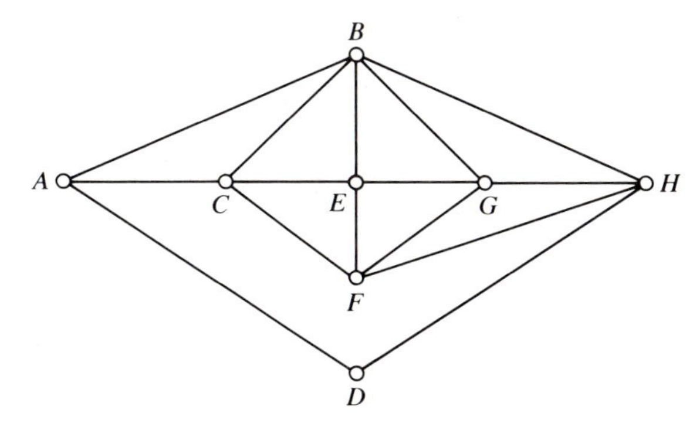

FIGURE 7.10 A graph of Freedonia ■

Exercises 7.2

1. Draw a graph of order 6 that has one vertex of degree 1, two vertices of degree 2, and three vertices of degree 3. How many edges must such a graph have?

2. If a graph has no adjacent edges, what can be said about the degrees of its vertices?

3. If G is a graph and $V(G) = \{v_1, v_2, \ldots, v_p\}$, then the sequence of numbers $\deg v_1, \deg v_2, \ldots, \deg v_p$ is called the *degree sequence* of G. For each of the following sequences, draw a graph with that degree sequence, or show that no such graph exists.

a. 4,4,3,3,2,2 **b.** 3,2,2,2,1,1
c. 4,3,2,1 **d.** 5,4,3,3,2,1

4. **a.** Show that a graph of order 5 cannot have degree sequence 4,4,3,2,1.

b. Give an example of a multigraph of order 5 that has degree sequence 4,4,3,2,1. (First define the terms "degree" and "degree sequence" for multigraphs.)

5. Show that, in any group of two or more people, there are always two people who know the same number of people inside the group.

★6. Let $d_1 \geq d_2 \geq \cdots \geq d_p \geq 0$, $p \geq 2$, and $d_1 \geq 1$ be given. Show that there exists a graph G (of order p) having degree sequence $d_1, d_2, \ldots, d_p$ if and only if there exists a graph H (of order $p - 1$) having degree sequence $d_2 - 1, d_3 - 1, \ldots, d_{d_1+1} - 1, d_{d_1+2}, \ldots, d_p$.

7. Given isomorphic graphs G_1 and G_2 and an isomorphism $\phi: V(G_1) \to V(G_2)$, prove the following facts.

a. G_1 and G_2 have the same order.

b. For each vertex v of G_1, $\deg \phi(v) = \deg v$.

c. G_1 and G_2 have the same size.

8. Show that "is isomorphic to" is an equivalence relation on the set of graphs.

9. For each of the following parts, give an example (preferably of smallest possible order) of nonisomorphic graphs G_1 and G_2 satisfying the stated conditions.

a. G_1 and G_2 have the same order.

b. G_1 and G_2 have the same order and the same size.

c. G_1 and G_2 have the same order and the same degree sequence (see Exercise 3).

10. Let n be an arbitrary positive integer. For each part, give an example of a graph having the stated order and size.

a. order n, size $n(n - 1)/2$ **b.** order $2n$, size n
c. order n, size n $(n \geq 2)$ **d.** order $2n$, size $3n$ $(n \geq 2)$

11. For each set of three graphs shown in Figure 7.11, (i) find the two graphs that are isomorphic, (ii) exhibit an isomorphism, and (iii) argue why the third graph is not isomorphic to the other two.

12. Given a graph G of order p, size q, minimum degree δ, and maximum degree Δ, prove that $p\delta \leq 2q \leq p\Delta$.

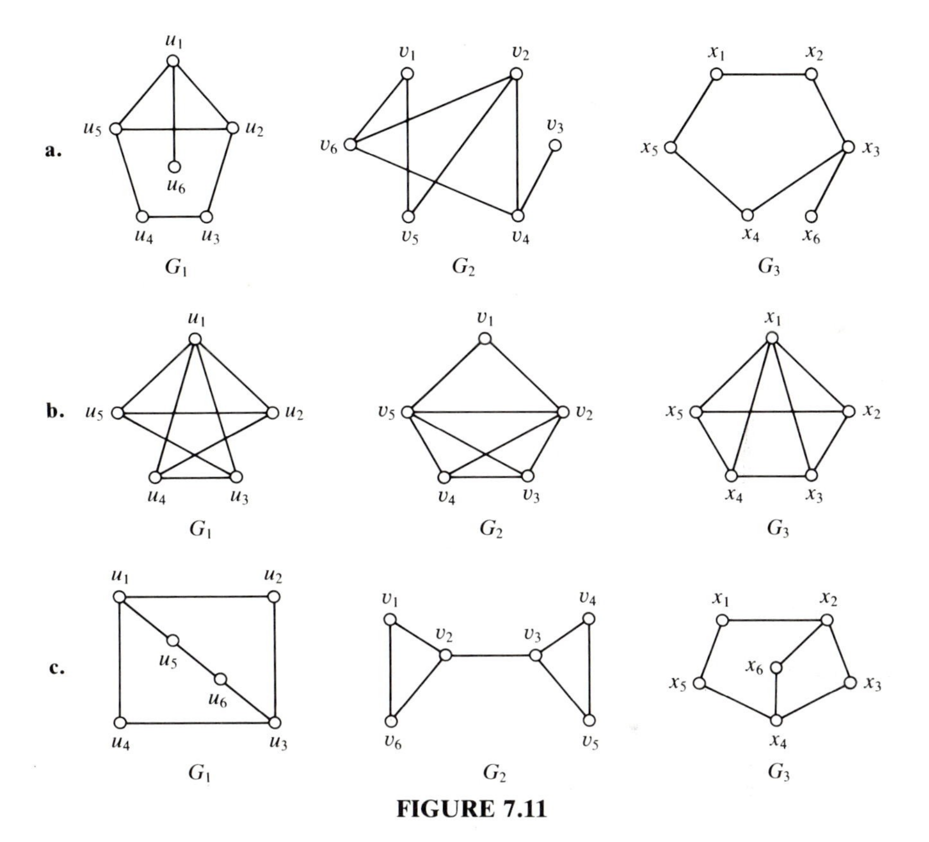

FIGURE 7.11

13. Let G be a k-regular graph of order p.
 a. What is the size of G?
 b. Show that either k or p must be even.

14. a. Show that the graph G_1 of Figure 7.12(a) is bipartite.
 b. Show that the graph G_2 of Figure 7.12(b) is not bipartite.

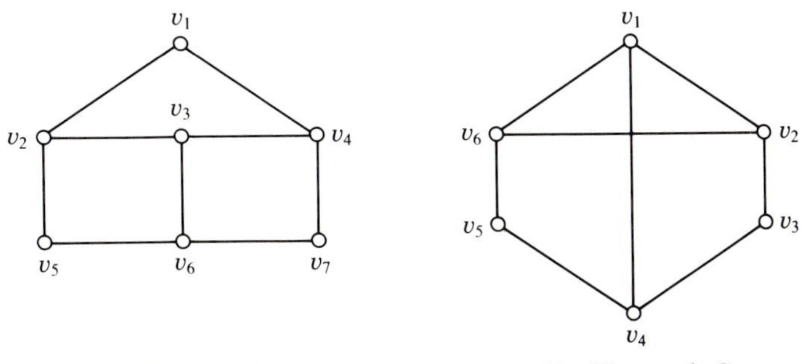

(a) The graph G_1 (b) The graph G_2

FIGURE 7.12

15. Prove: If G is a k-regular bipartite graph, then $V(G) = (V_1, V_2)$, where $n(V_1) = n(V_2)$ and $n(V_1) \geq k$.

16. Let G be a bipartite graph of order p and size q.
a. Prove that $4q \leq p^2$. **b.** If $4q = p^2$, what is G?

17. Give examples of each of the following graphs.
a. a 2-regular bipartite graph of order 6
b. a 3-regular bipartite graph of order 8
c. an r-regular bipartite graph of order $2r + 2$

18. Prove: If G_1 and G_2 are isomorphic graphs and G_1 is bipartite, then G_2 is bipartite.

19. Given that G is a k-regular graph of order p and size q, what can be said about G if
a. $k = 3$ and $q = p + 3$; **b.** $k = 4$ and $q = 3p - 5$.

20. Determine all nonisomorphic graphs of order 5.

21. A graph G is *n-partite*, $n \geq 2$, if it is possible to partition $V(G)$ into n subsets $V_1, V_2, \ldots, V_n$ (called partite sets) such that every edge of G joins a vertex of V_i to a vertex of V_j, $i \neq j$. Find the minimum n for which each of the graphs in Figure 7.13 is n-partite.

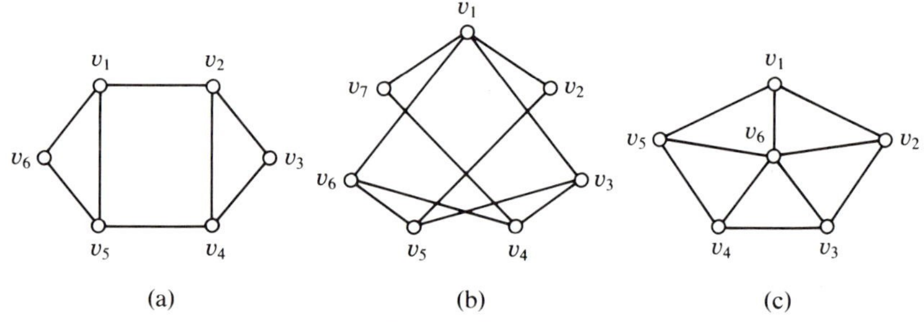

FIGURE 7.13

22. Give a definition of "G_1 is isomorphic to G_2" if G_1 and G_2 are multigraphs.

23. A *complete n-partite graph* G is an n-partite graph in which vertices u and v are adjacent if and only if u and v belong to different partite sets. If the n-partite sets have cardinalities $p_1 \leq p_2 \leq \cdots \leq p_n$, then the graph G is denoted $K_{p_1, p_2, \ldots, p_n}$. Draw each of the following complete n-partite graphs.
a. $K_{1,4}$ **b.** $K_{1,2,3}$ **c.** $K_{2,2,2}$ **d.** $K_{1,2,2,3}$ **e.** $K_{2,2,2,2}$

★**24.** Determine the size of the complete n-partite graph $K_{p_1, p_2, \ldots, p_n}$.

25. The men's intramural hockey league at a small college has seven teams: A, B, C, D, E, F, and G. Each team is scheduled to play four games over a five-week season. During each of the first four weeks of the season, six of the teams will play a game while the remaining team has a bye. For the last week, the three teams that have already played four games will have a bye, while the remaining four teams play.

a. Design a schedule for the hockey league, using a graph to indicate which teams each team will play.

b. Suppose the season is extended to six weeks. Is it now possible to have each of the seven teams play five games?

7.3 SUBGRAPHS AND CONNECTEDNESS

As is usually the case when a particular kind of mathematical structure is studied, we encounter the important notion of "substructure." In this regard, graphs are no exception.

DEFINITION 7.4

A graph H is called a *subgraph* of the graph G provided $V(H) \subseteq V(G)$ and $E(H) \subseteq E(G)$. We write $H \subseteq G$.

Example 7.7 Given the graph G in Figure 7.14(a), exhibit

(a) all subgraphs of G of size 3;

(b) all nonisomorphic subgraphs of size 3.

Solution The graphs H_1, H_2, H_3, and H_4 shown in Figure 7.14(b) through (e), respectively, are the four subgraphs of G of size 3. There are three nonisomorphic subgraphs of size 3, for observe that $H_3 \cong H_4$.

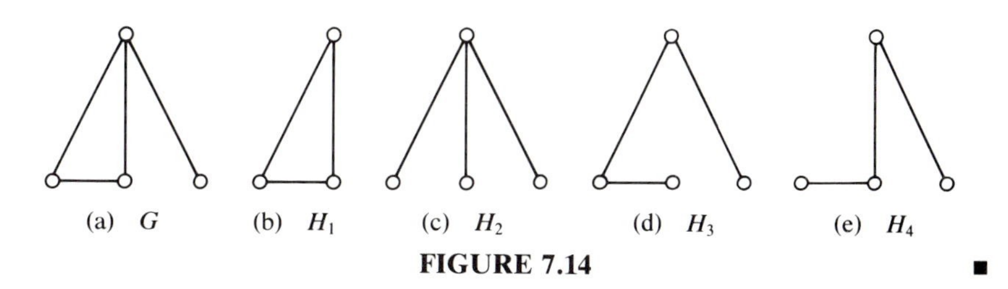

(a) G (b) H_1 (c) H_2 (d) H_3 (e) H_4

FIGURE 7.14 ■

If v is a vertex of the graph G, then by $G - v$ we mean the subgraph of G obtained by deleting v and all edges of G incident with v. More precisely,

$$V(G - v) = V(G) - \{v\}$$

and

$$E(G - v) = \{e \in E(G) \mid e \text{ is not incident with } v\}$$

Similarly, if $e \in E(G)$, then $G - e$ denotes the subgraph of G whose vertex set is $V(G)$ and whose edge set is $E(G) - \{e\}$. In general, if H is a subgraph of G and $V(H) = V(G)$, then H is called a *spanning subgraph* of G.

Example 7.8 Consider the graphs G, H_1, H_2, and H_3 shown in Figure 7.15. Note that $H_1 = G - u$, $H_2 = G - e$, and $H_3 = (G - f) - e$. Both H_2 and H_3 are spanning subgraphs of G.

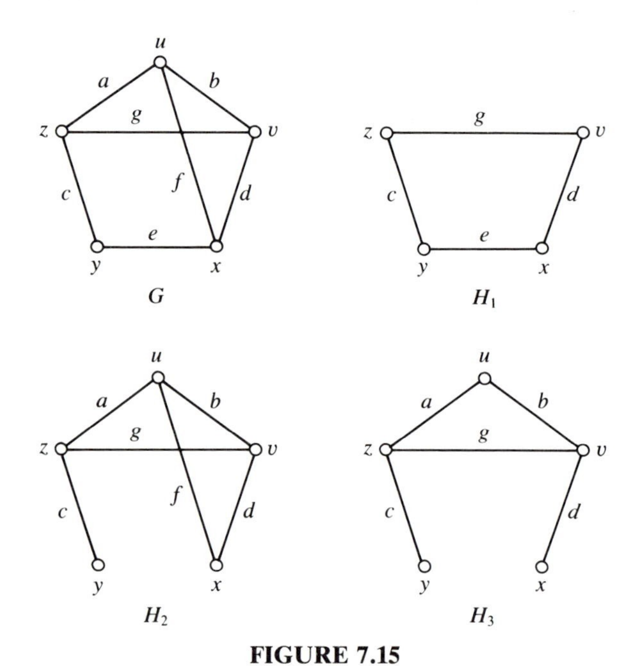

FIGURE 7.15 ■

DEFINITION 7.5

Let G be a graph and let U be a nonempty subset of $V(G)$. The *subgraph of G induced by U*, denoted by $\langle U \rangle$, is that subgraph with vertex set U such that two vertices of U are adjacent in $\langle U \rangle$ if and only if they are adjacent in G. A subgraph H of G is called an *induced subgraph* if $H = \langle U \rangle$ for some subset U of $V(G)$.

Example 7.9 Given the graph G of Figure 7.15, find induced subgraphs of G isomorphic to K_3 and to $K_{2,2}$.

Solution Both of the sets $\{u, v, x\}$ and $\{u, v, z\}$ induce a subgraph of G isomorphic to K_3. The graph H_1 of Figure 7.15 is an induced subgraph isomorphic to

$K_{2,2}$, as is $\langle\{u, x, y, z\}\rangle$. Note that $(G - y) - b$ is also isomorphic to $K_{2,2}$, but this is not an induced subgraph of G. ■

Let u and v be vertices (not necessarily distinct) in a graph G. A *u-v walk* in G is any finite, alternating sequence of vertices and edges,

$$u, uu_1, u_1, u_1u_2, \ . \ . \ . \ , u_{n-1}, u_{n-1}v, v$$

beginning with u and ending with v, such that each edge in the sequence is incident with the vertices that immediately precede and follow it. A *trivial walk* is one containing no edges (and hence exactly one vertex). A u-v walk is *closed* if $u = v$, *open* otherwise. A u-v walk is called a *u-v trail* if no edge of the walk is repeated; a *u-v path* is a u-v trail in which no vertex is repeated. A trivial walk is also called a *trivial path*. For the graph G of Figure 7.16, we have the following u-v walks:

R_1: $u, ux, x, xy, y, yx, x, xw, w, wv, v$
R_2: $u, ut, t, tw, w, wx, x, xy, y, yz, z, zw, w, wv, v$
R_3: $u, ut, t, tw, w, wx, x, xy, y, yz, z, zv, v$

The walk R_1 is not a trail due to the repeated edge xy, and R_2 is a trail but not a path due to the repeated vertex w. The walk R_3 is a path.

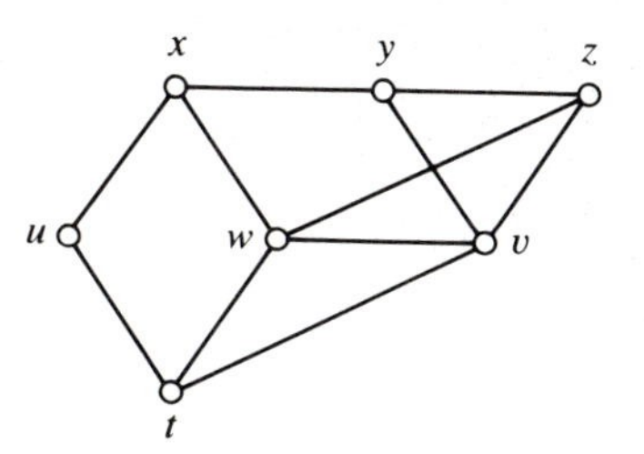

FIGURE 7.16 A graph G

Note that it is really superfluous to include the edges when listing the vertices and edges of a walk, because it is assumed that two consecutive vertices in a walk are adjacent. Thus we shall agree to list only the vertices in a walk. Thus the walks R_1, R_2, and R_3 become

R_1: u, x, y, x, w, v
R_2: u, t, w, x, y, z, w, v
R_3: u, t, w, x, y, z, v

Of special significance is the fact that the walk R_1 actually contains a u-v path (a path such that each edge belongs to R_1). We can obtain such a path as follows. Starting with u, trace R_1 vertex by vertex and note the first time a vertex is repeated; in this case it will be x. Delete from R_1 all vertices and edges that immediately follow the first occurrence of x, up to and including the next occurrence of x. We then obtain the walk

$$R_1': \quad u, x, w, v$$

The walk R_1' turns out to be a u-v path; however, if it were not a path we could repeat the above process to obtain a walk R_1'' that is properly contained in R_1'. Eventually, because R_1 contains a finite number of vertices, this process must yield a u-v path P that is contained in R_1.

We may refine the above process into a proof that every u-v walk in a graph G contains a u-v path. Before stating and proving this result, however, we require the additional notion of the length of a walk. The *length* of a walk W is the number of edges in W, with repetitions counted. Thus in the above discussion, R_1 has length 5, R_2 has length 7, and R_3 has length 6. We also remark that a trivial walk has length zero. Since we are listing only the vertices in a walk, it is also helpful to note that the length of a walk will be one less than the number of vertices listed.

THEOREM 7.2 Let u and v be vertices in a graph G and let W be a u-v walk in G. Then W contains a u-v path.

Proof Among all u-v walks of G that are contained in W, pick one of minimum length, say,

$$P: \quad u = u_0, u_1, \ldots, u_{r-1}, u_r = v$$

Such a walk exists by the principle of well-ordering. We claim that P is a path.

In order to establish this claim, we proceed by contradiction and suppose that P is not a path. This means that there is some vertex, say u_i, that is repeated in P. Thus P has the form

$$u_0, u_1, \ldots, u_{i-1}, u_i, \ldots, u_{j-1}, u_j, \ldots, u_r$$

where $u_i = u_j$. Hence the u-v walk

$$P': \quad u_0, u_1, \ldots, u_{i-1}, u_j, \ldots, u_r$$

is contained in W. But the length of P' is less than the length of P, which contradicts the choice of P. Hence P is a u-v path that is contained in W. ■

If u and v are vertices of a graph G and there is a u-v path in G, then it is common to say that *u is connected to v* in G. Of course, it is certainly possible that there exist vertices x and y in G such that x is not connected to y. For example, in the graph G of Figure 7.17, u_1 is connected to u_3, but u_1 is not connected to u_5.

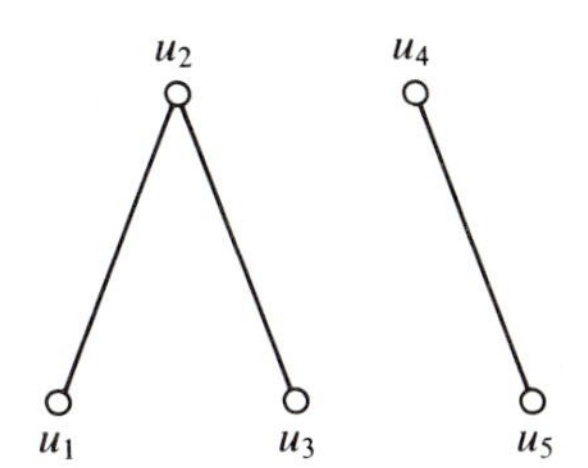

FIGURE 7.17 A disconnected graph G

DEFINITION 7.6

A graph G is termed *connected* if for every pair of vertices u and v of G, u is connected to v. Otherwise G is said to be *disconnected.*

The graph of Figure 7.16 is connected, but the graph of Figure 7.17 is disconnected.

The relation "is connected to" is an equivalence relation on the vertex set of a graph G (see Exercise 6). Hence, if $u, v \in V(G)$, we can simply ask whether u and v are "connected" in G. This equivalence relation induces a partition $\{V_1, V_2, \ldots, V_k\}$ of $V(G)$, where u and v belong to the same V_i if and only if u and v are connected in G. Let $C_i = \langle V_i \rangle$ for $i = 1, 2, \ldots, k$. Then each of the subgraphs C_i is connected and satisfies the property that there is no connected subgraph H of G that properly contains C_i (in the sense that $V(C_i) \subset V(H)$ or $E(C_i) \subset E(H)$). We call $C_1, C_2, \ldots, C_k$ the *components* of the graph G.

Example 7.10 A connected graph has exactly one component, namely, itself. The graph of Figure 7.17 has two components: $C_1 = \langle\{u_1, u_2, u_3\}\rangle$ and $C_2 = \langle\{u_4, u_5\}\rangle$. Give an example of a graph of order 5 that has (a) three, (b) four, (c) five components.

Solution (a) In Figure 7.18 are the three nonisomorphic graphs of order 5 that have three components.

(b) If a graph of order 5 has four components, then necessarily it has exactly one edge, and up to isomorphism there is only one such graph.

(c) The only graph of order 5 having five components is the graph of size zero.

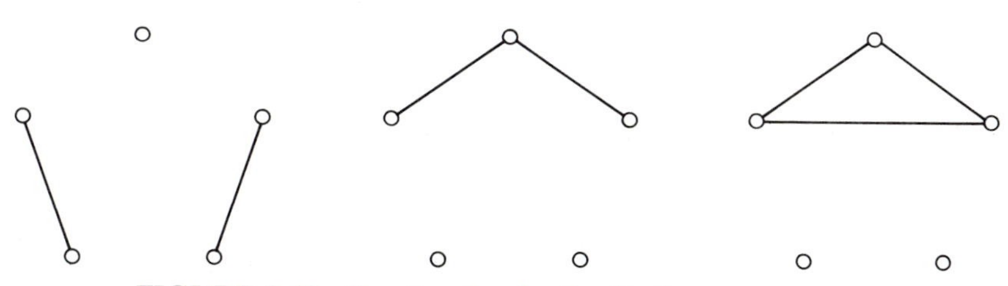

FIGURE 7.18 Graphs of order 5 with three components ■

If G is a connected graph, then for any two vertices u and v there is a u-v path of minimum length. The length of such a path is called the *distance* between u and v and is denoted by $d(u, v)$. This distance function satisfies the same properties as the ordinary distance function for points in a Euclidean plane, namely,

1. $d(u, v) \geq 0$, and $d(u, v) = 0$ if and only if $u = v$
2. $d(u, v) = d(v, u)$
3. $d(u, v) \leq d(u, w) + d(w, v)$

for all $u, v, w \in V(G)$. (See Exercise 20.)

A nontrivial path is necessarily an open walk. Next we define two special kinds of closed walks. A *circuit* in a graph is a nontrivial closed trail. If $v_1, v_2, \ldots, v_n$ are distinct vertices of a graph G, a circuit $v_1, v_2, \ldots, v_n, v_1$ is called a *cycle* of G. For example, in the graph of Figure 7.16,

$$W_1: \quad u, x, w, v, z, w, t, u$$

is a circuit, but it is not a cycle due to the repeated vertex w. However,

$$W_2: \quad u, x, y, z, v, t, u$$

is a cycle. The lengths of W_1 and W_2 are 7 and 6, respectively.

If C is a circuit and u is a vertex on C, then it is a direct consequence of Theorem 7.2 that C contains a cycle C' that contains the vertex u. To see this, first consider C as a u-u walk, say,

$$C: \quad u, u_1, u_2, \ldots, u_n, u$$

The walk W: $u, u_1, \ldots, u_n$ contains a u-u_n path P by Theorem 7.2, say,

$$P: \quad u, v_1, v_2, \ldots, u_n$$

Then C': $u, v_1, v_2, \ldots, u_n, u$ is a cycle contained in C and containing u.

Example 7.11 A traveling salesman lives and works in the country of Freedonia. (See Example 7.6; the graph representing the road map of Freedonia is repeated for convenience in Figure 7.19.) The salesman must go on a sales trip, beginning at his hometown H, visiting each of the cities A, B, C, D, E, F, and G, and then returning home to H. For obvious reasons, the

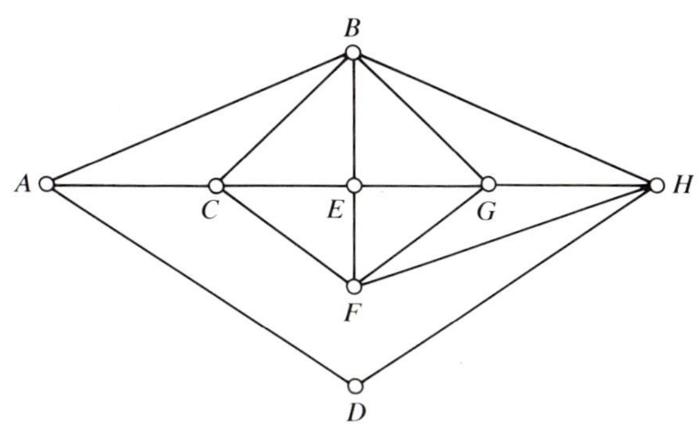

FIGURE 7.19

salesman would like to schedule a trip that allows him to visit each city exactly once; that is, he desires an *H*-*H* cycle in the graph that includes every vertex. Find such a cycle.

Solution By observation we find that one such cycle is

$$R: \quad H, D, A, B, C, F, E, G, H$$ ■

Notice that the vertices and edges of the cycle R in Example 7.11 form a spanning subgraph of the graph. In general, such a cycle is called a *spanning* or *hamiltonian cycle,* and a graph that possesses such a cycle is called a *hamiltonian graph.*

The problem of determining whether a given graph is hamiltonian is a special instance of a more general problem known as the *traveling salesman problem.* In general, no efficient algorithm is known for solving this problem, although several useful necessary conditions that a graph be hamiltonian are known. A couple of these will be explored in the exercises and problems.

Exercises 7.3

1. Determine each of the following subgraphs of the graph G of Figure 7.16.

a. $G - w$ **b.** $(G - w) - z$

c. $\langle\{u, x, y, v, t\}\rangle$ **d.** $(G - v) - xw$

2. Show that an induced subgraph H of a graph G can be obtained by successively deleting vertices from G.

3. For the graph G of Figure 7.16,

a. find a subgraph H_1 isomorphic to $K_{2,3}$;

b. find an induced subgraph H_2 isomorphic to $K_{1,3}$;

c. find a spanning subgraph H_3 that is regular of degree 2;

d. show that G does not contain a subgraph isomorphic to $K_{1,1,3}$.

4. Let G be a graph and let F be a nonempty subset of $E(G)$. The *subgraph of G induced by F*, denoted by $\langle F \rangle$, is that subgraph whose edge set is F and whose vertex set consists of those vertices of G incident with at least one element of F. A subgraph H of G is called *edge-induced* if $H = \langle F \rangle$ for some $F \subseteq E(G)$. For the graph G of Figure 7.16,

a. determine the subgraph H_1 induced by the set of edges incident to the vertex w;

b. find an edge-induced subgraph H_2 isomorphic to $K_{2,2}$.

Let G be a graph and let H be an edge-induced subgraph of G.

c. Show that $\delta(H) \geq 1$. (Thus not all subgraphs are edge-induced.)

d. Give an example to show that it is not always possible to obtain H by successively removing edges of G.

5. Find all nonisomorphic connected, spanning subgraphs of the graph H_2 in Figure 7.15.

6. Let G be a graph. Show that the relation "is connected to" is an equivalence relation on $V(G)$.

7. Let G be a graph of order p and size q.

a. Give a lower bound on q as a function of p if G is connected.

b. Give an upper bound on q as a function of p if G is disconnected.

c. Give examples to show that the bounds found in parts a and b are "sharp" (cannot be improved in general).

★d. Give an upper bound on q as a function of p if G has m components.

8. a. Characterize those graphs having the property that every induced subgraph is connected.

★b. Characterize those graphs having the property that every induced subgraph on three or more vertices is connected.

★c. Generalize parts a and b.

(Hint for parts b and c: see Exercise 23 in Section 7.2.)

9. a. Give an example of a $(p - 1)$-regular graph of order $2p$ that is disconnected.

b. Let G be a graph of order n. Prove: If $\delta(G) \geq (n - 1)/2$, then G is connected.

10. Let u and v be vertices of a connected graph G. Show that there is a u-v walk containing every vertex of G.

11. Let G_1 and G_2 be isomorphic graphs and let $\phi: V(G_1) \to V(G_2)$ be an isomorphism.

a. Prove: If W: $u_0, u_1, \ldots, u_n$ is a walk in G_1, then W': $\phi(u_0), \phi(u_1), \ldots, \phi(u_n)$ is a walk in G_2.

b. Prove: If G_1 is connected, then so is G_2. (In fact, G_1 and G_2 have the same number of components.)

12. a. Show that every subgraph of a bipartite graph is bipartite.

b. Show that every induced subgraph of a complete n-partite graph is complete m-partite for some $m \leq n$. (See Exercise 23 of Section 7.2.)

13. For the graph G of Figure 7.15, give an example of each of the following.

a. a u-y walk that is not a trail

b. a u-y trail that is not a path

c. a u-y path of length $d(u, y)$

d. a u-y path of maximum length

e. a cycle of length 5 that does not include the edge a

f. a cycle of length 4 that does not pass through the vertex y

14. Let G be a graph such that $\Delta(G) \leq 3$. Show that every circuit of G is a cycle.

15. Consider the graph G of Figure 7.16.

a. Find a circuit in G that is not a cycle.

b. Find a hamiltonian cycle in G.

c. Add the edge xt to G and call the resulting graph G'. Find a trail in G' that includes all edges of G'.

★**d.** Show that G does not have a trail that includes all of its edges.

16. Let G be a graph having vertex set V. Prove: G is connected if and only if for every partition of V into two nonempty subsets V_1 and V_2, there is an edge of G that joins a vertex of V_1 and a vertex of V_2.

17. Find the distance between each pair of vertices in the graph G of Figure 7.16.

18. Let G be a graph having minimum degree $\delta(G) = \delta$.

a. Prove that G has a path of length δ.

b. Prove: If $\delta \geq 2$, then G contains a cycle of length at least $\delta + 1$.

19. Let G be a connected graph. A trail in G that contains every edge is called an *eulerian trail;* if such a trail is a circuit then it is called an *eulerian circuit*.

a. Find an eulerian trail in the graph of Figure 7.19. (Hint: such a trail must begin and end at vertices of odd degree. Why?)

b. Remove the edge AB from the graph of Figure 7.19; find an eulerian circuit in the resulting graph.

c. Why might the Freedonia Highway Patrol be interested in knowing the eulerian trail you found in part a?

d. Let G be a connected (multi)graph. Prove: If G contains an eulerian circuit, then every vertex of G has even degree. (In fact, the converse also holds: see Chapter Problem 6.)

20. Let G be a connected graph. Show that the distance function satisfies the following properties for all $u, v, w \in V(G)$.

a. $d(u, v) \geq 0$, and $d(u, v) = 0$ if and only if $u = v$.

b. $d(u, v) = d(v, u)$

c. $d(u, v) \leq d(u, w) + d(w, v)$

21. The traveling salesman of Example 7.11 would probably be interested in knowing the "shortest" hamiltonian cycle in the graph of Figure 7.19, in terms of actual mileage. Assume that the graph is drawn to some scale, so that the length of an edge is proportional to the actual distance between the cities corresponding to its incident vertices. Let $\bar{e}$ denote the length of the edge e. Assume that $\overline{AD} = \overline{DH} > \overline{AB} = \overline{BH} > \overline{FH} > \overline{BC} = \overline{BG} > \overline{AC} = \overline{CF} = \overline{FG} = \overline{GH} > \overline{BE} = \overline{CE} = \overline{EG} > \overline{EF}$. Find a shortest hamiltonian cycle.

7.4 TREES

One of the more important and widely applied graphical structures is a "tree." In this section we shall introduce trees and discuss some of their basic properties.

DEFINITION 7.7

A graph that has no cycles is said to be *acyclic*. An acyclic graph is called a *forest* and a connected acyclic graph is called a *tree*.

Some examples of trees are shown in Figure 7.20. It is not surprising that the term tree is used to describe such graphs. Notice that $T_1 \cong K_{1,6}$ and that T_4 is a path of order 5. In general, we shall denote the path of order n by P_n; thus $T_4 \cong P_5$.

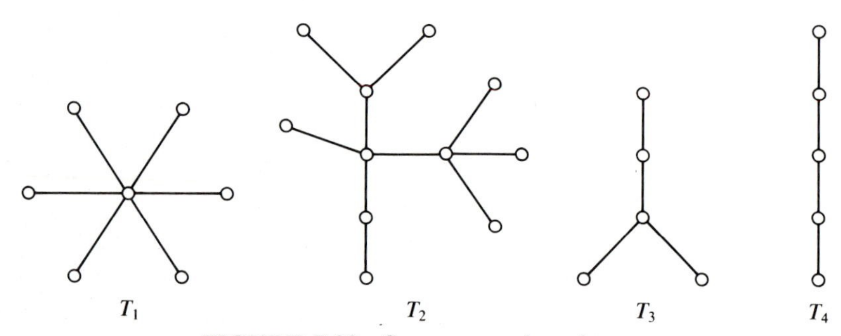

FIGURE 7.20 Some examples of trees

For $n = 1, 2, 3$, there is a unique tree of order n (up to isomorphism), namely P_n. The nonisomorphic trees of order 4 are P_4 and $K_{1,3}$. The graph $K_{1,n-1}$ is often called the *star* of order n. In many ways the path and star are the "extreme" trees. For example, if T is a tree of order $n \geq 3$, then $T \cong P_n$ if and only if $\Delta(T) = 2$, while $T \cong K_{1,n-1}$ if and only if $\Delta(T) = n - 1$ (see Exercise 3).

Example 7.12 Find (a) the nonisomorphic trees of order 5, and (b) the nonisomorphic trees of order 6 having maximum degree 3.

Solution (a) The nonisomorphic trees of order 5 are P_5, $K_{1,4}$, and the tree T_3 of Figure 7.20.

(b) The trees shown in Figure 7.21 are the three nonisomorphic trees of order 6 having maximum degree 3.

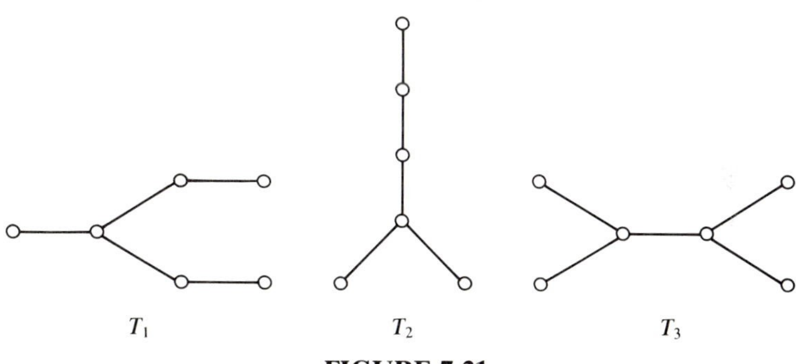

FIGURE 7.21 ■

An important characterization of a tree is that it is a connected graph in which every two vertices are joined by a unique path. In order to prove this result we shall employ the following lemma.

LEMMA 7.3 Let G be a graph and let u and v be distinct vertices of G. If P and Q are distinct u-v paths in G, then G contains a cycle.

Proof Assume that the paths P and Q are given as follows:

$$P: \quad u = u_1, u_2, \ldots, u_r = v \qquad Q: \quad u = v_1, v_2, \ldots, v_t = v$$

Since P and Q are not the same path, there must be an edge e that belongs to one path and not to the other, say $e \in E(P) - E(Q)$. We can in fact assume that e is the first occurrence in P of an edge that is not in Q; that is, if $e = u_iu_{i+1}$, then we may express the paths P and Q as follows:

$$P: \quad u = u_1, \ldots, u_i, u_{i+1}, \ldots, u_r = v$$
$$Q: \quad u = u_1, \ldots, u_i, v_{i+1}, \ldots, v_t = v$$

where $u_{i+1} \neq v_{i+1}$. Since both paths terminate in v, there must also be a least subscript k such that $k > i + 1$ and u_k is common to both paths. If $u_k = v_m$, then the paths

$$P': \quad u_i, u_{i+1}, \ldots, u_k \qquad \text{and} \qquad Q': \quad u_i, v_{i+1}, \ldots, v_m$$

are vertex-disjoint except for u_i and u_k. Hence it follows that

$$u_i, u_{i+1}, \ldots, u_k (= v_m), v_{m-1}, \ldots, v_{i+1}, u_i$$

is a cycle in G. ■

THEOREM 7.3 A graph T is a tree if and only if for all $u, v \in V(T)$ there is a unique u-v path in T.

Proof Assume first that T is a tree, and let u and v be vertices of T. We wish to show that there is a unique u-v path in T. Suppose, to the contrary, that there are two distinct u-v paths in T. Then, by the lemma, T would contain a cycle, contradicting the definition of a tree.

Next assume that T is a graph with the property that for all $u, v \in V(T)$, there is exactly one u-v path in T. Clearly then T is connected. To show that T is acyclic, suppose, on the contrary, that T contains a cycle, say,

$$C: \quad u_1, u_2, \ldots, u_n, u_1$$

Then $n \geq 3$, and therefore P: $u_1, u_2, \ldots, u_n$ and Q: u_1, u_n are different u_1-u_n paths in T, a contradiction. Hence T is a tree. ■

As a consequence of the above result, we note that if T is a tree and e is any edge of T, then the graph $T - e$ is disconnected. In particular, if $e = uv$, then there is no u-v path in $T - e$ and, in fact, $T - e$ has exactly two components, one containing u and the other containing v. To see this, let C_1 and C_2 be those components of $T - e$ that contain u and v, respectively. Let $x \in V(T) - \{u,v\}$, and let P: $u = u_1, u_2, \ldots, x$ be the unique u-x path in T. If v is on P, then it must be that $u_2 = v$, and hence Q: $v, u_3, \ldots, x$ is a path in $T - e$. Thus $x \in C_2$. On the other hand, if v is not on P, then P is a u-x path in $T - e$ and we obtain that $x \in C_1$. This shows that C_1 and C_2 are the components of $T - e$.

In general, if e is an edge of a connected graph G and the graph $G - e$ is disconnected, then e is called a *bridge* of G.

Another useful characterization of trees is provided by the following result.

THEOREM 7.4 A graph G of order p and size q is a tree if and only if G is acyclic and $q = p - 1$.

Proof We first prove the "only if" portion of the theorem. Thus we assume that G is a tree of order p and size q. Since G is a tree it is acyclic by definition. In order to prove that $q = p - 1$, we employ the strong form of mathematical induction. First, if $p = 1$, then $G \cong K_1$ and $q = 0$. Assume that the result holds for all trees of order less than p, where $p \geq 2$. Let T be any tree of order p. Consider $T - e$, where $e = uv$ is an edge of T. By the preceding remarks, the graph $T - e$ has exactly two components, T_1 and T_2, both of which are trees. If T_i has order p_i and size q_i for $i = 1, 2$, then by the induction hypothesis, $q_1 = p_1 - 1$ and $q_2 = p_2 - 1$. Thus

$$q = q_1 + q_2 + 1 = (p_1 - 1) + (p_2 - 1) + 1 = (p_1 + p_2) - 1 = p - 1$$

Thus by induction the result $q = p - 1$ holds for any tree of order p and size q.

Conversely, assume that G is an acyclic graph of order p and size q with $q = p - 1$. Let $T_1, T_2, \ldots, T_r$ be the components of G; we wish to show that $r = 1$ and hence that G is connected. Since G is acyclic, each of $T_1, T_2, \ldots, T_r$ is a tree; assume T_i has order p_i and size q_i for $i = 1, 2, \ldots, r$. Then, by the first part of the proof, each $q_i = p_i - 1$, and hence

$$\begin{aligned} p - 1 = q &= q_1 + q_2 + \cdots + q_r \\ &= (p_1 - 1) + (p_2 - 1) + \cdots + (p_r - 1) \\ &= p_1 + p_2 + \cdots + p_r - r \\ &= p - r \end{aligned}$$

Hence $r = 1$ and we conclude that G is connected. Since G is a connected acyclic graph, G is a tree. ■

COROLLARY 7.4.1 If F is a forest of order p that has r components, then F has size $p - r$. ■

COROLLARY 7.4.2 If T is a tree of order $p \geq 2$, then T has at least two vertices of degree 1. ■

The proofs of these two corollaries are left to the exercises.

It has already been noted that if T is a tree, then every edge of T is a bridge. In fact, the converse of this result holds, providing us with another characterization of trees.

THEOREM 7.5 A graph T is a tree if and only if T is connected and every edge of T is a bridge.

Proof We need only prove sufficiency, so assume that T is a connected graph in which every edge is a bridge. If T is not a tree, then T contains a cycle, say C. If e is an edge of C, then $T - e$ is a connected graph (see Exercise 16), which contradicts the hypothesis that every edge of T is a bridge. Therefore, T is a tree. ■

Using Theorem 7.5, we can prove that every connected graph G contains a spanning tree. Recall that this would be a subgraph T of G such that T is a tree and $V(T) = V(G)$.

COROLLARY 7.5 If G is a connected graph, then G contains a spanning tree.

Proof Since G is connected, it is a connected spanning subgraph of itself. It thus makes sense to choose a connected spanning subgraph of G of minimum size; call it T. We wish to show that T is a tree. Certainly it must be the case that for each edge e of T, the graph $T - e$ is disconnected. For otherwise $T - e$ would be a connected spanning subgraph of G of size smaller than the size of T. Thus every edge e of T is a bridge, and we conclude from Theorem 7.5 that T is a tree. ■

The next theorem is a consequence of Corollary 7.5 and provides us with another characterization of a tree.

THEOREM 7.6 A graph of order p and size q is a tree if and only if it is connected and $q = p - 1$.

Proof Let G be a graph of order p and size q. If G is a tree, then we know that G is connected and $q = p - 1$. Conversely, assume G is connected and $q = p - 1$. By Corollary 7.5, G contains a spanning tree T. Then T has order p, and since T is a tree, it must have size $p - 1$. Since G also has size $p - 1$, it must be that $T = G$, which shows that G is a tree. ■

To summarize, then, a graph G of order p and size q is a tree if and only if it satisfies any one of the following:

1. G is connected and acyclic (Definition 7.7).
2. There is a unique path joining any two vertices of G (Theorem 7.3).
3. G is acyclic and $q = p - 1$ (Theorem 7.4).
4. G is connected and $q = p - 1$ (Theorem 7.6).
5. G is connected and every edge of G is a bridge (Theorem 7.5).

This is a good point in our discussion of trees to describe an application of trees to transportation problems. In order to facilitate the description, we need to define a special type of graph. Given a graph G with vertex set $V = \{x_1, x_2, \ldots, x_p\}$, we associate with each edge x_ix_j of G a positive number w_{ij} and call w_{ij} the *weight* of the edge x_ix_j. If x_i and x_j are nonadjacent vertices, then we set $w_{ij} = \infty$. The resulting structure is called a *weighted graph*, and its *total weight* is the sum of the weights of all the edges of G.

Suppose that we are presented with the problem of connecting p cities $c_1, c_2, \ldots, c_p$ by a network of highways. The cost of building a highway between c_i and c_j is w_{ij}, with the agreement that $w_{ij} = \infty$ if no highway can be built between c_i and c_j (or if such a highway is not feasible). The problem is to build such a network so that

1. it is possible to get from one city to any other;
2. the cost of construction is a minimum.

We can model this problem as a weighted graph G in which vertices represent cities, and two vertices are joined by an edge of weight w_{ij} provided the cost of highway construction between the associated cities is the positive number w_{ij} (not ∞). To meet condition 1, we must find a connected spanning subgraph of G. To satisfy both conditions 1 and 2, we must find, among all connected spanning subgraphs T of G, one of minimum total weight. We denote the total weight of a subgraph H of G by $w(H)$. Clearly a connected spanning subgraph of G of minimum total weight will be a spanning tree of G. Such a subgraph is termed a *minimum spanning tree* of G.

In Section 7.8 we shall present an algorithm for finding a minimum spanning tree in a given connected weighted graph; this will provide a general solution to the highway problem.

Example 7.13 Find a minimum spanning tree for the weighted graph G of Figure 7.22. (The weights of the edges are indicated in the figure.)

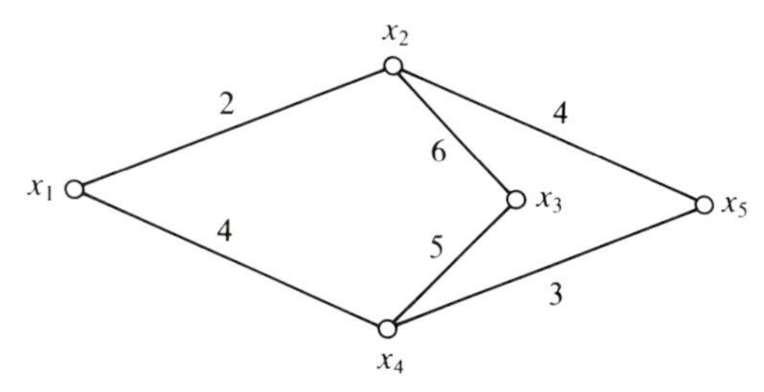

FIGURE 7.22

Solution The graph G has order 5 and size 6, so to obtain a spanning tree, two edges of G must be removed. We would like to remove the two edges of largest weight, namely, x_2x_3 and x_3x_4; however, the resulting subgraph is not connected. The best we can do is to remove x_2x_3 and one of the edges having weight 4. For instance, $H = (G - x_2x_3) - x_2x_5$ is a minimum spanning tree of G with $w(H) = 14$. ■

Exercises 7.4

1. Find the nonisomorphic trees of order 6 having maximum degree 4.
2. Find the nonisomorphic trees of order 7.
3. Let T be a tree of order $n \geq 4$ and maximum degree Δ.
   a. Show that T is a path if and only if $\Delta = 2$.
   b. Clearly $T \cong K_{1,n-1}$ if and only if $\Delta = n - 1$. Show that, up to isomorphism, there is a unique tree with $\Delta = n - 2$.
   c. Show that, for $n \geq 6$, there are three nonisomorphic trees with $\Delta = n - 3$.
4. Prove Corollary 7.4.1.
5. Give an example of two nonisomorphic trees having the same order and the same degree sequence.

6. Prove Corollary 7.4.2
   a. using the fact that $q = p - 1$;
   b. by considering the first and last vertices on a longest path in T.

7. Show that if a tree T has maximum degree Δ, then T contains at least Δ vertices of degree 1.

8. Let ε denote the number of vertices of degree 1 in a tree T, where T has order $n \geq 3$. Show:
   a. $2 \leq \varepsilon \leq n - 1$;
   b. $\varepsilon = 2$ if and only if $T \cong P_n$;
   c. $\varepsilon = n - 1$ if and only if $T \cong K_{1,n-1}$;
   d. there are $(n - 2)$ DIV 2 nonisomorphic trees with $\varepsilon = n - 2$;
   e. $\varepsilon = 1 + \frac{1}{2}\sum_{i=1}^{n} |\deg v_i - 2|$, where $V(T) = \{v_1, v_2, \ldots, v_n\}$.

9. Find all nonisomorphic spanning trees of the graph G shown in Figure 7.23.

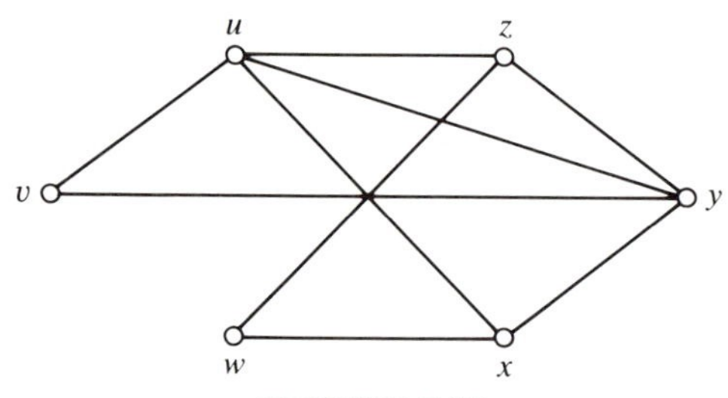

FIGURE 7.23

10. Use induction to prove that every tree is a bipartite graph.

11. Let G be a connected graph and let $v \in V(G)$. A spanning tree T of G is said to be *distance-preserving from* v if the distance $d_T(u, v)$ from u to v in T equals the distance $d_G(u, v)$ from u to v in G for all $u \in V(G)$.
   a. For the graph G of Figure 7.23, find a spanning tree that is distance-preserving from v.
   b. Prove: If G is a connected graph and $v \in V(G)$, then G contains a spanning tree T that is distance-preserving from v.

12. Prove that a graph G is a forest if and only if every induced subgraph of G contains a vertex of degree at most 1.

13. Show that every tree T of order $n \geq 3$ contains a vertex v such that $\deg v \geq 2$ and every vertex adjacent to v, with at most one exception, has degree 1.

14. Prove that a graph G is a forest if and only if every connected subgraph of G is an induced subgraph.

15. A connected graph G is said to be *traceable* if G contains a spanning path.
  a. Show that the graph of Figure 7.22 is traceable. (Note that this graph is isomorphic to $K_{2,3}$.)
  b. Show that the graph $K_{2,4}$ is not traceable.
  c. Give a necessary and sufficient condition for $K_{m,n}$ to be traceable.
  ★**d.** Prove: If G is a graph of order $p \geq 3$ such that for all distinct nonadjacent vertices u and v, $\deg u + \deg v \geq p - 1$, then G is traceable.
  e. Give an example to show that the bound $p - 1$ in the result of part d cannot be lowered.

16. Let G be a connected graph and let $e \in E(G)$. Show that e is a bridge if and only if no cycle of G contains e.

17. Let G be a connected graph that contains a bridge e. Prove that G contains at least one vertex of odd degree.

18. **a.** Develop an algorithm for finding a spanning tree T of a connected graph G.
  b. Apply the algorithm to find a spanning tree of the graph of Figure 7.19.

19. Under the assumptions of Exercise 21 of Section 7.3, find a minimum spanning tree of the graph of Figure 7.19, where the weight of each edge is equal to its length.

20. Prove that an edge e of a connected graph G is a bridge if and only if there exist vertices u and v such that e is on every u-v path of G.

7.5 DIRECTED GRAPHS

In many applications of graph theory, it may be necessary (in a certain sense) to assign directions to the edges in a graph. The structure that results can no longer be called a graph; it is called a "directed graph."

Consider the problem of traffic flow in a given city, in which some streets are one-way and others are two-way. This situation can be modeled using a directed graph as follows. Associate a vertex with each intersection, and then join two vertices a and b with an edge directed from a to b if it is possible to travel from a to b without passing through a third intersection. Note that it is possible for both the directed edge from a to b and the directed edge from b to a to be included in the directed graph. (When will this happen?)

We now proceed to the formal definition of a directed graph. It should be noted that this definition agrees with the notion of directed graph as presented in Section 4.2.

DEFINITION 7.8

A *directed graph* (or *digraph*) D consists of a finite nonempty set V, together with a subset A of the product set $V \times V$. We call V the *vertex set* of D and A the *arc set* of D. The digraph D is denoted by the ordered pair (V, A).

Given the digraph $D = (V, A)$, each $u \in V$ is called a *vertex* of D, while each $e \in A$ is called an *arc* or *directed edge* of D. As usual, we represent the vertices of D by points and represent an arc $(u,v) \in A$ by a directed line segment or curve from u to v.

Example 7.14 The digraph $D = (V, A)$, with $V = \{u, v, x, y, z\}$ and $A = \{(v,u), (v,x), (x,y), (y,x), (y,u), (z,u), (z,z)\}$, is represented in Figure 7.24. The arc (z,z) is called a *loop*.

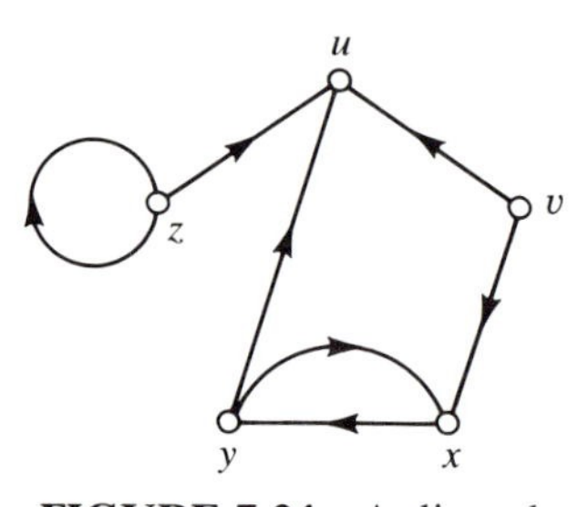

FIGURE 7.24 A digraph ■

From the definition it is clear that a digraph $D = (V, A)$ determines a relation on the set V, namely A. If this relation is symmetric and irreflexive (no element is related to itself), then we see that $(v,u) \in A$ whenever $(u,v) \in A$, and $(u,u) \notin A$ for all $u \in V$. Thus D has no loops, and we may as well replace the two arcs joining u and v by the simple edge uv. What results is a graph.

In almost all instances, the digraphs we will encounter will be "loopless." Since this is the case, we will simply use the term digraph in place of loopless digraph hereafter.

Suppose now that $D = (V, A)$ is a digraph and $(x,y) \in A$. We then say that x is *adjacent to* y and that y is *adjacent from* x. If both x is not adjacent to y and y is not adjacent to x, then we say that x and y are *nonadjacent*.

An x-y *semiwalk* in D is a finite alternating sequence of vertices and arcs,

$$W: \quad x = x_0, a_1, x_1, a_2, \ldots, x_{n-1}, a_n, x_n = y$$

such that $a_i = (x_{i-1}, x_i)$ or $a_i = (x_i, x_{i-1})$ for $i = 1, 2, \ldots, n$. We call n the *length* of the semiwalk W. If $a_i = (x_{i-1}, x_i)$ for $i = 1, 2, \ldots, n$, then W is called an *x-y walk*; in this case we can conveniently exhibit the walk W by listing only its vertices:

$$x = x_0, x_1, x_2, \ldots, x_n = y$$

An x-y walk is called *open* or *closed* according as $x \neq y$ or $x = y$, respectively. An x-y walk in D with no repeated arcs is called an *x-y trail,* and an x-y trail with no repeated vertices is an *x-y path*. An *x-x circuit* is a closed trail, while an *x-x cycle* is an x-x circuit in which only the vertex x is repeated.

The *distance* from x to y in D, denoted $d(x, y)$, is the minimum length among all x-y paths in D. If D contains no x-y path, then we write $d(x, y) = \infty$.

Example 7.15 Find the distance between each pair of vertices in the digraph D shown in Figure 7.25. Also, find a longest cycle and a longest circuit in D.

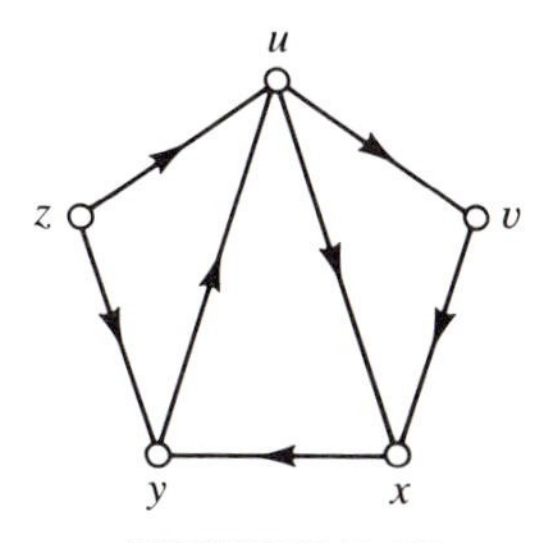

FIGURE 7.25

Solution We find that

$$d(u, v) = d(u, x) = d(v, x) = d(x, y) = d(y, u) = d(z, u) = d(z, y) = 1$$
$$d(u, y) = d(v, y) = d(x, u) = d(y, x) = d(y, v) = d(z, v) = d(z, x) = 2$$
$$d(v, u) = d(x, v) = 3$$
$$d(u, z) = d(v, z) = d(x, z) = d(y, z) = \infty$$

(Also, of course, the distance from any vertex to itself is zero.) A longest cycle in D is u, v, x, y, u, which has length 4; this is also a longest circuit in D. Note that u, v, x, y, u, x, y, u is not a circuit, although it is a closed walk of length 7. ■

If a digraph D contains an x-y walk, then we say that the vertex y is *reachable* from the vertex x. We now state two basic but important results concerning walks, paths, and reachability in a digraph.

THEOREM 7.7 If W is an x-y walk in a digraph D, then W contains an x-y path. ■

The proof of this theorem is very much like that of Theorem 7.2; that is, begin by selecting, among all x-y walks contained in W, one of minimum length. This will turn out to be an x-y path in D. The details are left to Exercise 4.

THEOREM 7.8 In a digraph D, if y is reachable from x and z is reachable from y, then z is reachable from x and

$$d(x, z) \leq d(x, y) + d(y, z)$$ ■

The proof of this result is also not difficult and is left to Exercise 6.

What about the other properties of the distance function? Well, it is easy to see that if y is reachable from x, then $d(x, y) \geq 0$, with $d(x, y) = 0$ if and only if $y = x$. However, the distance function is not symmetric for digraphs. We leave it to the reader to find an example where x is reachable from y and y is reachable from x, but $d(x, y) \neq d(y, x)$.

As contrasted with graphs, there are different degrees of connectedness that can exist in a digraph. For our purposes it is sufficient to consider just one type of connectedness for digraphs, one that we will apply further in Section 7.9.

DEFINITION 7.9

A digraph D is called *strongly connected* or *strong* if, for all $x, y \in V$, both x is reachable from y and y is reachable from x.

Example 7.16 The digraph of Figure 7.25 is not strong, for vertex z is not reachable from any of the other vertices. Determine whether the digraphs D_1 and D_2 in Figure 7.26 are strong.

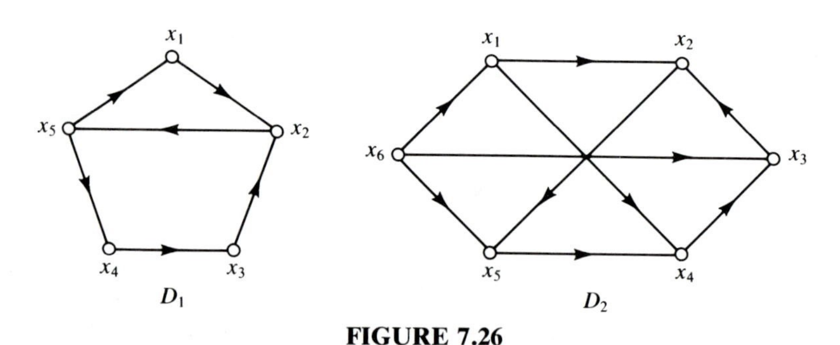

FIGURE 7.26

Solution It is not difficult to check that D_1 is strong. However, an easy way to see that D_1 is strong is to note that the walk

$$W: \quad x_1, x_2, x_5, x_4, x_3, x_2, x_5, x_1$$

is a spanning closed walk of D_1. Thus, given any two distinct vertices u and v of D_1, we may locate u before v on W, which yields a u-v walk contained in W. This shows that v is reachable from u.

The digraph D_2 is not strong; by inspection note that x_1 is not reachable from x_4. ■

Example 7.17 As seen in Example 7.16, the digraph D_2 is not strong. However, show that it is possible to reverse the direction of a single arc of D_2 so that the resulting digraph D_2' contains a spanning cycle. Must then D_2' be strong? Explain.

Solution If D_2' is to have a spanning cycle, then D_2 must contain a spanning path. One such path is

$$P: \quad x_6, x_1, x_2, x_5, x_4, x_3$$

(Note that P cannot be extended to a spanning cycle since x_6 is not adjacent from x_3.) Let D_2' be obtained from D_2 by reversing the direction of the arc (x_6, x_3). In other words, replace the arc (x_6, x_3) with the arc (x_3, x_6). The digraph D_2' is shown in Figure 7.27.

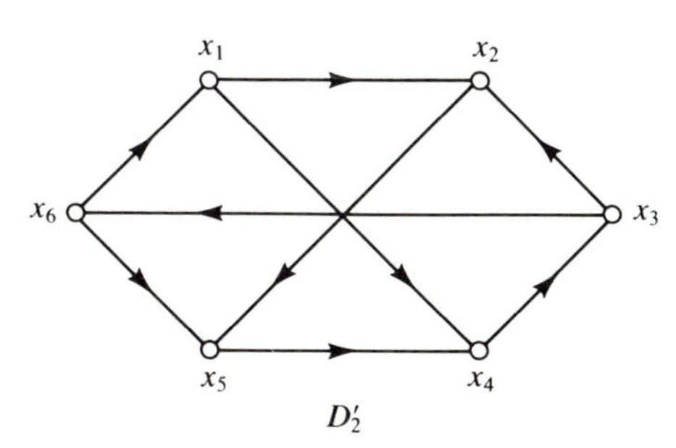

FIGURE 7.27

So D_2' contains the spanning cycle

$$C: \quad x_6, x_1, x_2, x_5, x_4, x_3, x_6$$

It follows that D_2' is strong, for we may use C in much the same way that the spanning closed walk W in the last example was used to show that D_1 is strong. Given distinct vertices u and v of D_2', the cycle C is composed of a u-v path and a v-u path. For instance, with $u = x_2$ and $v = x_4$, we see that C is composed of the paths

$$P_1: \quad x_2, x_5, x_4 \qquad \text{and} \qquad P_2: \quad x_4, x_3, x_6, x_1, x_2$$ ■

The next result provides us with a useful characterization of strongly connected digraphs.

THEOREM 7.9 A digraph D is strongly connected if and only if D has a spanning closed walk.

Proof Assume first that D has a spanning closed walk W and let $x, y \in V(D)$. We can represent W as a walk with x as the initial and terminal vertex:

$$W: \quad x, x_2, \ldots, x_r \,(= y), \ldots, x_n, x$$

As noted, since W is a spanning walk, the vertex y is on W, say $x_r = y$. Then W_1: $x, x_2, \ldots, x_r = y$ is an x-y walk in D and W_2: $x_r = y, x_{r+1}, \ldots, x_n, x$ is a y-x walk in D, which shows that D is strongly connected.

Next assume that D is strongly connected and, among all closed walks in D, choose one that includes a maximum number of distinct vertices. Call this walk W; then we claim that W is a spanning closed walk of D. We proceed by contradiction and suppose that there is some vertex u of D that is not on W. Let x be any vertex of W. Since D is strongly connected, we know both that u is reachable from x and that x is reachable from u. Let W_1 be an x-u walk and let W_2 be a u-x walk in D. We can now describe a closed walk in D that includes more distinct vertices than does W, as follows: take W_1 from x to u, followed by W_2 from u to x, followed by W from x to x. This contradicts the choice of W as a closed walk of D having a maximum number of distinct vertices. Thus W must be a spanning closed walk of D. ■

We close this section with a discussion of two special types of digraphs, "tournaments" and "rooted trees." These have some nice applications.

DEFINITION 7.10

A digraph D is called a *tournament* if, for any two distinct vertices u and v of D, exactly one of (u,v) and (v,u) is an arc of D.

We observe that if D is a tournament, then every pair of distinct vertices of D is joined by exactly one arc. In other words, we can obtain a tournament of order p simply by assigning directions to the edges of the complete graph K_p.

Three tournaments of order 4 are shown in Figure 7.28. In fact, once the notion of "isomorphism" has been defined for digraphs (see Exercise 5), then it can be shown that these three tournaments are mutually nonisomorphic.

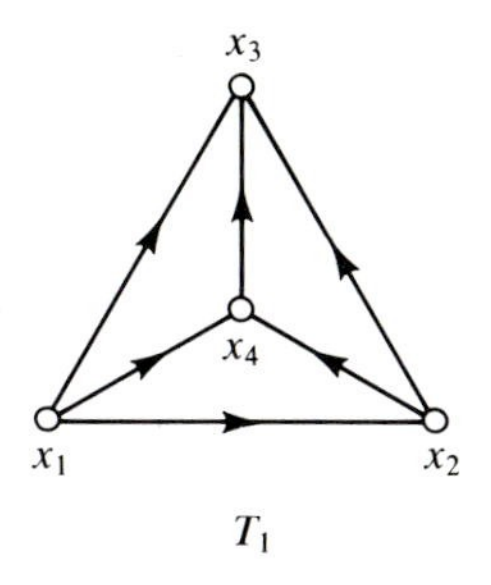

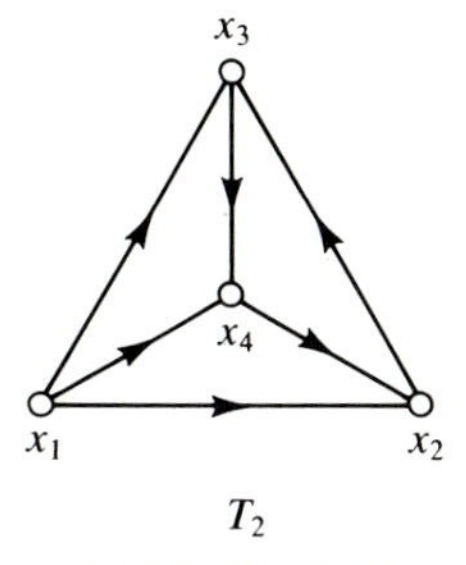

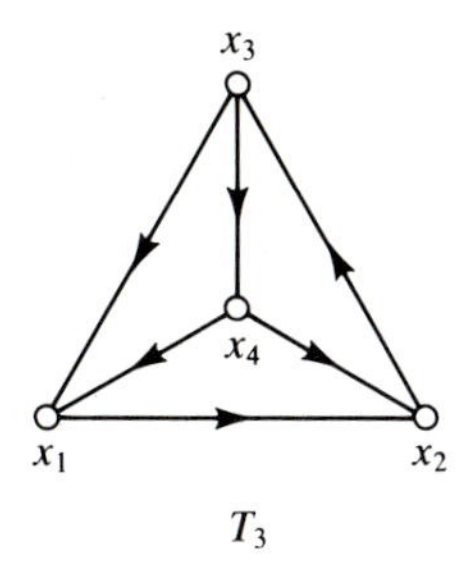

FIGURE 7.28

If G is a graph and v is a vertex of G, then we know that the degree of v is the number of vertices of G to which v is adjacent. If v is a vertex of a digraph D, then possibly there are vertices of D adjacent to v and vertices of D adjacent from v. Thus it makes sense to assign two different degree values to v. The *indegree* of v, denoted id v, is the number of vertices of D adjacent to v, while the *outdegree* of v, denoted od v, is the number of vertices of D that are adjacent from v. For the tournament T_1 of Figure 7.28, we have the following degrees:

$$\begin{array}{ll} \text{id } x_1 = 0 & \text{od } x_1 = 3 \\ \text{id } x_2 = 1 & \text{od } x_2 = 2 \\ \text{id } x_3 = 3 & \text{od } x_3 = 0 \\ \text{id } x_4 = 2 & \text{od } x_4 = 1 \end{array}$$

Notice that for each vertex x_i of T_1, id x_i + od $x_i = 3$, which is the degree of each vertex in the underlying complete graph K_4. In general, if T is any tournament of order p, then id x + od $x = p - 1$ for all vertices x of T.

A more interesting result concerning indegrees and outdegrees is provided by the next theorem, whose proof is left to Exercise 10.

THEOREM 7.10 If D is a digraph with vertex set $V = \{u_1, u_2, \ldots, u_p\}$ and size q, then

$$\sum_{i=1}^{p} \text{id } u_i = \sum_{i=1}^{p} \text{od } u_i = q \qquad \blacksquare$$

As already mentioned, tournaments provide us with some interesting and natural applications. For instance, a *round robin tournament* is a contest involving a finite number of players, in which each player competes against every other player exactly once. If we associate with each player a vertex, then we join vertices u and v by the arc (u,v) if and only if the player corresponding to u defeats the player corresponding to v (we assume no contest ends in a tie). The resulting digraph is clearly seen to be a tournament. Here are some questions that might be asked about round robin tournaments:

1. Is there a unique player who wins a maximum number of games? More generally, how can the players be ranked?
2. If there are p players, is it possible to label the players, say, $x_1, x_2, \ldots, x_p$, where x_1 defeated x_2, x_2 defeated $x_3, \ldots, x_{p-1}$ defeated x_p?

In digraph terminology, notice that question 2 asks if a tournament has a spanning path. This is indeed the case.

THEOREM 7.11 Every tournament has a spanning path.

Proof Let T be a tournament and let

$$P: \quad u_0, u_1, \ldots, u_n$$

be a longest path in T. We claim that P is a spanning path. If not, let v be a vertex of T that is not on P. Since P is a longest path, it must be that (u_0, v) and (v, u_n) are arcs of T, as shown in Figure 7.29. It then follows that there must be a vertex u_k on P such that (u_k, v) and (v, u_{k+1}) are both arcs of T. But then the path

$$P': \quad u_0, \ldots, u_k, v, u_{k+1}, \ldots, u_n$$

is longer than P, which is a contradiction. Therefore P is a spanning path of T.

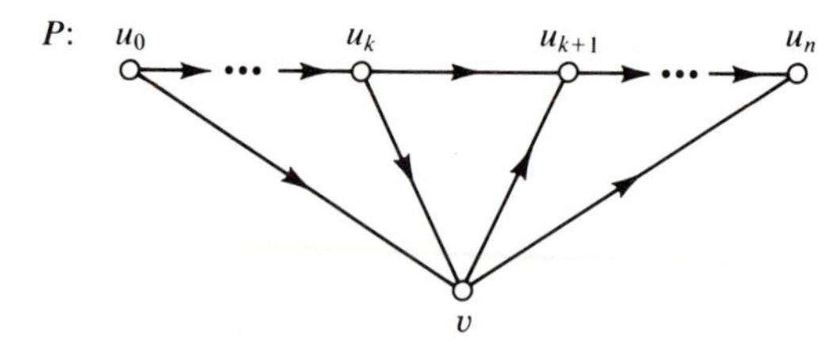

FIGURE 7.29

■

DEFINITION 7.11

A digraph D is called a *rooted tree* if D contains a vertex u, called the *root*, such that for every vertex v of D there is a unique u-v path in D.

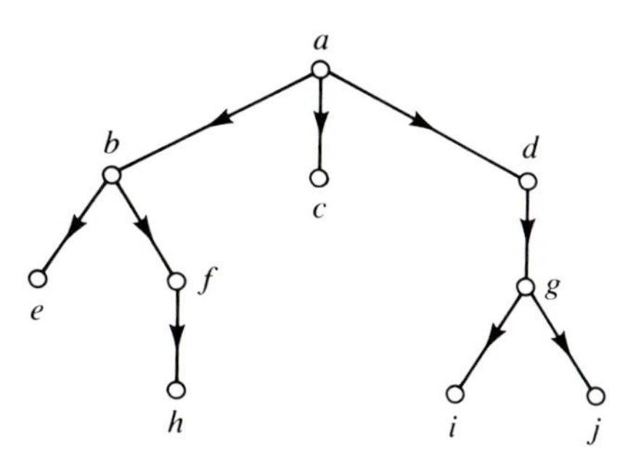

FIGURE 7.30 A rooted tree

We remark that a rooted tree necessarily has a unique root. An example of a rooted tree is the digraph D shown in Figure 7.30; here vertex a is the root.

Note that if D is a rooted tree, then the underlying graph is a tree. In fact, a rooted tree T' can be obtained from a tree T by choosing some vertex u of T as the root of T', and then directing all edges "away from" u. In other words, if v is a vertex of T and P is the unique u-v path in T, then direct the edges of P so that it becomes a u-v path in the digraph T'.

There is quite a bit of standard terminology associated with rooted trees. Let v be a vertex of the rooted tree D. Those vertices adjacent from v are called *children* of v; if v has no children, then v is called a *leaf* of D. If v is not the root of D, then it follows from the definition that there is a unique vertex x of D from which v is adjacent; we call x the *parent* of v. Vertices of D that have the same parent are called *siblings*. The notions of "child" and "parent" extend naturally to the notions of "descendant" and "ancestor."

As mentioned, there is a unique path in D from the root to the vertex v. The length of this path is the *level* of v in D. Thus the root has level zero, the children of the root have level 1, and so on. The *height* of D is the maximum level of all the vertices.

Example 7.18 For the rooted tree D of Figure 7.30, find (a) the children of b, (b) the ancestors of i, (c) the vertices having level 2, and (d) the height of D.

Solution
(a) The children of b are e and f.
(b) The ancestors of i are g, d, and a.
(c) The vertices at level 2 are e, f, and g.
(d) The height of D is 3. ■

Let m be a positive integer. A rooted tree D is said to be an *m-ary rooted tree* (or simply an *m-ary tree*) if each vertex of D has at most m children. Note that if $m = 1$, then D is a path. If $m = 2$, then D is called a *binary tree*. The rooted tree shown in Figure 7.30 is a 3-ary tree.

One of the many uses made of rooted trees in the study of computer science is as a convenient method of representing and working with expressions. For instance, consider the expression

$$(a + b) * (c \,/\, \mathrm{SQRT}(d))$$

where SQRT denotes the square root function. This expression can be represented by the rooted tree of Figure 7.31. Note that each operator corresponds to a non-leaf vertex of the rooted tree, in such a way that its subtrees (the subtrees of which its children are the roots) represent the operands for that operator. Note that a rooted tree represents an expression in an unambiguous manner, whereas the expression itself may require parentheses, or use implicit precedence rules, to make clear the order in which the operations are to be applied.

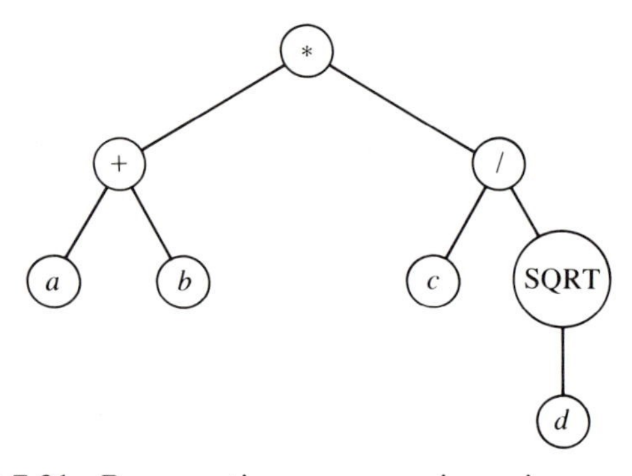

FIGURE 7.31 Representing an expression using a rooted tree

Exercises 7.5

1. For the digraph D_2' of Figure 7.27, find each of the following.
   a. a cycle of length 4
   b. a spanning path whose first vertex is adjacent to its last vertex
   c. a trail of maximum length
   d. a walk that contains every arc of D_2'
2. Prove or disprove: If the digraph D is strong, then D has a walk that includes every arc of D.
3. Find the distance between each pair of vertices in the following digraphs.
   a. D_1 of Figure 7.26
   b. D_2' of Figure 7.27
4. Prove Theorem 7.7.
5. **a.** Define the notion of "isomorphism" for digraphs.
   b. Figure 7.32 shows two digraphs D_1 and D_2 of order 5. Show that D_1 is isomorphic to D_2.

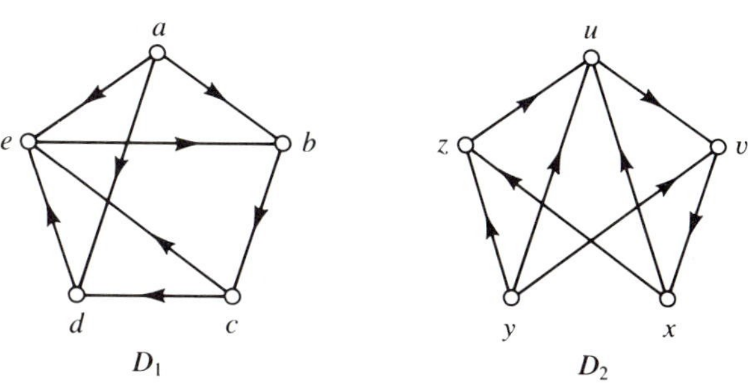

FIGURE 7.32

6. Prove Theorem 7.8.

7. Give an example of a strongly connected digraph D that does not contain a spanning circuit.

8. Let $D = (V, A)$ be a digraph and let $U \subseteq V$.
  a. Define the notion of "subdigraph" of D.
  b. Define the notion of "subdigraph of D induced by U."

9. Find id x_i and od x_i for each vertex x_i of the digraph D_2 of Figure 7.26.

10. Prove Theorem 7.10.

11. A digraph D is called *r-regular* if od v = id v = r for each v in the vertex set of D.
  a. Give an example of a 2-regular digraph on five vertices.
  b. Prove or disprove: If r and p are integers with $0 \leq r < p$, then there exists an r-regular digraph having p vertices.

12. **a.** Prove or disprove: In any digraph, the number of vertices having odd outdegree is even. Compare with Corollary 7.1.
  b. Give an example of a digraph D having p vertices whose outdegrees are $0, 1, \ldots, p - 1$, respectively. Compare with Exercise 5 of Section 7.2.

13. Let D be a digraph having p vertices such that whenever u and v are distinct vertices of D and u is not adjacent to v, then od u + id $v \geq p - 1$. Show that D is strongly connected.

14. Let D be a digraph such that every vertex of D has outdegree at least k.
  a. Prove that D contains a path of length k.
  b. If $k > 0$, prove that D contains a cycle of length at least $k + 1$.

15. A digraph D is said to be *unilaterally connected* or *unilateral* if, for all vertices x and y of D, the vertex x is reachable from y or y is reachable from x.
  a. Show that the digraph D_2 of Figure 7.26 is unilateral.
  b. Give an example of a digraph that contains a spanning semiwalk but is not unilateral.
  c. Prove: A digraph D is unilateral if and only if D contains a spanning walk.

d. Give an example of a unilateral digraph that does not contain a spanning trail.

e. Let D be a unilateral digraph. Prove: Either D is strong, or there exist vertices u and v of D such that the addition of the arc (u,v) to D results in a strong digraph.

16. Let $D = (V, A)$ be a digraph and define a relation $\sim$ on V by $u \sim v$ if and only if there is a closed walk of D containing both u and v.

a. Show that $\sim$ is an equivalence relation on V.

Let $U \subseteq V$ be an equivalence class for this equivalence relation.

b. Show that $\langle U \rangle$, the subdigraph of D induced by U, is strongly connected.

c. If $U \subset X \subseteq V$, show that $\langle X \rangle$ is not strongly connected.

Let $U_1, U_2, \ldots, U_n$ be the equivalence classes, and let $S_i = \langle U_i \rangle$ for $i = 1, 2, \ldots, n$. The S_i are called the *strong components* of D.

d. Find the strong components of each of the digraphs D_1 and D_2 shown in Figure 7.33.

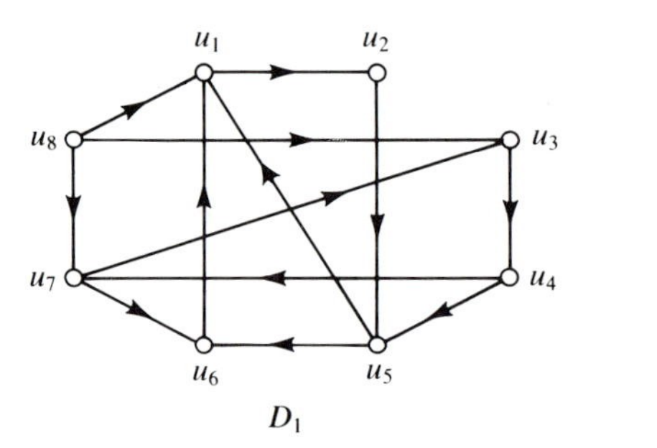

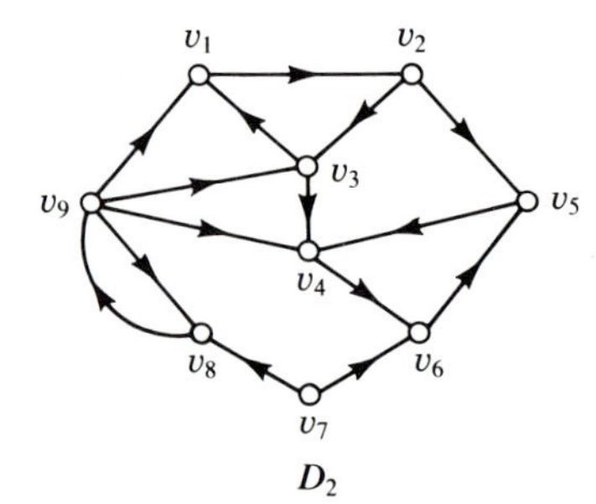

FIGURE 7.33

17. a. Find all nonisomorphic tournaments of order 3.

b. There are four nonisomorphic tournaments of order 4. Figure 7.28 shows three of them; find the fourth.

c. Find six nonisomorphic tournaments of order 5.

18. Let T be a tournament of order $n \geq 3$. Prove that T is strongly connected if and only if T contains a spanning cycle. (Hint: If T is strongly connected, let C be a cycle of maximum length in T. If C is not a spanning cycle, show that for every vertex u not on C, either u is adjacent to every vertex of C, or u is adjacent from every vertex of C.)

19. Let T be a tournament with vertex set $V = \{v_1, v_2, \ldots, v_n\}$, where od $v_1 \leq$ od $v_2 \leq \cdots \leq$ od v_n. The sequence od v_1, od $v_2, \ldots,$ od v_n is called the *score sequence* of T (od v is often termed the *score* of vertex v).

a. Find the score sequence of each of the nonisomorphic tournaments of order 4.

b. Give an example of two nonisomorphic tournaments having the same score sequence.

c. Show that, up to isomorphism, there is a unique tournament of order n having score sequence $0, 1, \ldots, n-1$. This tournament is called the *transitive tournament* of order n.

Let S: $s_1, s_2, \ldots, s_n$ $(n > 2)$ be a nondecreasing sequence of nonnegative integers, with $s_n < n$, and let S' denote the sequence $s_1, \ldots, s_{s_n}, s_{s_n+1} - 1, \ldots, s_{n-1} - 1$.

d. Prove: If S' is the score sequence of some tournament T', then S is the score sequence of some tournament T.

★**e.** Prove: If S is the score sequence of some tournament T, then S' is the score sequence of some tournament T'.

f. Use the results of parts d and e to decide whether the following are possible score sequences:
(i) 1,2,2,2,3 (ii) 0,1,2,3,3,5,6,7

20. Discuss some ways in which the participants in a tournament can be ranked. In particular, come up with a ranking system that reflects the popular notion that a player's ranking should reflect the strength of the opponents defeated by that player.

21. Draw binary trees representing each of the following expressions.

a. $x + (y * z)$ **b.** $(x + y) * z$

c. $(-x)/(y + (u \uparrow z))$ **d.** $(u * v) + ((w \uparrow (x/y)) + z)$

22. Find all nonisomorphic rooted trees whose underlying tree is the tree of Figure 7.34. Give the root and the height of each.

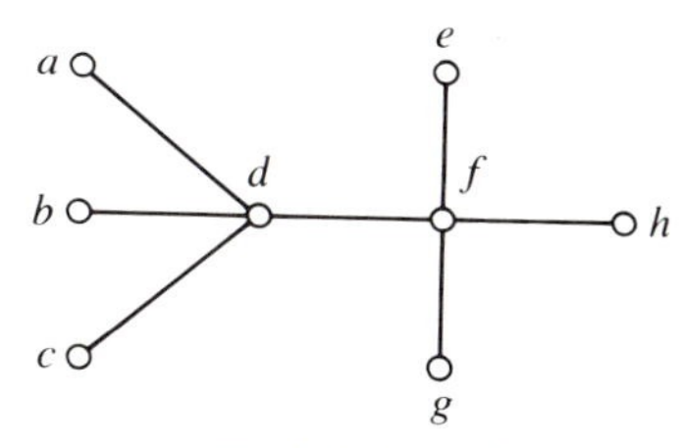

FIGURE 7.34

23. Let T be an m-ary rooted tree having height h, and let v be a vertex of T whose level is n.

a. How many ancestors does v have?

b. Give an upper bound for the number of descendants of v.

24. An m-ary rooted tree is said to be *full* if each non-leaf vertex has m children.

a. Draw a full binary tree of height 3 having as many vertices as possible.

b. Draw a full 3-ary tree of height 3 having as few vertices as possible.

c. Give a lower bound for the number of vertices in a full m-ary rooted tree having height h.

25. A common operation on a rooted tree is to *traverse* it, that is, to process or visit the vertices of the tree in some definite order. One method of traversing a tree is known as a *preorder traversal*. Let T be a tree having root r, and let $T_1, T_2, \ldots, T_k$ be the subtrees of T whose roots are the children of r. To traverse T in preorder, first visit the root r, and then traverse each of the subtrees $T_1, \ldots, T_k$ in preorder. (We assume that the children of each vertex of T are ordered in some way, say from left to right.) For example, a preorder traversal of the rooted tree of Figure 7.30 would visit the vertices in the order $a, b, e, f, h, c, d, g, i, j$.

a. List the vertices of the rooted tree T of Figure 7.35 in the order a preorder traversal would visit them.

b. A preorder traversal of the rooted tree of Figure 7.31 would yield $* + a\ b\ /\ c$ SQRT d. What does this represent?

c. Perform preorder traversals of each of the binary trees found in Exercise 21.

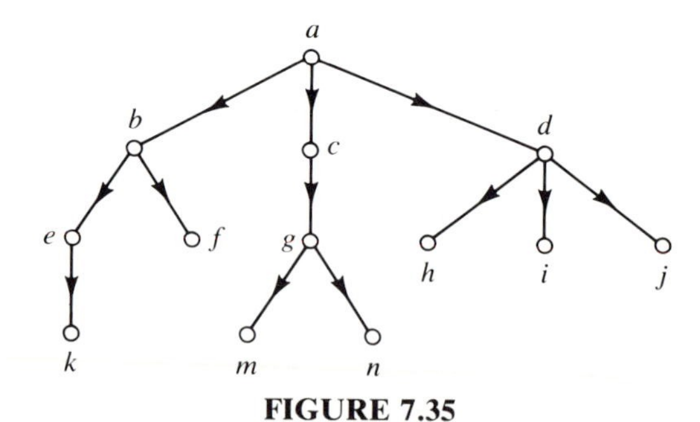

FIGURE 7.35

26. A *complete m-ary rooted tree of height h* is a full m-ary rooted tree in which each leaf is at level h; thus, each vertex at levels $0, 1, \ldots, h - 1$ has m children. How many vertices are there in such a rooted tree?

7.6 VERTEX COLORING AND PLANAR GRAPHS

Given a graph G and k different colors, is it possible to assign one of the k colors to each vertex of G in such a way that adjacent vertices have different colors? If so, then we speak of "coloring the vertices" of the graph using k colors, and such an assignment is termed a *k-coloring* of G. We also say that G is *k-colorable*.

The idea of coloring the vertices of a graph arises naturally in many problems concerned with "scheduling." For example, suppose that the department of mathematics at a particular university is concerned with scheduling its courses for the next semester. There are a number of courses to be offered and, no doubt, a smaller number of time periods during which courses may be given. Certain courses may be intended for roughly the same audience of students and thus should not be offered during the same time period. This suggests modeling the situation with a graph whose vertices correspond to the courses to be offered, such that two vertices are adjacent if the two courses should not be offered during the same time period. Suppose now that a k-coloring of the graph exists. Then all the vertices of a given color are mutually nonadjacent, which means that the corresponding courses could all be offered during the same time period; hence all the courses offered by the department can be accommodated using k time periods.

The minimum k for which a graph G is k-colorable is termed the *chromatic number* of G and is denoted $\chi(G)$. Note that if G has order p, then $1 \leq \chi(G) \leq p$. Also, $\chi(G) = 1$ if and only if G is empty, and $\chi(G) = p$ if and only if $G \cong K_p$.

The chromatic number of a cycle is also easily determined. If C_p denotes the cycle of order p, then we call C_{2n} an *even cycle* and C_{2n+1} an *odd cycle*. It is readily verified that $\chi(C_{2n}) = 2$ and $\chi(C_{2n+1}) = 3$.

Example 7.19 Find the chromatic number of the "wheel of order 6," W_6, which is shown in Figure 7.36.

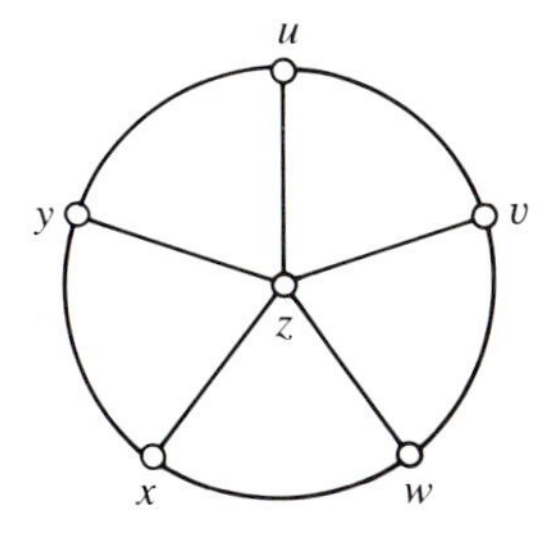

FIGURE 7.36 The wheel of order 6

Solution Notice that the graph $W_6 - z$ is isomorphic to the odd cycle C_5. Also, z is adjacent to each of the five vertices on this cycle. Thus, since $\chi(C_5) = 3$ and z requires its own fourth color, $\chi(W_6) = 4$. (In the exercises, the reader will be asked to show that for $n \geq 2$, $\chi(W_{2n}) = 4$ and $\chi(W_{2n+1}) = 3$, where W_p denotes the wheel of order p.) ■

A standard technique for showing that $\chi(G) = k$ is to exhibit a k-coloring of G, and then to argue that G is not $(k - 1)$-colorable. This technique is illustrated in the following example.

Example 7.20 Find $\chi(G)$ for the graph G shown in Figure 7.37.

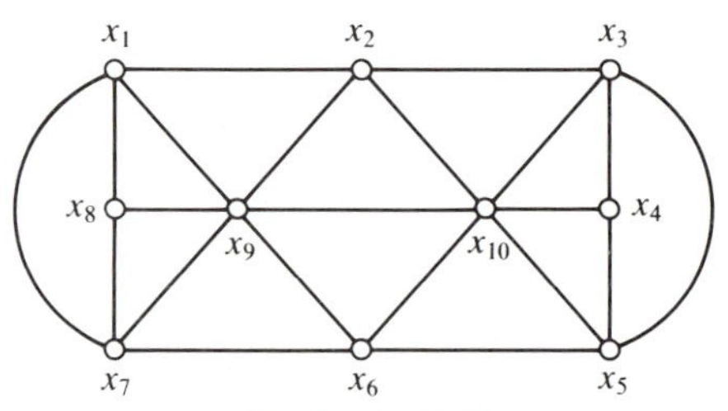

FIGURE 7.37

Solution We first note that $\langle\{x_1, x_7, x_8, x_9\}\rangle$ is an induced subgraph of G isomorphic to K_4. This shows that $\chi(G) \geq 4$. To show that $\chi(G) = 4$, we must exhibit a 4-coloring of G. One possibility is to color x_1 and x_{10} with one color, say red, x_2, x_4, x_6, and x_8 blue, use green for x_3 and x_9, and yellow for x_5 and x_7. ■

Suppose that a graph G is k-colorable, and a particular k-coloring of G is given. Define a relation on $V(G)$ as follows: for $x, y \in V(G)$, the vertex x is related to y provided x and y are the same color. This relation is an equivalence relation on $V(G)$ and, as such, partitions $V(G)$ into equivalence classes, each class consisting of all vertices of G of a particular color. These classes are called the *color classes* of the k-coloring of G. If the color classes are $V_1, V_2, \ldots, V_k$, then for each $i = 1, 2, \ldots, k$, V_i has the property that if $x, y \in V_i$, then x and y are nonadjacent in G.

DEFINITION 7.12

If G is a graph and U is a nonempty subset of $V(G)$, then U is called an *independent set* of vertices provided no two vertices in U are adjacent in G. An independent set U is called a *maximal independent set* provided U is not properly contained in another independent set.

Thus the color classes for a given k-coloring of G are independent sets in G. It should be noted that the chromatic number of G is the minimum number of independent sets into which $V(G)$ may be partitioned.

Example 7.21 The sets $V_1 = \{x_1, x_{10}\}$, $V_2 = \{x_2, x_4, x_6, x_8\}$, $V_3 = \{x_3, x_9\}$, and $V_4 = \{x_5, x_7\}$, which are the color classes for the 4-coloring found in Example 7.20, are independent sets in the graph G of Figure 7.37. In fact, V_1, V_2, and V_3 are maximal independent sets in G, but V_4 is not maximal, since V_4 is properly contained in the independent set $V_4 \cup \{x_2\}$. Also notice that if we let $V_2' = V_2 - \{x_2\}$ and $V_4' = V_4 \cup \{x_2\}$, then V_1, V_2', V_3, V_4' are independent sets in G whose union is $V(G)$. This shows that color classes may vary from one k-coloring of G to another. ■

As was mentioned earlier, a graph has chromatic number 1 if and only if the graph is empty. The following theorem characterizes those graphs having chromatic number 2.

THEOREM 7.12 A graph G has chromatic number 2 if and only if G is a nonempty bipartite graph.

Proof If G is a nonempty bipartite graph then clearly $\chi(G) \geq 2$. Moreover, if V_1 and V_2 are the partite sets of G, then V_1 and V_2 are independent sets in G whose union is $V(G)$. Thus $\chi(G) = 2$.

Conversely, if $\chi(G) = 2$, then G is nonempty. Let G be 2-colored, and let B_1 and B_2 be the resulting color classes. Then each of B_1 and B_2 is an independent set in G, and so every edge of G must join a vertex of B_1 with a vertex of B_2. It follows that G is bipartite. ■

In Exercise 2, the reader will be asked to show that a graph G is bipartite if and only if G contains no odd cycles. With this result, we can restate Theorem 7.12 as follows: A graph G has chromatic number 2 if and only if G is nonempty and contains no odd cycles. Thus, if G has an odd cycle, then $\chi(G) \geq 3$, and conversely, if $\chi(G) \geq 3$, then G must contain an odd cycle.

Unfortunately, for $n \geq 3$, no nice characterization is known of those graphs with chromatic number n. Even so, several interesting and important results are known regarding the chromatic number of a graph.

DEFINITION 7.13

A graph G is called *critical* if $\chi(G - v) < \chi(G)$ for all $v \in V(G)$. Specifically, a graph G is called *k-critical* if G is critical and $\chi(G) = k$.

As a consequence of this definition, we can readily show that if G is k-critical, then $\chi(G - v) = k - 1$ for all $v \in V(G)$.

It is not difficult to verify that K_2 is the only 2-critical graph, while the only 3-critical graphs are the odd cycles (see Exercise 4). The wheel W_{2n}, $n \geq 2$, is a 4-critical graph; we illustrate the verification of this fact using the wheel W_6 of Figure 7.36. In Example 7.19 we showed that $\chi(W_6) = 4$. Since $W_6 - z$ is the odd cycle C_5, we have that $\chi(W_6 - z) = 3$. Now consider any one of the vertices on the "rim" of the wheel, such as u. The graphs $W_6 - u$ and $(W_6 - u) - z$ are shown in Figure 7.38(a) and (b). Since $(W_6 - u) - z$ is a nontrivial path, it has chromatic number 2. Since z is adjacent to each of y, x, w, and v, a third color is required for z; hence $\chi(W_6 - u) = 3$. It follows that W_6 is 4-critical.

For $n \geq 4$, the n-critical graphs have not as yet been determined. However, we can obtain the following useful result.

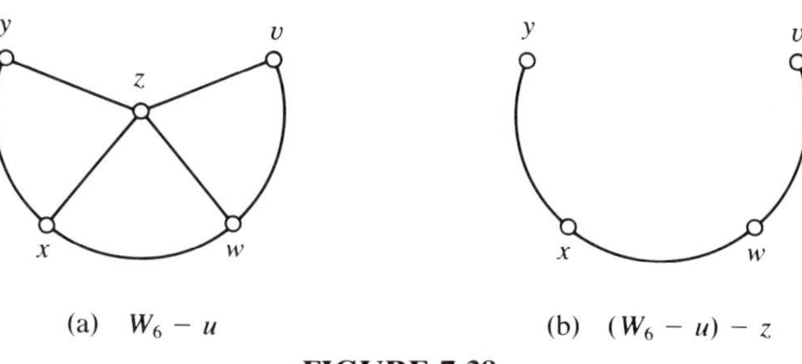

(a) $W_6 - u$ (b) $(W_6 - u) - z$

FIGURE 7.38

THEOREM 7.13 If G is a k-critical graph, then $\delta(G) \geq k - 1$.

Proof Let G be a k-critical graph. We proceed by contradiction and suppose there is a vertex u of G such that $\deg u < k - 1$. We know that $\chi(G - u) = k - 1$, so let a $(k - 1)$-coloring of $G - u$ be given. Since $\deg u < k - 1$, one of the $k - 1$ colors used, say α, is such that none of the vertices adjacent to u in G is colored α. We may then color u with α to obtain a $(k - 1)$-coloring of G. This contradicts the fact that $\chi(G) = k$ and thus establishes the result. ∎

If $\chi(G) = k$, then G may or may not be k-critical, but G does contain a k-critical subgraph. For if G is not k-critical, then there is some vertex u of G such that $\chi(G - u) = k$. If $G - u$ is k-critical, then we have the sought-after subgraph; if not, then we can repeat the foregoing process and, after a finite number of steps, arrive at a k-critical subgraph of G. (In fact, notice that we actually obtain a k-critical induced subgraph of G.)

By applying Theorem 7.13 to a $\chi(G)$-critical subgraph of G, we may obtain the following upper bound on the chromatic number of G.

COROLLARY 7.13 For any graph G, the chromatic number $\chi(G) \le \Delta(G) + 1$.

Proof Let $\chi(G) = k$ and let H be a k-critical subgraph of G. By Theorem 7.13, $\delta(H) \ge k - 1$, and hence $\Delta(H) \ge k - 1$. But certainly $\Delta(H) \le \Delta(G)$, and so we have that $k \le \Delta(H) + 1 \le \Delta(G) + 1$. Therefore, $\chi(G) \le \Delta(G) + 1$. ■

There is a stronger result than Corollary 7.13, proved by R. L. Brooks in 1941, which we now state.

THEOREM 7.14 (Brooks) If G is a connected graph that is neither a complete graph nor an odd cycle, then $\chi(G) \le \Delta(G)$. ■

It should be mentioned that the bound provided by Theorem 7.14 is not particularly satisfying for certain classes of graphs. For example, $\chi(K_{1,n}) = 2$ while $\Delta(K_{1,n}) = n$.

Example 7.22 Applying Brooks' theorem to the graph G in Figure 7.37, we find that $\chi(G) \le \Delta(G) = 6$. For a lower bound, we observe that G contains a subgraph isomorphic to K_4; thus $4 \le \chi(G)$. Another useful observation is the following: if v is a vertex of degree less than k in G, then G is k-colorable if and only if $G - v$ is k-colorable. Let us apply this observation to the graph G, using $k = 4$. Note that x_4 has degree 3 in G, so G is 4-colorable if and only if $G_1 = G - x_4$ is 4-colorable. Next, x_8 has degree 3 in G_1, so G_1 is 4-colorable if and only if $G_2 = G_1 - x_8$ is 4-colorable. Applying Brooks' theorem to G_2, we find that $\chi(G_2) \le \Delta(G_2) = 5$, so let's continue. Note that both x_1 and x_3 have degree 3 in G_2, so G_2 is 4-colorable if and only if $G_3 = G_2 - \{x_1, x_3\}$ is 4-colorable. But $\chi(G_3) \le \Delta(G_3) = 4$, which shows that G_3 is 4-colorable. Therefore, $\chi(G) = 4$. ■

In certain applications, we may be interested in an algorithmic procedure for partitioning the vertex set $V(G)$ of a graph G into k color classes $V_1, V_2, \ldots, V_k$. It is not hard to show (Exercise 6) that if $k = \chi(G)$, then this can be done so that V_1 is a maximal independent set in G, and V_2 is a maximal independent set in $G - V_1$, and so on. However, in general, no "good" algorithm, in terms of efficiency, is known for finding the chromatic number of a general graph.

At this point it is appropriate to return to a discussion of the four-color problem, which was described in Section 7.1. In particular, we shall attempt to explain how the "map-coloring problem" translates into a vertex-coloring problem for a particular class of graphs. A rigorous and formal treatment would take considerable time and would undoubtedly cloud the issue. Since it is our intent to provide only an intuitive description, we shall necessarily be somewhat crude and imprecise, but hopefully clear.

DEFINITION 7.14

A graph G is called a *planar graph* if it can be represented in the plane so that edges intersect only at incident vertices. We refer to this representation of G as a *plane embedding* of G or as a *plane graph*.

Example 7.23 The graph G of Figure 7.37 is a plane graph, as is the wheel W_6 shown in Figure 7.36. The graph H shown in Figure 7.39(a) is isomorphic to K_5 minus an edge. Show that H is a planar graph by representing H as a plane graph.

Solution The representation of H as a plane graph is given in Figure 7.39(b).

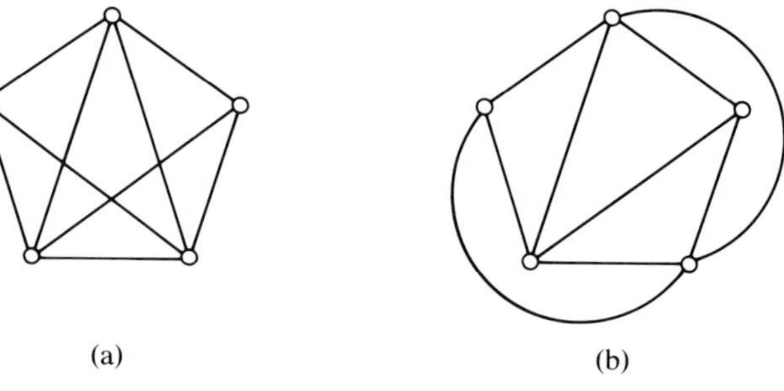

(a) (b)

FIGURE 7.39 A planar graph ■

Two noteworthy and interesting examples of nonplanar graphs are K_5 and $K_{3,3}$; the reader will be asked to show that these graphs are nonplanar in the exercises. Indeed, planar graphs can be characterized in terms of K_5 and $K_{3,3}$. In order to state this result, we need some additional terminology. An *elementary subdivision* of a nonempty graph G is obtained by first removing an edge $e = uv$ of G, and then adding a new vertex w to G, along with the edges uw and vw. For example, in Figure 7.40(b) we show an elementary subdivision of the graph in Figure 7.40(a). A *subdivision* of G is a graph obtained from G by a finite number of elementary subdivisions, and a graph H is said to be *homeomorphic from* a graph G provided H is isomorphic to a subdivision of G. We may now give the chief characterization of planar graphs, due to K. Kuratowski (1930).

THEOREM 7.15 (Kuratowski) A graph G is planar if and only if G does not contain a subgraph H homeomorphic from K_5 or $K_{3,3}$. ■

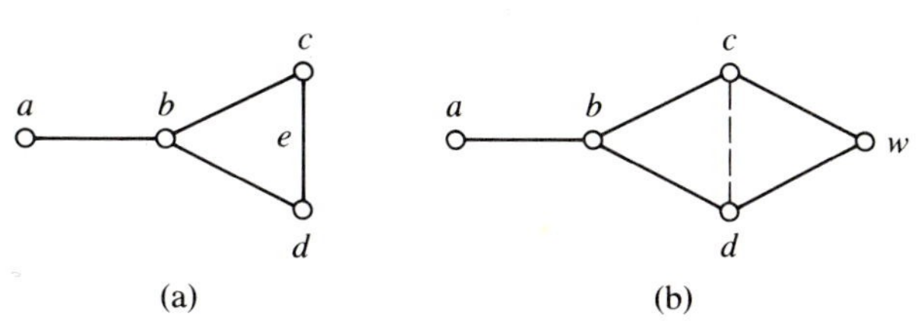

FIGURE 7.40 An elementary subdivision of a graph

Example 7.24 The graph G in Figure 7.41 is known as "Petersen's graph." Show that G is nonplanar by finding a subgraph of G homeomorphic from $K_{3,3}$.

Solution One possibility is the subgraph H shown in Figure 7.42, which is easily seen to be a subdivision of $K_{3,3}$.

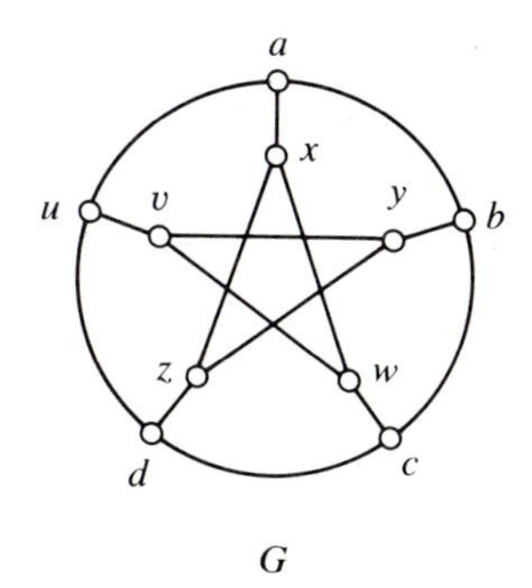

FIGURE 7.41 Petersen's graph

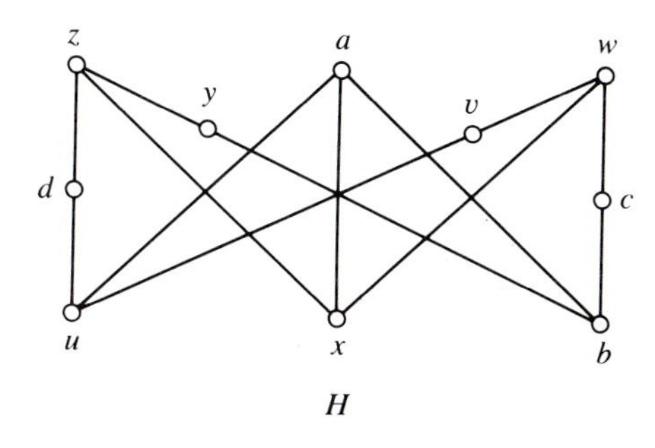

FIGURE 7.42 A subgraph of Petersen's graph homeomorphic from $K_{3,3}$ ∎

If G is a plane graph, then G partitions the plane into a finite number of regions, where each is bounded by a closed walk of G. For example, the plane graph H shown in Figure 7.43 partitions the plane into four regions, which are numbered 1 through 4 as shown. Note that region 2 is bounded by the cycle C: t, y, z, t, while the infinite region, region 4, is bounded by the closed walk W: $s, t, y, x, u, v, w, x, y, z, s$. In general, we refer to the

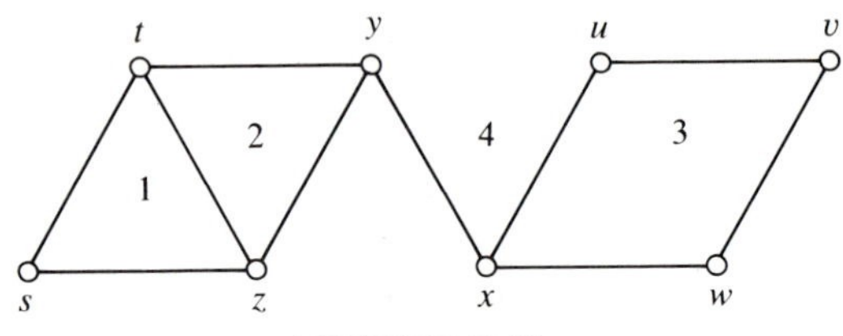

FIGURE 7.43

regions of a plane graph G as its *faces;* the infinite region is called the *exterior face*. The closed walk that encloses or bounds a given face f is called the *boundary* of f.

We now describe how the four-color problem for plane maps is equivalent to the problem of finding the maximum chromatic number among all planar graphs. Let a plane map M, such as the one in Figure 7.44(a), be given. We construct a planar graph G from M as follows. In each region of M insert a vertex of G, and join two vertices with an edge provided the associated regions have a nontrivial segment of their boundaries in common. Figure 7.44(b) shows the planar graph G corresponding to the map M of Figure 7.44(a).

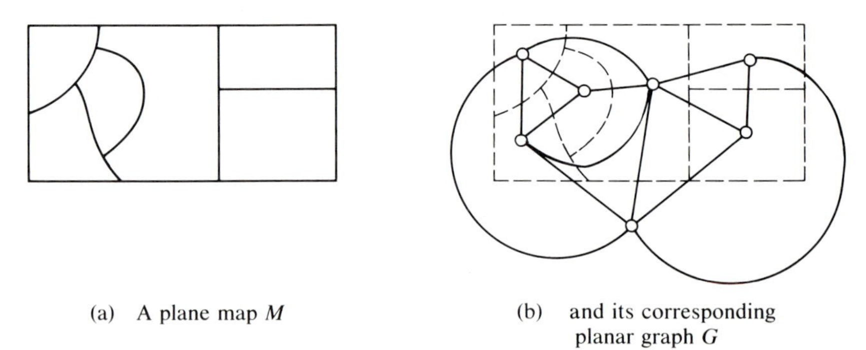

(a) A plane map M

(b) and its corresponding planar graph G

FIGURE 7.44

Our problem is to color the "countries" of a plane map M so that countries sharing a common boundary receive different colors. If this can be done with k colors, then we say that M is *k-region colorable*. Note that this problem is equivalent to that of coloring the vertices of the associated planar graph so that adjacent vertices are colored differently. Thus the four-color conjecture—that any plane map is 4-region colorable—is true provided that any planar graph is 4-colorable. We may now restate the celebrated result of Appel, Haken, and Koch.

THEOREM 7.16 **The four-color theorem** If G is a planar graph, then $\chi(G) \leq 4$. ■

Exercises 7.6

1. Find the chromatic number of each of the graphs shown in Figure 7.45.

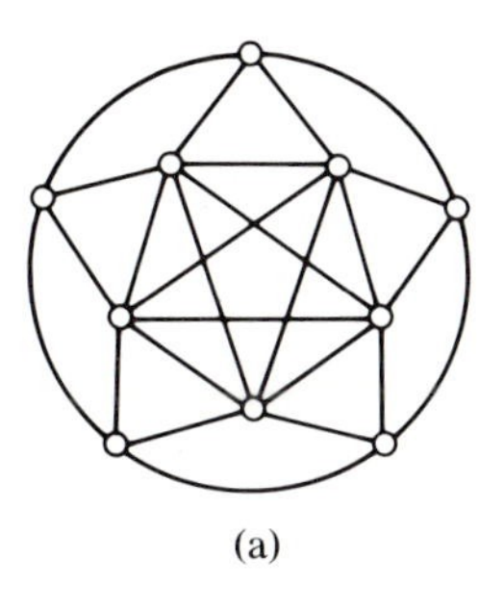

(a)

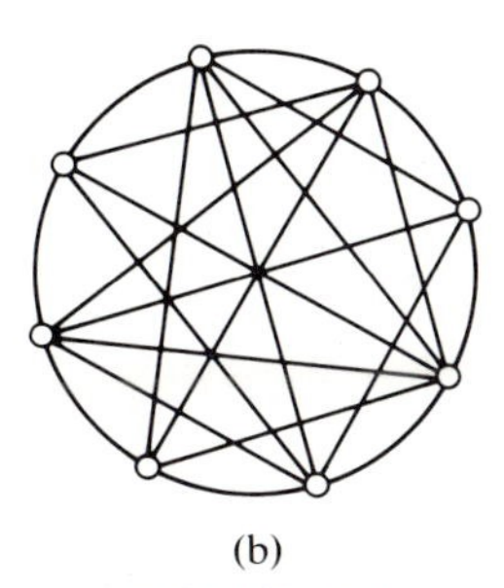

(b)

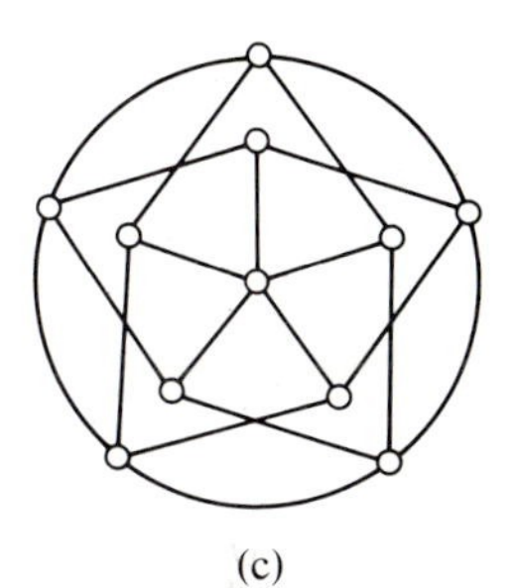

(c)

FIGURE 7.45

2. Prove that a graph G is bipartite if and only if G contains no odd cycles. (Hint: it suffices to prove the result for connected graphs.)

3. Find the chromatic number of the Petersen graph (Figure 7.41).

4. a. Show that the 3-critical graphs are precisely the odd cycles.

b. Give an example of a 4-critical graph that is not isomorphic to any wheel.

c. Give an example of a 5-critical graph.

5. Find the chromatic number of any nontrivial tree.

6. Show that if $\chi(G) = k$, then it is possible to partition $V(G)$ into k color classes $V_1, V_2, \ldots, V_k$ such that the set V_1 is maximally independent in G, the set V_2 is maximally independent in $G - V_1$, and so on.

7. Suppose we have n students $s_1, s_2, \ldots, s_n$, each taking some subset of the courses $c_1, c_2, \ldots, c_p$. We wish to schedule a final exam for each of the courses in such a way that no student has the conflict of two finals scheduled for the same time period. What is the minimum number of time periods required?

a. Formulate this as a problem in determining the chromatic number of a graph.

b. Suppose the set of students is {Al, Fred, Joe, Nancy, Steve}, where Al is taking MA 332, MA 337, and MA 351; Fred is taking MA 332, MA 351, and MA 423; Joe is taking MA 423 and MA 457; Nancy is taking MA 332, MA 351, and MA 423; and Steve is taking MA 332 and MA 457. Construct the graph and solve the problem.

8. Given two graphs G_1 and G_2 with $V(G_1) \cap V(G_2) = \phi$, the *join* of G_1 and G_2 is the graph G defined as follows:

$$V(G) = V(G_1) \cup V(G_2)$$
$$E(G) = E(G_1) \cup E(G_2) \cup \{uv \mid u \in V(G_1) \wedge v \in V(G_2)\}$$

We write $G = G_1 + G_2$ to denote that G is the join of G_1 and G_2.

a. The wheel W_p of order p is that graph isomorphic to $K_1 + C_{p-1}$. For $n \geq 2$, show that $\chi(W_{2n}) = 4$ and $\chi(W_{2n+1}) = 3$.

b. More generally, show that $\chi(G_1 + G_2) = \chi(G_1) + \chi(G_2)$.

9. Let $U = \{1, 2, \ldots, 10\}$ and consider the following subsets of U: $A = \{1, 2, 9\}$, $B = \{3, 6\}$, $C = \{1, 4, 8\}$, $D = \{2, 5, 10\}$, $E = \{3, 7, 9\}$, $F = \{6, 8, 10\}$, $G = \{4, 5, 7\}$. Form a graph H whose vertex set is $V(H) = \{A, B, C, D, E, F, G\}$ and whose edges join those pairs of sets having a nonempty intersection.

a. Find $\chi(H)$.

b. Given a $\chi(H)$-coloring of H, what properties do the color classes have?

c. Is there a subset of $V(H)$ that is a partition of U?

10. Show that if G is k-critical, then $G + K_1$ is $(k + 1)$-critical.

11. Show that if H is a subgraph of G, then $\chi(H) \leq \chi(G)$.

12. Show that $\chi(G)$ is equal to the minimum n for which the graph G is n-partite.

13. a. Prove or disprove: If G contains a unique odd cycle, then $\chi(G) = 3$.

b. Prove: If any two odd cycles of the graph G have a vertex in common, then $\chi(G) \leq 5$. (Hint: prove the contrapositive.)

14. Show that $\chi(G) \leq 1 + \max \delta(H)$, where the maximum is taken over all induced subgraphs H of G.

15. Let G be a graph of order p and let $\beta(G)$ denote the maximum cardinality of an independent set of vertices in G. Show that

$$\frac{p}{\beta(G)} \leq \chi(G) \leq p + 1 - \beta(G)$$

16. Let G be a graph of order p and size q. Show that $\chi(G) \geq p^2/(p^2 - 2q)$.

17. A *cut-vertex* in a connected graph G is a vertex whose removal results in a disconnected graph. Prove: If G is a k-critical graph, $k \geq 2$, then G is connected and contains no cut-vertices.

18. Prove Euler's formula: If G is a connected plane graph with p vertices, q edges, and r faces, then $p - q + r = 2$. (Hint: Use induction on q. If $q = p - 1$, then G is a tree, so $r = 1$ and the formula holds. If $q \geq p$, let e be a cycle edge of G and apply the induction hypothesis to $G - e$.)

19. Prove: If G is a planar graph of order $p \geq 3$ and size q, then $q \leq 3p - 6$. (Hint: Let r be the number of faces in a plane embedding of G. Count the number of edges on the boundary of each face, and sum over all faces. Since each edge is on the boundary of exactly two faces, and each face boundary contains at least three edges, we have that $3r \leq 2q$. Now apply Euler's formula.)

20. Modify the proof of the result of the preceding exercise to show that, if G is a planar graph of order $p \geq 3$ and size q, and G does not contain a subgraph isomorphic to K_3, then $q \leq 2p - 4$.

21. **a.** Use the result of Exercise 19 to show that K_5 is nonplanar.
b. Use the result of Exercise 20 to show that $K_{3,3}$ is nonplanar.

22. **a.** Show that if G is a planar graph, then $\delta(G) \leq 5$.
b. Construct the plane graph corresponding to the plane map of Figure 7.2(b). What is the minimum degree of this graph?
c. Use the result of part a to prove that if G is a planar graph, then $\chi(G) \leq 6$.

23. Determine whether each of the following graphs is planar or nonplanar.
a. $K_{2,4}$
b. $K_{2,2,2}$
c. $K_{1,1,2,2}$
d. the Petersen graph of Figure 7.41 minus edge ax
e. the "Grötzsch graph" of Figure 7.45(c)
f. the 4-regular subgraph of K_6

24. Find the maximal independent sets in the graph G of Figure 7.46. Try to do this in a systematic way.

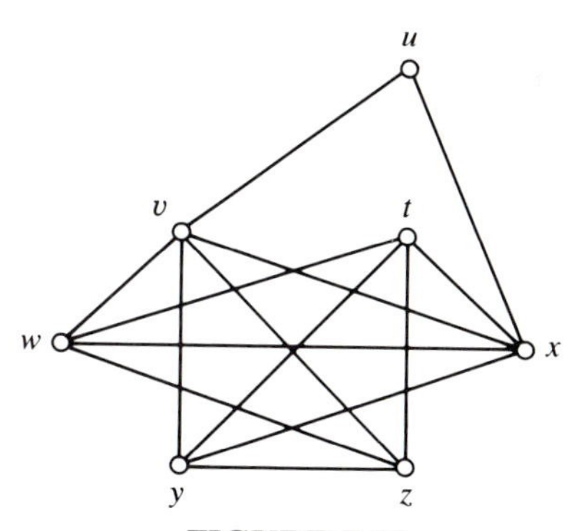

FIGURE 7.46

7.7 THE SHORTEST-PATH ALGORITHM

In many instances it is possible to apply graph theory to solve problems in communications and transportation, as well as in other areas. For example, suppose that a given company has plants in six different cities, c_1, c_2, . . . , c_6, and has its headquarters in city c_0. From time to time the company must send representatives from c_0 to one of the cities c_1, c_2, . . . , c_6. Assume the cost of a direct flight from c_i to c_j ($i \neq j$) is given by the (i,j) entry of the table shown below. (The entry ∞ means that there is no direct flight.)

City	c_0	c_1	c_2	c_3	c_4	c_5	c_6
c_0	0	185	75	68	∞	190	∞
c_1	185	0	101	∞	90	84	51
c_2	75	101	0	93	∞	∞	77
c_3	68	∞	93	0	72	∞	45
c_4	∞	90	∞	72	0	48	∞
c_5	190	84	∞	∞	48	0	∞
c_6	∞	51	77	45	∞	∞	0

The company would very much like to determine the cheapest routes from c_0 to each of $c_1, c_2, \ldots, c_6$. This problem can be conveniently modeled by a weighted graph if we associate a vertex with each city, where two vertices are adjacent provided there is a direct flight between those cities. In addition, define the weight of an edge in this graph to be the cost of the associated direct flight. The resulting structure is the weighted graph G depicted in Figure 7.47.

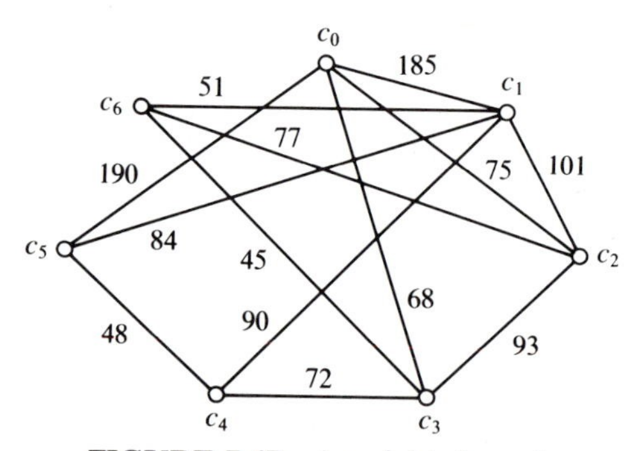

FIGURE 7.47 A weighted graph

With this interpretation, a cheapest route from c_0 to c_i would then correspond to a c_0-c_i path P such that the sum of the weights of the edges of P is a minimum among all c_0-c_i paths in G. In general, suppose that G is a weighted graph and $u, v \in V(G)$. The weights are assumed to be positive integers and the weight of the edge xy is denoted $w(xy)$. We define the *length* of a u-v path P in G to be the sum of the weights of the edges of P; the minimum length among the u-v paths in G will be called the *distance* between u and v and will be denoted by $d(u, v)$. If u and v are not connected in G then $d(u, v) = \infty$.

Given a vertex u_0 in a weighted graph G, an algorithm can be devised for finding the shortest paths from u_0 to all other vertices of G. We shall present an algorithm that was discovered by E. W. Dijkstra in 1959.

In order to understand the algorithm clearly, it is necessary to point out a very pertinent property of the distance function. Suppose that A is a proper subset of $V(G)$ and $u_0 \in A$. If $A' = V(G) - A$, then the distance from u_0 to A' is defined to be the length of a shortest path from u_0 to a vertex of A'. If P: $u_0, u_1, \ldots, u_r, u_{r+1}$ $(r \geq 0)$ is such a path, then it is not hard to show the following (see Exercise 2):

1. $u_{r+1} \in A'$
2. $u_0, u_1, \ldots, u_r \in A$
3. P': $u_0, u_1, \ldots, u_r$ is a shortest u_0-u_r path.

It follows that the distance $d(u_0, A')$ from u_0 to A' is given by

$$d(u_0, A') = \min_{\substack{u \in A \\ v \in A'}}\{d(u_0, u) + w(uv)\} \tag{1}$$

If the minimum occurs at $u = x$ and $v = y$, then

$$d(u_0, y) = d(u_0, x) + w(xy)$$

and the distance from u_0 to y is determined.

Formula (1) is the gist of the algorithm. At stage i of the process ($i = 0, 1, 2, \ldots$), A_i will denote the set of those vertices in the weighted graph G whose distances from u_0 have been determined. Initially, $A_0 = \{u_0\}$. At the $(i + 1)$st stage we will find a vertex v in A_i' such that

$$d(u_0, A_i') = d(u_0, v)$$

Then $A_{i+1} = A_i \cup \{v\}$. The process continues until $A_i' = \phi$; in other words, until $A_i = V(G)$.

Before formally describing the algorithm, we shall illustrate its use with the weighted graph G of Figure 7.47. We begin by setting $A_0 = \{c_0\}$ and finding

$$\min_{\substack{c \in A_0 \\ v \in A_0'}}\{d(c_0, c) + w(cv)\} = \min_{v \in A_0'}\{w(c_0 v)\}$$

Clearly this minimum is 68 and a shortest path from c_0 to A_0' is P_1: c_0, c_3. Thus $A_1 = \{c_0, c_3\}$ and $d(c_0, A_1')$ must now be determined. In this case we find that $d(c_0, A_1') = d(c_0, c_2) = w(c_0 c_2) = 75$, and a shortest path from c_0 to c_2 is P_2: c_0, c_2. At the next stage, then, set $A_2 = \{c_0, c_2, c_3\}$ and find $d(c_0, A_2')$. This occurs with $d(c_0, c_3) + w(c_3 c_6) = 68 + 45 = 113$, and a shortest path from c_0 to c_6 is P_3: c_0, c_3, c_6. Moreover, $d(c_0, c_6) = 113$

and now $A_3 = \{c_0, c_2, c_3, c_6\}$. We leave it to the reader to check the following:

$$d(c_0, A_3') = d(c_0, c_3) + w(c_3c_4) = 68 + 72 = 140$$

P_4: c_0, c_3, c_4 is a shortest c_0-c_4 path

$$A_4 = \{c_0, c_2, c_3, c_4, c_6\}$$

$$d(c_0, A_4') = d(c_0, c_6) + w(c_6c_1) = 113 + 51 = 164$$

P_5: c_0, c_3, c_6, c_1 is a shortest c_0-c_1 path

$$A_5 = \{c_0, c_1, c_2, c_3, c_4, c_6\}$$

$$d(c_0, A_5') = d(c_0, c_4) + w(c_4c_5) = 140 + 48 = 188$$

P_6: c_0, c_3, c_4, c_5 is a shortest c_0-c_5 path

$$A_6 = V(G)$$

The actual algorithm involves a labelling of the vertices at each stage. We begin by assigning to each $v \in V(G)$ a label $L(v)$; initially u_0 is labelled zero and the other vertices of G are labelled ∞. (In actual practice, we would replace ∞ by some number greater than any distance $d(u_0, v)$ in the graph, assuming the graph is connected.) The labels $L(v)$, $v \neq u_0$, will change as the algorithm progresses, and at the end of each stage, some vertex v will receive a permanent label, equal to $d(u_0, v)$. Here, then, is the algorithm.

ALGORITHM 7.1 (Dijkstra)

Step 0 (Initialize) Set $i = 0$, $A_0 = \{u_0\}$, $L(u_0) = 0$, and $L(v) = \infty$ for $v \neq u_0$.

Step 1 For each $v \in A_i'$, set

$$L(v) = \min_{u \in A_i}\{L(v), L(u) + w(uv)\}$$

Let y be a vertex in A_i' that now has the minimum label, that is

$$L(y) = \min_{v \in A_i'}\{L(v)\}$$

(Then $d(u_0, y) = L(y)$.) Set $A_{i+1} = A_i \cup \{y\}$.

Step 2 Replace i by $i + 1$. If $A_i = V(G)$, stop; otherwise repeat Step 1. ■

When the algorithm terminates, each vertex of the graph will be labelled with its distance from u_0. However, it should be pointed out that the actual shortest paths are not determined by Algorithm 7.1. Nevertheless, it is possible to determine the shortest paths from the labelled graph. Suppose we desire a shortest path from u_0 to some other vertex v. Starting at v, choose a vertex u such that its label differs from $L(v)$ by the weight $w(uv)$; that is, find u such that $L(u) = L(v) - w(uv)$. Then u is the predecessor of v on a shortest u_0-v path. If $u \neq u_0$, repeat this process to find a shortest path from u_0 to u.

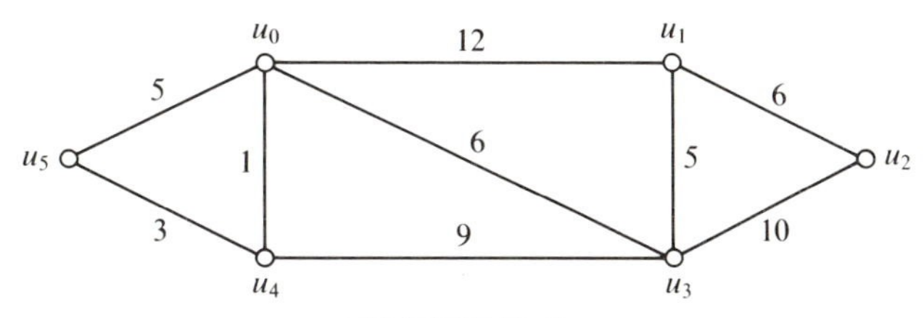

FIGURE 7.48

Example 7.25 Apply Dijkstra's algorithm to the graph G shown in Figure 7.48. Also, determine a shortest path from u_0 to u_2.

Solution In Step 0, set $i = 0$, $A_0 = \{u_0\}$, $L(u_0) = 0$, and $L(v) = \infty$ for $v \neq u_0$.

Step 1. Set $L(u_1) = 12$, $L(u_3) = 6$, $L(u_4) = 1$, and $L(u_5) = 5$. Among the vertices in A_0', the minimum label is $L(u_4) = 1$. Thus $A_1 = \{u_0, u_4\}$.

Step 2. Set $i = 1$. Since $A_1 \neq V(G)$, repeat Step 1.

Step 1. Set $L(u_5) = 4$ and $A_2 = \{u_0, u_4, u_5\}$.

Step 2. Set $i = 2$. Since $A_2 \neq V(G)$, repeat Step 1.

Step 1. No labels change. Since the minimum label among the vertices in A_2' is $L(u_3) = 6$, we have $A_3 = \{u_0, u_3, u_4, u_5\}$.

Step 2. Set $i = 3$. Since $A_3 \neq V(G)$, repeat Step 1.

Step 1. Set $L(u_1) = 11$ and $L(u_2) = 16$. Thus $A_4 = \{u_0, u_1, u_3, u_4, u_5\}$.

Step 2. Set $i = 4$. Since $A_4 \neq V(G)$, repeat Step 1.

Step 1. The label on u_2 does not change since $L(u_1) + w(u_1u_2) = 17$. Now $A_5 = V(G)$, so in Step 2 we set $i = 5$ and the algorithm terminates. The resulting labelled graph is shown in Figure 7.49.

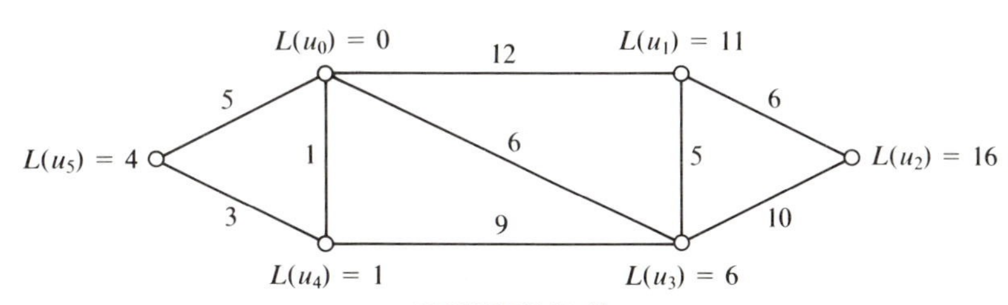

FIGURE 7.49

Now suppose we wish to find a shortest u_0-u_2 path. Since $L(u_3) = 6 = L(u_2) - w(u_2u_3)$, the vertex u_3 must precede u_2 on such a path. Hence we must determine a shortest u_0-u_3 path. Noting that $L(u_0) = 0 = L(u_3) - w(u_0u_3)$, this path is P': u_0, u_3. Therefore a shortest u_0-u_2 path is P: u_0, u_3, u_2. ■

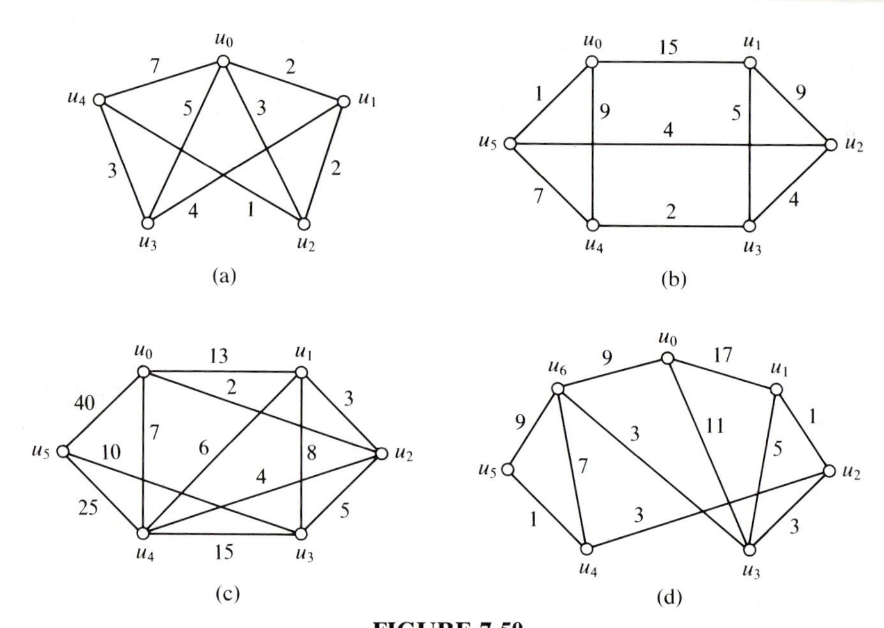

FIGURE 7.50

Exercises 7.7

1. Apply Dijkstra's algorithm to each of the graphs in Figure 7.50.
2. Let A be a proper subset of $V(G)$ and let $u_0 \in A$. If P: $u_0, \ldots, u_r, u_{r+1}$ is a shortest path from u_0 to A', show the following.
   a. $u_{r+1} \in A'$
   b. $u_0, u_1, \ldots, u_r \in A$
   c. P': $u_0, u_1, \ldots, u_r$ is a shortest u_0-u_r path.
3. For each of the graphs in Figure 7.50,
   a. find shortest paths from u_0 to the other vertices;
   b. find the subgraph induced by those edges that belong to one of these shortest paths. What kind of a subgraph is it?
4. During application of Dijkstra's algorithm, explain precisely the condition(s) under which the label on some vertex changes. Can this observation be used to make the algorithm more efficient?
5. Algorithm 7.1 may be modified so that it finds, in a sense, shortest paths from u_0 to the other vertices. In addition to the label $L(v)$, we may also associate with each vertex $v \neq u_0$ another vertex, Pred(v), which is the predecessor of v on the (current) shortest u_0-v path. When the algorithm terminates, a shortest v-u_0 path is then P: v, Pred(v), Pred(Pred(v)), $\ldots, u_0$.
   a. Modify Algorithm 7.1 so that it computes Pred(v) for each vertex $v \neq u_0$. (Hint: Pred(v) changes whenever $L(v)$ changes.)
   b. Apply the modified algorithm to each of the graphs of Exercise 1.

6. Let G be a weighted graph with vertex set $V(G) = \{u_0, u_1, \ldots, u_n\}$, such that no two edges of G have the same weight.

  a. Show that it is not always the case that the edge of smallest weight is used in some shortest u_0-u_i path.

  b. Give a sufficient condition under which the edge of smallest weight must be used in some shortest u_0-u_i path.

  c. For each $i = 1, 2, \ldots, n$, is it true that there is a unique shortest u_0-u_i path in G?

7.8 MINIMUM SPANNING TREES

We now return to the problem of finding a minimum spanning tree in a connected weighted graph G. Recall that this means that we are seeking, among all spanning trees of G, one of minimum total weight. Minimum spanning trees provide optimal solutions to problems in transportation networks, as was explained in Section 7.4.

In working with spanning trees, there is a particularly useful graphical operation. Let G be a graph and let u and v be nonadjacent vertices of G. Then $G + uv$ denotes the graph obtained by adding the edge uv to the graph G. Note that G is a proper spanning subgraph of $G + uv$. Now let T be a spanning tree of G and let $e \in E(G) - E(T)$. It can be shown (see Exercise 2) that the graph $T + e$ contains a unique cycle. If f is an edge of this cycle, $f \neq e$, then $T' = (T + e) - f$ is another spanning tree of G. By repeating this operation a finite number of times, it is possible to transform a given spanning tree T into any other spanning tree T_1 of G (see Exercise 4).

Example 7.26 Find all minimum spanning trees of the weighted graph G shown in Figure 7.51.

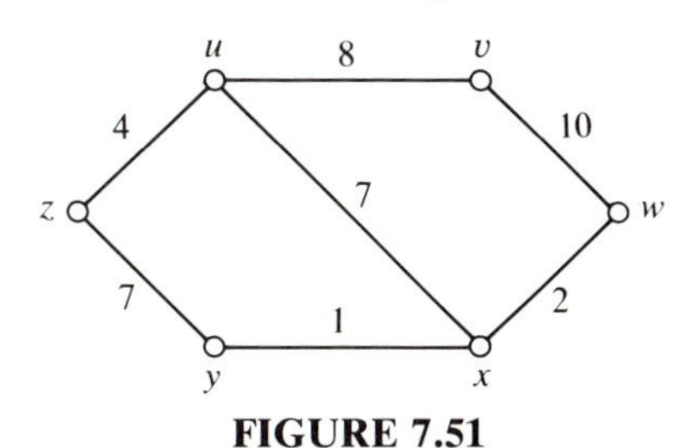

FIGURE 7.51

Solution Note that G has seven edges, while a spanning tree of G has five edges. Thus to obtain a spanning tree, we must remove two edges of G. We would like to remove the two edges of largest weight; however, $H = (G - uv) - vw$ is not acyclic. Next try $T = (G - vw) - yz$. This is a spanning tree and it is not hard to see that T is, in fact, a minimum spanning tree with $w(T) = 22$. Now, if we could replace an edge of T with one of equal weight, in such a way that the resulting subgraph is acyclic, then we would obtain another minimum spanning tree T'. Note that the graph $T + yz$ contains the unique cycle C: u, x, y, z, u. On this cycle, the edge ux has the same weight as the edge yz. Thus $T' = (T + yz) - ux$ is also a minimum spanning tree of G. Finally, it is not hard to verify that T and T' are the only minimum spanning trees of G. ■

Suppose that we are given a connected weighted graph G of order p and size q, and we wish to find a minimum spanning tree of G. There exist two well-known algorithms for this problem; one was discovered by J. B. Kruskal, Jr., in 1956 and the other by R. C. Prim in 1957. Since Prim's algorithm involves steps that closely resemble those used in Dijkstra's algorithm for finding shortest paths, we shall present Kruskal's algorithm.

ALGORITHM 7.2 (Kruskal) To find a minimum spanning tree T of a weighted connected graph G of order $p > 2$, perform the following steps:

Step 1 Choose, as the first edge of T, an edge e_1 in $E(G)$ of minimum weight. Set $t = 1$ and $A = \{e_1\}$.

Step 2 Let $A = \{e_1, \ldots, e_t\}$ be the set of edges of T that have been chosen so far, and let $A' = E(G) - A$. Let B denote the set of those edges e in A' such that $A \cup \{e\}$ induces an acyclic subgraph of G. For the next edge e_{t+1} of T, choose an edge in B of minimum weight. Replace A by $A \cup \{e_{t+1}\}$ and t by $t + 1$.

Step 3 If $t = p - 1$, stop; otherwise repeat Step 2. ■

Example 7.27 Apply Kruskal's algorithm to the graph G shown in Figure 7.52.

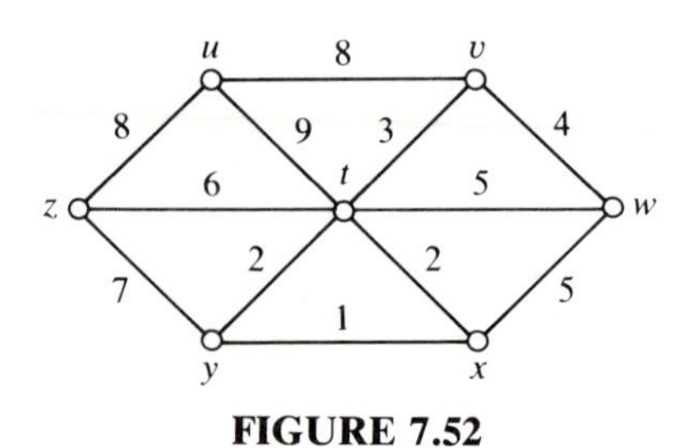

FIGURE 7.52

Solution At Step 1 we choose $e_1 = xy$. At Step 2 we have a choice: we may choose either ty or tx. Suppose we choose $e_2 = ty$. At the next iteration, the edge of smallest weight not yet chosen is tx, but $\langle\{xy, ty, tx\}\rangle$ is a cycle. So we must instead choose $e_3 = tv$. At the next stage we choose $e_4 = vw$. Neither tw nor wx can be chosen as e_5, for including either of these edges would produce a cycle. We are forced to choose $e_5 = tz$. Now the edge yz would complete a cycle, so there are two possible choices for e_6: uz or uv. Let us choose uz. This produces the minimum spanning tree $T_1 = \langle\{xy, ty, tv, vw, tz, uz\}\rangle$. Remembering the choices that we made earlier, we can at this stage note that other minimum spanning trees are $T_2 = (T_1 + tx) - ty$, $T_3 = (T_1 + uv) - uz$, and $T_4 = (T_2 + uv) - uz$. ■

Some pertinent questions arise in connection with Kruskal's algorithm. For one, how do we check that the edge-induced subgraph $T = \langle\{e_1, e_2, \ldots, e_{t+1}\}\rangle$ is acyclic at each stage? For another, how do we know that the algorithm does indeed produce a minimum spanning tree?

To check that T is acyclic, given that $T' = T - e_{t+1}$ is acyclic, consider the incident vertices of e_{t+1}, say $e_{t+1} = uv$. Bear in mind that the components of T' are trees. Thus if u and v belong to the same component of T', then $T' + uv$ contains a cycle. However, if u and v belong to different components of T', then $T' + uv$ is acyclic (see Exercise 6). So the problem of checking that T is acyclic reduces to checking that u and v belong to different components of T'. This is accomplished by labelling the vertices of G so that, after each execution of Step 2, two vertices have the same label if and only if they belong to the same component of T.

If $V(G) = \{u_1, u_2, \ldots, u_p\}$, we initially give u_i the label i for $i = 1, 2, \ldots, p$. In Step 1 an edge e_1 of minimum weight is chosen, and its incident vertices are relabelled with the same label. We agree to use the smaller of the two labels as the common label. Suppose now that after t iterations we have an acyclic graph $T = \langle\{e_1, e_2, \ldots, e_t\}\rangle$. In Step 3 the check on t is made, and if $t < p - 1$, then Step 2 is repeated. In Step 2 we choose an edge $e \in A'$, say $e = uv$. If u and v have the same label (so u and v belong to the same component of T), then $e \notin B$, and so we consider another edge $f \in A'$. In a finite number of steps we will have determined B, and can then choose an edge $e_{t+1} = xy$ in B of minimum weight. Note that x and y do not have the same label. At this point the edge e_{t+1} is added to T, and the vertices in that component of T containing x and y are relabelled with the smaller of the two labels on x and y.

That Kruskal's algorithm produces a spanning tree is clear, since the resulting graph has order p, size $p - 1$, and is acyclic. The next result states that the tree T produced by the algorithm is actually a minimum spanning tree.

THEOREM 7.17 Let G be a connected weighted graph. If T is a spanning tree of G that is produced by Kruskal's algorithm, then T is a minimum spanning tree of G.

Proof Let T_0 be a minimum spanning tree of G. If $T = T_0$ we are done, so assume $T \neq T_0$. Then there is an edge e in T such that e is not an edge of T_0. It follows that the subgraph $T_0 + e$ contains a unique cycle, say C. Since T is a tree there must be some edge of C, say f, such that $f \notin E(T)$. Now consider the subgraph $T_1 = (T_0 + e) - f$. By our earlier remarks, T_1 is also a spanning tree of G and its weight is

$$w(T_1) = w(T_0) + w(e) - w(f)$$

Since T_0 is a minimum spanning tree, it must be that $w(e) \geq w(f)$. In addition, $T' = (T + f) - e$ is a spanning tree of G. If $w(f) < w(e)$, then this would contradict the choice of e during the application of Kruskal's algorithm that produced T. Hence we must conclude that $w(e) = w(f)$, so that $w(T_1) = w(T_0)$. Therefore, T_1 is a minimum spanning tree of G. Moreover, the trees T and T_1 have one more edge in common, namely e, than do T and T_0. If $T = T_1$ we are done; otherwise the above replacement process may be repeated to produce a minimum spanning tree T_2 that has an additional edge in common with T. Thus, after a finite number of steps, we shall obtain a minimum spanning tree T_n such that $T_n = T$. Therefore, T is a minimum spanning tree of G. ■

Exercises 7.8

1. Apply Kruskal's algorithm to find a minimum spanning tree of the graph G shown in Figure 7.48.
2. Let T be a spanning tree of the connected graph G and let $e \in E(G) - E(T)$. Show that the graph $T + e$ contains a unique cycle.
3. Apply Kruskal's algorithm to each of the graphs of Figure 7.50.
4. Let T be a spanning tree of the connected graph G, let $e \in E(G) - E(T)$, and let $f \neq e$ be an edge on the unique cycle in $T + e$. Let $T' = (T + e) - f$.
   a. Show that T' is a spanning tree of G.
   b. If G has order p, how many edges do T and T' have in common?
   We call the above operation, which produced T' from T, *edge replacement*.
   c. Let T_1 and T_2 be two different spanning trees of G. Show that it is possible to obtain T_2 from T_1 by performing a finite number of edge replacements.
5. True or false: If no two edges of the connected weighted graph G have the same weight, then G contains a unique minimum spanning tree.
6. Let F be a forest and let u and v be nonadjacent vertices of F.
   a. Show that the graph $F + uv$ is a forest if and only if u and v belong to different components of F.
   b. If F has order p and size q, how many edges must be added to F to produce a tree T?
7. Show that, in Kruskal's algorithm, once an edge $e \in A'$ is such that $e \notin B$, then it need not be given further consideration (as a possible

edge of the minimum spanning tree to be produced). Incorporate this observation into Algorithm 7.2.

8. What would happen if Kruskal's algorithm were applied to a disconnected graph?

9. Here is Prim's algorithm to find a minimum spanning tree T of the connected weighted graph G, where $V(G) = V$:

Step 0. Initially, set $E(T) = \phi$ and $U = \{u_0\}$, where u_0 is any vertex of G.

Step 1. Find the edge $e = uv$ of smallest weight such that $u \in U$ and $v \in V - U$. Replace $E(T)$ by $E(T) \cup \{e\}$ and U by $U \cup \{v\}$.

Step 2. If $U = V$, stop; otherwise repeat Step 1.

a. Show that Prim's algorithm produces a spanning tree.

★**b.** Show that Prim's algorithm produces a minimum spanning tree.

c. Apply Prim's algorithm to the graph G of Figure 7.48.

d. Apply Prim's algorithm to the graph G of Figure 7.52.

10. Let G be a connected weighted graph and let T be a minimum spanning tree of G. Is it always possible, by making the proper choices of edges, for Kruskal's algorithm to produce T? In other words, is every minimum spanning tree of G obtainable from Kruskal's algorithm?

7.9 TRAFFIC FLOW

We now return to the traffic flow problem, in which we are presented with a network of two-way streets in a given city and faced with the task of making one-way street assignments for each street in such a way that it is possible to travel from any one intersection to any other. As was mentioned in Section 7.5, this situation can be modeled by a graph G if we represent each intersection by a vertex and join two vertices a and b with an edge provided it is possible to travel between a and b without passing through a third intersection. The problem then becomes one of assigning directions to the edges of G so that the resulting digraph is strongly connected. Such a digraph is termed a *strong orientation* of G. In what follows, we will use the term "orient" in place of "assign a direction to."

If the graph G is to have a strong orientation, then it is clear that G must be connected. More than this, G can have no bridges. For suppose G is connected and has a bridge $e = uv$, and the edges of G have been oriented so that e is directed from u to v. The resulting digraph will then contain no v-u path. The reason for this is that the graph $G - e$ is disconnected and therefore has no v-u path. Thus, if G is to possess a strong orientation, then G must be a connected graph with no bridges. What is indeed interesting is that the converse of this result also holds.

THEOREM 7.18 A graph G has a strong orientation if and only if G is connected and has no bridges.

Proof In view of the discussion preceding the theorem, we need only prove the "if" part of the theorem. Thus we assume that G is connected and has no bridges; we must show that G has a strong orientation.

Since the graph consisting of a single vertex, K_1, has a strong orientation, we may consider each subset U of $V(G)$ such that $\langle U \rangle$ has a strong orientation; let us choose such a U having maximum cardinality, and let $H = \langle U \rangle$. If $H = G$ we are done, so suppose, on the contrary, that $V(H)$ is a proper subset of $V(G)$. Choose vertices u and v of G so that $u \in V(H)$ and $v \notin V(H)$. Also, let D be a strong orientation of H. Now, since G is connected and contains no bridges, there exist two edge-disjoint u-v paths in G (see Exercise 2), say,

$$P_1: \quad u = u_0, u_1, \ldots, u_n = v$$

and

$$P_2: \quad v = v_0, v_1, \ldots, v_m = u$$

Let i be the largest subscript for which u_i is in $V(H)$ and let j be the smallest subscript for which v_j is in $V(H)$. Also, let Q_1 and Q_2 be the paths

$$Q_1: \quad u_i, u_{i+1}, \ldots, u_n = v$$

and

$$Q_2: \quad v = v_0, v_1, \ldots, v_j$$

Orient the edges of Q_1 from u_i to v and orient those of Q_2 from v to v_j, as shown in Figure 7.53. Let Q_1' and Q_2' be the resulting directed u_i-v and v-v_j paths, respectively. Let D_1 be the digraph whose vertex set is $V(D) \cup V(Q_1') \cup V(Q_2')$ and whose arcs are those of D, Q_1', and Q_2'. It is then a simple exercise to show that D_1 is strongly connected (see Exercise 4). Thus the graph $\langle V(D_1) \rangle$ has a strong orientation, which is a contradiction since $V(D_1)$ has cardinality greater than that of $U = V(D)$. Therefore, it must be that $V(D) = V(G)$, so that G has a strong orientation.

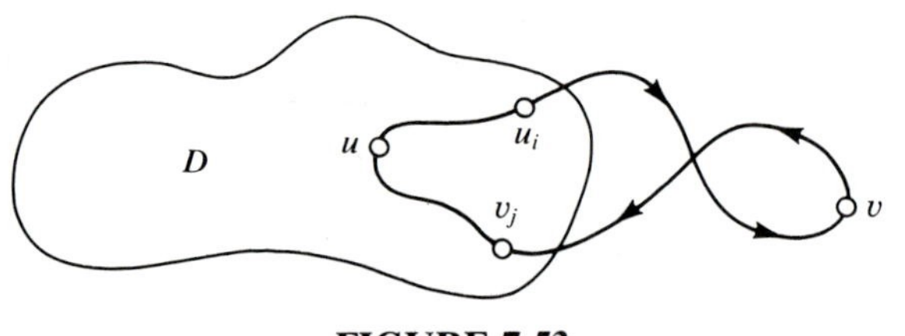

FIGURE 7.53 ∎

Given a bridgeless connected graph G, how do we obtain a strong orientation of G? The proof of Theorem 7.18 is an "existence proof" in that the proof concentrates on showing that a strong orientation of G

exists, rather than indicating a method for finding one. We shall now describe an algorithm, first discovered by J. E. Hopcroft and R. Tarjan in 1971, for finding a strong orientation of a connected bridgeless graph. This algorithm is an example of a more general class of algorithms, called *depth-first search* algorithms, that are widely used in studying graphs.

ALGORITHM 7.3 (Hopcroft and Tarjan) Let G be a connected bridgeless graph. To find a strong orientation $D = (V, A)$ of G, perform the following steps:

Step 1 (Initialize) Choose an arbitrary vertex x and give it the label $L(x) = 1$. (The vertex x is called the *root*.) Set $V = \{x\}$ and $U = V(G) - \{x\}$. (U is the set of unlabelled vertices.)

Step 2 Let v be that vertex in V having the highest label for which there is some vertex u in U such that uv is an edge of G. Set $L(u) = L(v) + 1$, replace V by $V \cup \{u\}$ and U by $U - \{u\}$. Also replace A by $A \cup \{(v,u)\}$, that is, orient the edge uv from v to u.

Step 3 If $V \subset V(G)$, repeat Step 2.

Step 4 (At this point, each vertex of G has been labelled and each arc of A is oriented from a vertex to a vertex having a higher label. In fact, it can be shown that the arcs of A form a rooted tree (see Exercise 3). Moreover, it can be shown that no edge of G joins two vertices having the same label (see Exercise 5).) Orient all remaining edges of G from the vertex of higher label to the one with lower label. ■

It is certainly not obvious that Algorithm 7.3 always works! Some aspects of the proof that Algorithm 7.3 does indeed produce a strong orientation of a connected bridgeless graph will be explored in the exercises.

Example 7.28 Apply Algorithm 7.3 to the graph G shown in Figure 7.54.

Solution Suppose we choose x_1 as the root and label it $L(x_1) = 1$. In Step 2, $v = x_1$ and we choose $u = x_2$. Set $L(x_2) = 2$ and orient the edge x_1x_2 from x_1 to x_2.

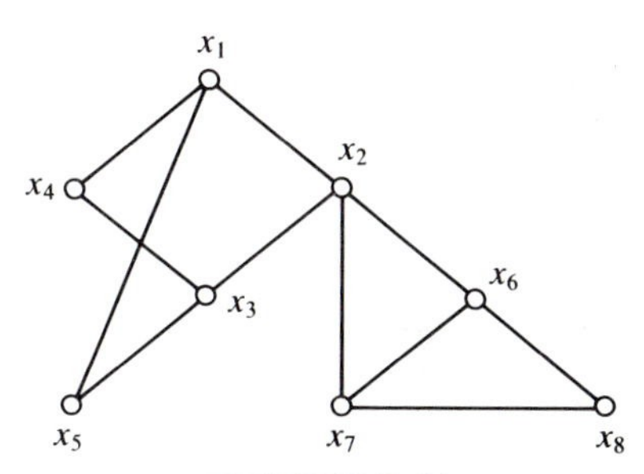

FIGURE 7.54

Then $V = \{x_1, x_2\}$ and $U = \{x_3, x_4, \ldots, x_8\}$. Since $V \neq V(G)$, we repeat Step 2. Now $v = x_2$ and u could be any one of x_3, x_6, or x_7. Choose $u = x_3$, set $L(x_3) = 3$, and orient x_2x_3 from x_2 to x_3. Then $V = \{x_1, x_2, x_3\}$ and $U = \{x_4, x_5, \ldots, x_8\}$. Again we repeat Step 2. Now $v = x_3$ and u could be either x_4 or x_5. Let $u = x_4$, set $L(x_4) = 4$, and orient x_3x_4 from x_3 to x_4. Thus $V = \{x_1, x_2, x_3, x_4\}$ and $U = \{x_5, x_6, x_7, x_8\}$. Repeating Step 2, since there are no unlabelled vertices adjacent to x_4, it must be that $v \neq x_4$. However, x_5 is unlabelled and x_3 is adjacent to x_5. Thus $v = x_3$ and $u = x_5$, with $L(x_5) = 4$, and x_3x_5 is oriented from x_3 to x_5. Now $V = \{x_1, x_2, \ldots, x_5\}$ and $U = \{x_6, x_7, x_8\}$. We repeat Step 2. At this point note that $v = x_2$ and u is either x_6 or x_7. Choose $u = x_6$, set $L(x_6) = 3$, and orient the edge x_2x_6 from x_2 to x_6. This gives $V = \{x_1, x_2, \ldots, x_6\}$ and $U = \{x_7, x_8\}$. Repeat Step 2. Now $v = x_6$ and u is either x_7 or x_8. Choosing $u = x_7$, we have $L(x_7) = 4$, with x_6x_7 oriented from x_6 to x_7, $V = \{x_1, x_2, \ldots, x_7\}$, and $U = \{x_8\}$. Then one final iteration of Step 2 will yield $L(x_8) = 5$, and the edge x_7x_8 will be oriented from x_7 to x_8. At this point we have the situation depicted in Figure 7.55. Since all vertices of G have been labelled, we now move on to Step 4. In this step the remaining edges of G, namely x_1x_4, x_1x_5, x_2x_7, and x_6x_8 are oriented from the vertex of higher label to the vertex having the lower label, resulting in the final digraph D that is shown in Figure 7.56. To see that D is strong, note that W: $x_1, x_2, x_6, x_7, x_8, x_6, x_7, x_2, x_3, x_5, x_1, x_2, x_3, x_4, x_1$ is a closed spanning walk of D.

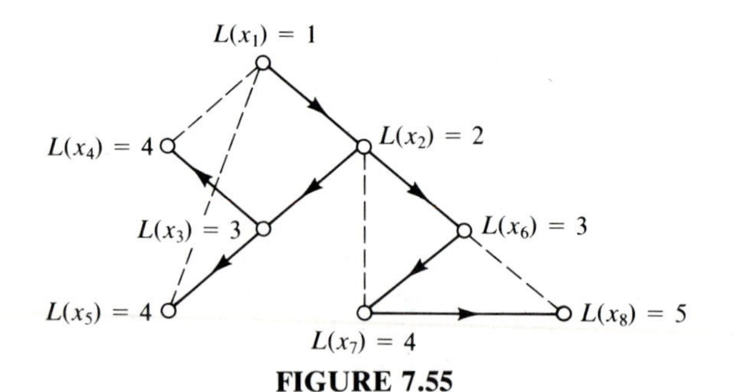

FIGURE 7.55

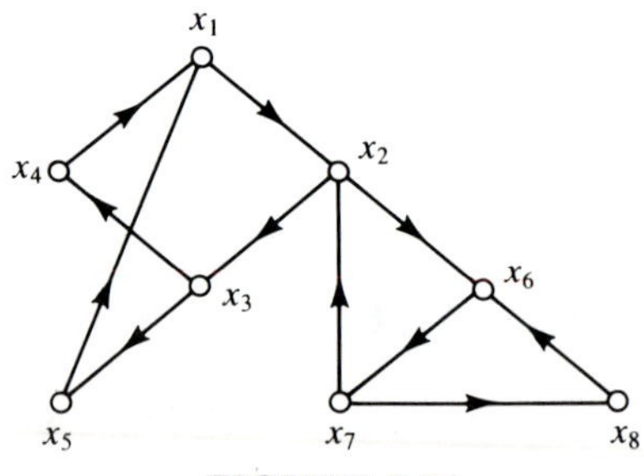

FIGURE 7.56 ■

Exercises 7.9

1. Apply Algorithm 7.3 to each of the graphs in Figure 7.57. (Show the label on each vertex and the orientation of each edge.)
2. Let G be a connected bridgeless graph and let u and v be distinct vertices of G. Show that G contains two edge-disjoint u-v paths.
3. Show that, in Steps 1 through 3 of Algorithm 7.3, a spanning rooted tree of the resulting digraph D is constructed.

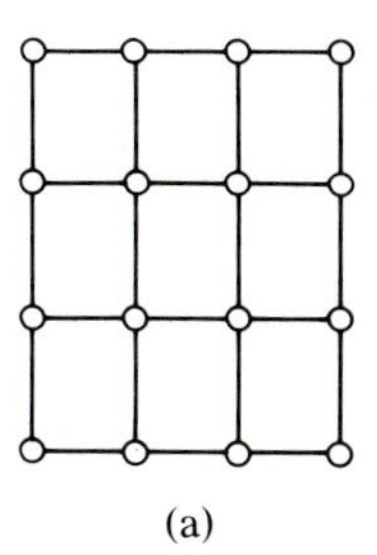

(a)

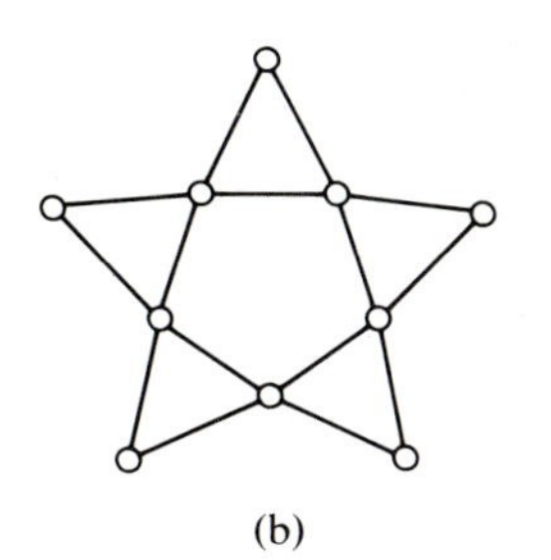

(b)

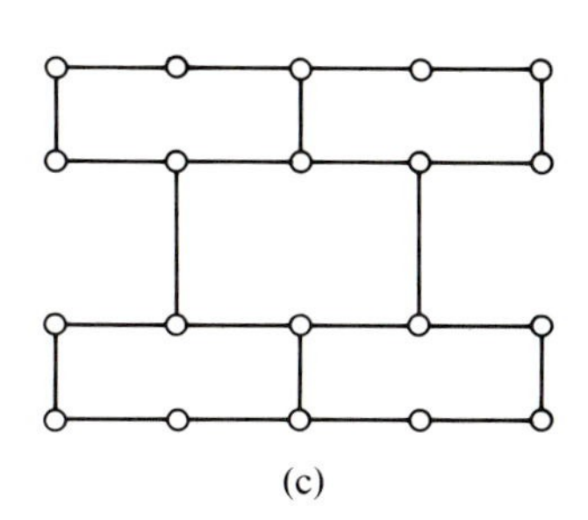

(c)

FIGURE 7.57

4. Show that the digraph D_1, constructed in the proof of Theorem 7.18, is strongly connected.

5. Show that, at the conclusion of Step 3 in Algorithm 7.3, no edge of G joins two vertices having the same label.

6. Prove or disprove: If the graph G is connected, then G has a unilateral orientation.

7. The *converse* of a digraph D is the digraph D^c obtained by reversing the orientation of each arc of D.

 a. Show that D is strong if and only if D^c is strong.

 b. It follows from part a that if G is a connected bridgeless graph and D is a strong orientation of G, then D^c is also a strong orientation of G. In this sense, every connected bridgeless graph has at least two strong orientations. Determine those graphs that have exactly two strong orientations.

8. Let G be a connected bridgeless graph.

 a. Develop several ways to define the ''efficiency'' of a given strong orientation of G.

 b. How efficient are the orientations found in Exercise 1?

9. In this exercise, we outline a proof that the digraph D constructed by Algorithm 7.3 is strongly connected.

 a. First, verify that every vertex of D is reachable from the root x.

 b. Thus, to complete the proof, it suffices to show that x is reachable from each vertex of D. This is equivalent to showing that every vertex of D other than x can reach one of its ancestors in the rooted tree T constructed by the first part of the algorithm. Why?

 c. Thus, let $v \neq x$ be a vertex of D and let u be the parent of v in T. Consider the subtree T_v of T formed by v and all of its descendants. Since the edge uv is not a bridge of G, there must be an edge wz of G joining a vertex w in T_v with a vertex z not in T_v. Why?

 d. To complete the proof, show that the arc (w,z) is in D and that z must be an ancestor of v in T. Hence, since v can reach w and w can reach z, v can reach z.

CHAPTER PROBLEMS

1. **a.** Given any graph G, show that there exists a graph H such that G is an induced subgraph of H and H is $\Delta(G)$-regular. (Hint: if G is not regular, consider two isomorphic copies of G; join a vertex with its "twin" if that vertex has degree less than $\Delta(G)$ in G.)
   b. Prove or disprove: Given any graph G, there exists a graph H such that G is a spanning subgraph of H and H is $\Delta(G)$-regular.
2. The *complement* $\bar{G}$ of a graph G is the graph whose vertex set is $V(G)$ and whose edge set is the complement of $E(G)$; in other words, two vertices are adjacent in $\bar{G}$ if and only if they are nonadjacent in G.
   a. Find the complements $\bar{G}_1$, $\bar{G}_2$, and $\bar{G}_3$ of the graphs in Figure 7.9.
   b. If G has order p and size q, determine the size of $\bar{G}$.
   c. If $d_1 \geq d_2 \geq \cdots \geq d_p$ is the degree sequence of G, find the degree sequence of $\bar{G}$. Thus, determine $\delta(\bar{G})$ and $\Delta(\bar{G})$ in terms of $\delta(G)$ and $\Delta(G)$.
   d. Show that it is not possible for both G and $\bar{G}$ to be disconnected.
3. A graph G is said to be *self-complementary* if G is isomorphic to its complement $\bar{G}$.
   a. Prove: If G is a self-complementary graph of order p, then $p \equiv 0$ or $1 \pmod 4$.
   ★b. Prove: If p is a positive integer and $p \equiv 0$ or $1 \pmod 4$, then there exists a self-complementary graph of order p.
4. Let m, n, and p be integers with $0 \leq m \leq n < p$. We are interested in the question, Does there exist a graph G of order p such that $\delta(G) = m$ and $\Delta(G) = n$? The answer is obviously no if $m = 0$ and $n = p - 1$, or if $m = n$ and m and p are both odd, so we shall exclude these cases.
   a. If $m = n$ and mp is even, then there exists an m-regular graph of order p. (See, for example, Section 8.2 of *Graphs and Digraphs*, by Behzad, Chartrand, and Lesniak-Foster.) Give examples of the following graphs.
      i. a 4-regular graph of order 9
      ii. a 5-regular graph of order 10
   b. Using the result stated in part a, answer the question in the case $m = n - 1$.
   c. Answer the question in the remaining cases.
5. Let $d_1 \geq d_2 \geq \cdots \geq d_p > 0$. Prove: If $d_1, d_2, \ldots, d_p$ is the degree sequence of some graph G of order p, then
   a. $d_1 + d_2 + \cdots + d_p$ is even;
   b. for $n = 1, 2, \ldots, p - 1$,

$$d_1 + \cdots + d_n \leq n(n-1) + \min\{n, d_{n+1}\} + \cdots + \min\{n, d_p\}.$$

   (Erdös and Gallai have shown that conditions a and b are also sufficient for $d_1 \geq d_2 \geq \cdots \geq d_p > 0$ to be the degree sequence of some graph.)

6. An *eulerian circuit* in a connected graph G is a circuit containing every edge of G. For example, the graph G of Figure 7.58 does not contain an eulerian circuit, but the graph $G + xv$ contains the eulerian circuit E: $x, z, s, w, t, v, w, y, u, x, t, z, v, x$. Prove that a connected graph G contains an eulerian circuit if and only if every vertex of G has even degree. (See Exercise 19 of Section 7.3.)

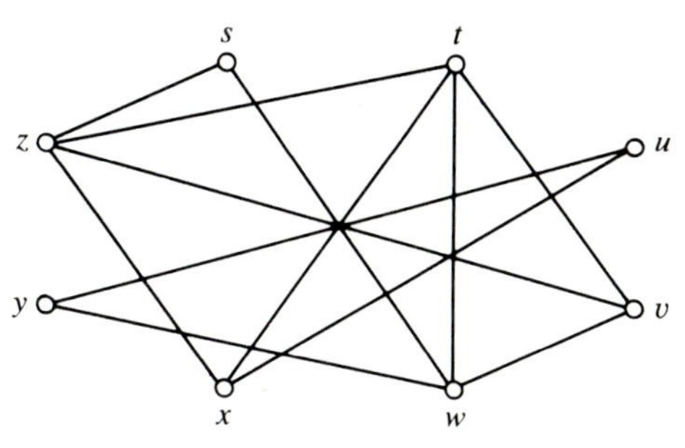

FIGURE 7.58

7. Let G be a connected graph. The *eccentricity* $e(v)$ of a vertex v is the maximum distance $d(u, v)$ from any vertex u to v in G.

a. Find the eccentricity of each vertex in the graph of Figure 7.59. The *radius* rad G and the *diameter* diam G of G are the minimum and maximum eccentricities among the vertices of G, respectively.

b. Find the radius and diameter of the graph G of Figure 7.59.

c. Show that for any connected graph G, rad $G \leq$ diam $G \leq 2$ rad G.

★**d.** Let m and n be positive integers such that $m \leq n \leq 2m$. Show that there exists a connected graph G with rad $G = m$ and diam $G = n$.

e. Let G be a nontrivial self-complementary graph. Prove that rad $G = 2$ and diam $G \leq 3$.

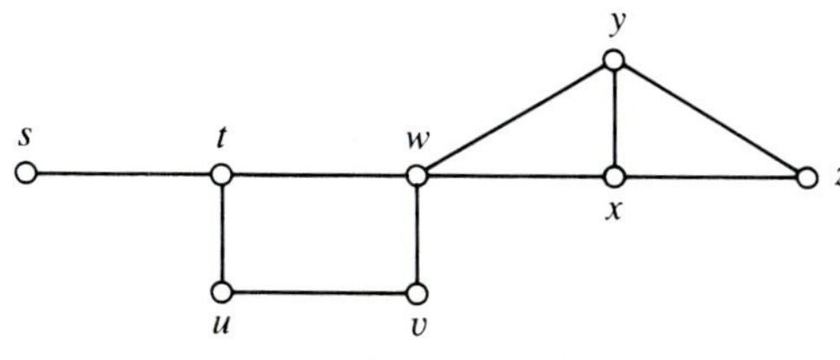

FIGURE 7.59

8. Let G be a graph of order $p \geq 3$, and let u and v be nonadjacent vertices of G such that $\deg u + \deg v \geq p$.

a. Prove that G is hamiltonian if and only if $G + uv$ is hamiltonian. (Hint: For the "if" part, assume $G + uv$ is hamiltonian, and let C be a spanning cycle of $G + uv$. If C does not contain the edge uv, then C is a spanning cycle of G. Thus we may assume that G contains a spanning u-v path, say P: $u = u_1, u_2, \ldots, u_p = v$. Use the fact that $\deg u + \deg v \geq p$ to show that there must be some i, $2 \leq i < p$, such that uu_i and $u_{i-1}v$ are both edges of G. Use these edges to construct a spanning cycle of G.)

Given G of order p, the *closure* of G is the graph $C(G)$ defined recursively as follows: if G contains nonadjacent vertices u and v such that $\deg u + \deg v \geq p$, then $C(G) = C(G + uv)$; otherwise, $C(G) = G$. (It can be shown that the operation of closure is well-defined.)

b. Show that G is hamiltonian if and only if $C(G)$ is hamiltonian. In particular, if $C(G)$ is complete, then G is hamiltonian. This result, due to J. A. Bondy and V. Chvátal, is one of the best sufficient conditions known for a graph to be hamiltonian.

c. Apply the result of part b to show that the graph G of Figure 7.58 is hamiltonian.

9. Suppose $V(G) = \{v_1, v_2, v_3, v_4, v_5\}$ and the subgraphs $G_i = G - v_i$, $1 \leq i \leq 5$, are isomorphic to the graphs shown in Figure 7.60. Deter-

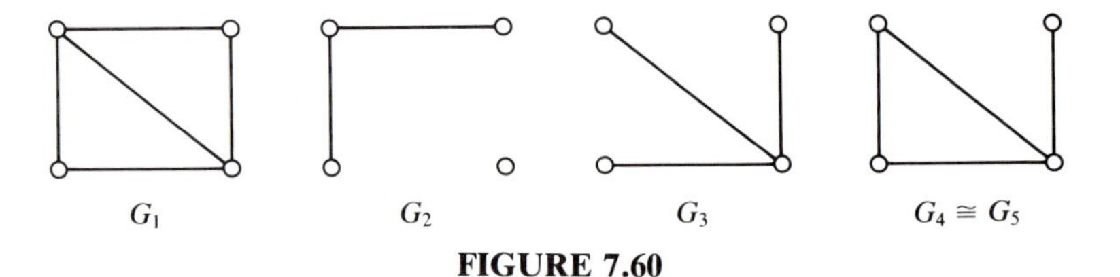

FIGURE 7.60

mine the graph G. (This is an example of a famous problem known as the reconstruction problem: Do the subgraphs $G - v$, where $v \in V(G)$, determine (up to isomorphism) the graph G?)

10. Prove that every tree of order $p \geq 4$ that is not a star is isomorphic to a subgraph of its complement.

11. Prove: If G is a graph with $\delta(G) \geq n - 1$ and T is any tree of order n, then G contains a subgraph isomorphic to T.

12. A graph G is said to be *unicyclic* if G is connected and contains a unique cycle. Prove that a graph G of order p and size q is unicyclic if and only if G is connected and $q = p$.

13. A tree T of order p is called *graceful* if the vertices of T can be assigned the numbers $1, 2, \ldots, p$ in such a way that the $p - 1$ absolute differences $|i - j|$ of the numbers assigned to adjacent vertices yield precisely the numbers $1, 2, \ldots, p - 1$. It is conjectured that every tree is graceful. Show that paths and stars are graceful.

14. When is a graph the graph of an equivalence relation?

15. Let G be a connected graph that is not a tree. The length of a shortest cycle in G is called the *girth* of G and is denoted $g(G)$. Consider the following problem: For $r \geq 2$ and $n \geq 3$, determine those graphs having the smallest order among all r-regular graphs with girth n; such graphs are called *(r,n)-cages*. (A result of Erdös and Sachs guarantees that there exists an (r,n)-cage for all $r \geq 2$ and $n \geq 3$.)

a. Show that the cycle C_n is the unique $(2,n)$-cage.
b. Show that the complete graph K_{r+1} is the unique $(r,3)$-cage.
c. Find the unique $(r,4)$-cage.
★d. Find a $(3,5)$-cage and a $(3,6)$-cage.

16. For positive integers m and n, the *ramsey number* $r(m,n)$ (named for the mathematician Frank Ramsey) is the least positive integer p such that every graph G of order p either contains a subgraph isomorphic to K_m (m mutually adjacent vertices) or contains a subgraph isomorphic to $\bar{K}_n$ (an independent set of n vertices). It can be remarked that $r(m,n) = r(n,m)$, $r(1,n) = 1$, and $r(2,n) = n$.

a. Show that $r(3,3) = 6$.
b. Prove, for $m \geq 2$ and $n \geq 2$, that $r(m,n) \leq r(m-1,n) + r(m,n-1)$. (This shows that the ramsey numbers exist.)
c. From part b it follows that $r(3,4) \leq r(2,4) + r(3,3) = 4 + 6 = 10$. Show, in fact, that $r(3,4) \leq 9$.
d. Show that $r(3,4) \geq 9$ by exhibiting a graph of order 8 that contains neither K_3 nor $\bar{K}_4$ as a subgraph.

From parts b and c it follows that $r(3,4) = 9$. The other known ramsey numbers are $r(3,5) = 14$, $r(3,6) = r(4,4) = 18$, and $r(3,7) = 23$.

17. For $n \geq 1$, the *nth power* of a connected graph G is that graph G^n with $V(G^n) = V(G)$ such that two vertices u and v are adjacent in G^n if and only if the distance from u to v in G is at most n.

a. Find G^2, G^3, and so on, for the graph G of Figure 7.59.
b. Prove: If G is a connected graph of order $p \geq 3$, then G^3 is hamiltonian. (Hint: it suffices to prove the following stronger result for trees: If T is a nontrivial tree and u and v are any two vertices of T, then T^3 contains a spanning u-v path.)
c. Give an example of a tree T of order 7 such that T^2 is not hamiltonian.

18. Prove: Every tournament T contains a vertex u with the property that $d(u, v) \leq 2$ for all $v \in V(T)$.

19. Consider expressions involving integer operands and the operators + (addition), * (multiplication), DIV, and MOD.

a. Find the expression represented by the binary tree of Figure 7.61. What is the value of the expression?

★ **b.** Given such a binary tree, develop an algorithm that finds the value of the expression.

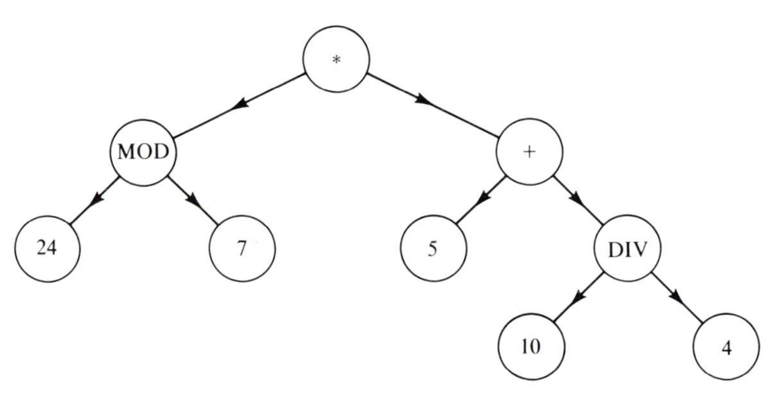

FIGURE 7.61

20. The 1984 NCAA basketball tournament began with 53 teams. In the first round, five games were played, thus eliminating five teams. In the second round, 16 games were played, eliminating 16 teams. The remaining 32 teams continued on to the next round of the tournament; from this point, half the remaining teams were eliminated at each round, until the winner of the tournament was determined.

a. Describe how such a single-elimination tournament can be represented as a binary tree.

b. What is the height of the binary tree representing the 1984 NCAA basketball tournament?

c. What was the total number of games played in the tournament?

A rooted tree of height h is called *balanced* if all leaves are at level h or at level $h - 1$.

d. Show that the binary tree that represents the 1984 NCAA basketball tournament is not balanced. Why do you think the tournament was designed in this way?

e. If T is a full m-ary tree of height h, such that T is balanced and has k leaves, show that $m^{h-1} < k \leq m^h$.

21. Let T be an m-ary tree, $m \geq 3$, with root x, and assume the children of each node of T are ordered from left to right. We may define a binary tree T' with $V(T') = V(T)$ as follows.

(i) The root of T' is x.
(ii) If q is a non-leaf vertex of T, then the leftmost child of q in T becomes the left child of q in T'.
(iii) If $q \neq x$ is not the rightmost child of its parent in T, then the sibling to the right of q in T becomes the right child of q in T'.

Using this method, it is possible to convert an m-ary tree T into a binary tree T' having the same vertex set.

a. Find the binary tree T' corresponding to the tree T of Figure 7.35.
b. When will a vertex of T be a leaf of T'?
c. If T has height h, give an upper bound on the height of T' (in terms of m and h).
d. Given T', is it possible to recover T?

22. Let G be a graph with $V(G) = \{v_1, v_2, \ldots, v_p\}$ such that $\deg v_1 \geq \deg v_2 \geq \cdots \geq \deg v_p$. Prove that

$$\chi(G) \leq \max_i\{\min\{i, \deg v_i + 1\}\}$$

23. **a.** Given integers k and p, $1 \leq k \leq p$, give an example of a graph G of order p such that $\chi(G) = k$ and $\chi(\bar{G}) = p + 1 - k$.
b. Prove the following result, due to Nordhaus and Gaddum: If G is a graph of order p, then $\chi(G) + \chi(\bar{G}) \leq p + 1$.

24. A graph G is called *k-minimal* if $\chi(G) = k$ and $\chi(G - e) = k - 1$ for every edge e of G.
a. Show that every connected k-minimal graph is k-critical.
b. Show that every k-critical graph, $k = 2$ or 3, is k-minimal.
c. Give an example of a 4-critical graph that is not 4-minimal.

25. Let $\omega(G)$ denote the order of the largest complete subgraph of G. Then clearly $\chi(G) \geq \omega(G)$. Interestingly, there exist graphs with $\omega = 2$ having arbitrarily large chromatic number. The following method for constructing such graphs is due to Mycielski. Let H be a graph having $V(H) = \{v_1, \ldots, v_p\}$, $\omega(H) = 2$, and $\chi(H) = k$. We construct a graph G from H by adding $p + 1$ new vertices $u, u_1, \ldots, u_p$; the vertex u is made adjacent to each u_i, and u_i is joined to each vertex v_j that is adjacent to v_i in H.
a. Show that $\omega(G) = 2$ and $\chi(G) = k + 1$.
b. Starting with $H \cong C_5$, apply the construction to obtain a graph G having $\omega(G) = 2$ and $\chi(G) = 4$.
c. Show that the graph G constructed in b is isomorphic to the Grötzsch graph of Figure 7.45(c).

26. Let G be a connected graph. Prove that $\chi(G) \leq k$ if and only if G has an orientation D such that the length of a longest path in D is at most $k - 1$.

27. Determine all values of $n \geq 2$ and $p_1, p_2, \ldots, p_n$ for which the complete n-partite graph $K_{p_1,p_2,\ldots,p_n}$ is planar.

28. Let G be a connected graph of order p and size q.

a. Prove: If $q \leq p + 2$, then G is planar.

b. For $p \geq 6$, give an example of a nonplanar connected graph of order p and size $p + 3$.

29. The *crossing number* $\nu(G)$ of a graph G is the minimum number of edge crossings among all representations of G in a plane (assuming that no more than two edges may cross at a point). Thus $\nu(G) = 0$ if and only if G is planar.

a. Show that $\nu(K_5) = 1$.

b. Show that $\nu(K_{3,3}) = 1$.

c. Show that $\nu(K_6) = 3$. (Hint: To show that $\nu(K_6) \leq 3$, find a representation of K_6 with three crossings. To show that $\nu(K_6) \geq 3$, suppose there is a representation of K_6 having k crossings. Form a plane graph G of order $k + 6$ and size $2k + 15$ by introducing a new vertex at each crossing. Now apply the result of Exercise 19 of Section 7.6.)

d. Show that $\nu(K_{4,4}) = 4$.

30. Let G be a planar graph having girth g.

a. Generalize the results of Exercises 19 and 20 of Section 7.6.

b. Prove: If $g \geq 4$, then $\chi(G) \leq 4$.

31. A *dominating set* of vertices in a graph G is a set J such that every vertex of G is either in J or is adjacent to a vertex in J. A *minimal dominating set* is a dominating set that does not properly contain any other.

a. Show that a maximal independent set in G is a minimal dominating set.

b. Give an example to show that a minimal dominating set in a graph may not be an independent set.

Let $\sigma(G)$ and $\beta(G)$ denote, respectively, the minimum cardinality of a dominating set and the maximum cardinality of an independent set in G. From part a it follows that $\beta(G) \geq \sigma(G)$.

c. For each nonnegative integer n, give an example of a graph G_n such that $\beta(G_n) - \sigma(G_n) = n$.

32. An *independent set of edges* in a graph is a set of mutually nonadjacent edges. Let $\beta_1(G)$ denote the maximum cardinality of an independent set of edges in the graph G.

a. Let G be a graph of order p with $\delta(G) \geq 1$. Prove that

$$\frac{p}{1 + \Delta(G)} \leq \beta_1(G) \leq \frac{p}{2}$$

★**b.** Prove the following result of König: If G is a bipartite graph, then $\beta_1(G) = \min(n(V_1))$, where the minimum is taken over all partitions (V_1, V_2) of $V(G)$ into partite sets.

c. Prove: If G is a k-regular bipartite graph of order p, then $\beta_1(G) = p/2$.

33. For each of the weighted graphs shown in Figure 7.62, find

a. shortest paths from x to the other vertices;

b. a minimum spanning tree;

c. a strong orientation (ignoring the weights).

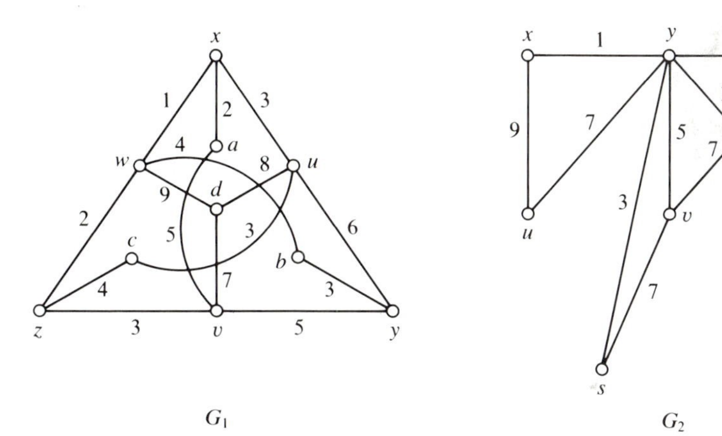

FIGURE 7.62

34. Adapt Kruskal's algorithm to solve the following problem: Let G be a connected weighted graph and let $E' \subseteq E(G)$; find the connected spanning subgraph of G of minimum weight that includes each edge of E'.

35. a. Describe an algorithm for finding a maximal independent set in a graph G.

b. Consider the following "algorithm for finding $\chi(G)$."

Step 1. Initially set $n = 0$ and $V = V(G)$.

Step 2. Find a maximal independent set U in $\langle V \rangle$. Replace n by $n + 1$ and V by $V - U$. If $V = \phi$, then $\chi(G) = n$; otherwise repeat Step 2.

Give an example to show that this algorithm will not find $\chi(G)$ in general.

c. What does the algorithm produce?

36. Here is an algorithm to find the maximal independent sets in a graph G. It employs two mutually recursive procedures, A and B.

Procedure A. To find the maximal independent sets in a graph G: If G is empty, then $V(G)$ is the only maximal independent set. If G is not empty, let uv be an edge of G. Then the maximal independent sets in G may be partitioned into two classes. Class I consists of those maximal independent sets in G that do not contain the vertex u. To find these, first call Procedure A recursively to find the maximal independent sets in $G - u$. If D is a maximal independent set in $G - u$, and u is adjacent to some vertex of D, then D belongs to Class I. Class II consists of those maximal independent sets in G that do contain the vertex u. To find these, call Procedure B to find the maximal independent sets in $G - v$ that contain u; each of these belongs to Class II.

Procedure B. To find the maximal independent sets in a graph H that include a specific vertex u:

If u has degree zero in H, call Procedure A to find the maximal independent sets in $H - u$; then add u to each of these. If u has degree at least 1 in H, let ux be an edge of H. Call Procedure B recursively to find the maximal independent sets in $H - x$ that contain u.

Apply the algorithm to find the maximal independent sets in the graph G of Figure 7.63.

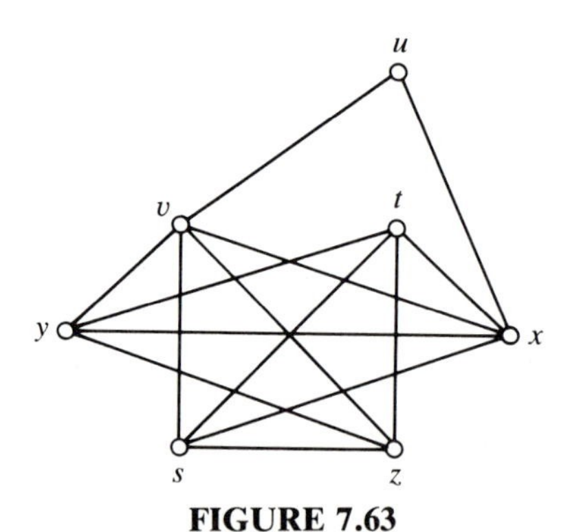

FIGURE 7.63

Answers to Selected Odd-Numbered Exercises and Problems

CHAPTER 1

Section 1.1

a, c, f, and **h** are statements, **d** and **g** are declarative sentences but are not statements, and **b** and **e** are not declarative sentences.

Section 1.2

1. a. $p \wedge q$ **b.** $p \vee r$ **c.** $p \to \sim q$ **d.** $p \leftrightarrow r$ **e.** $\sim(r \to p)$ **f.** $(q \vee r) \wedge \sim(q \wedge r)$

3. a. $p \vee q$, true **c.** $p \to q$, true **e.** $p \leftrightarrow q$, false

5. a. (triangle ABC is equilateral) $\to$ (triangle ABC is isosceles)

c. ($\pi^{\sqrt{2}}$ is real) $\to$ ($\pi^{\sqrt{2}}$ is rational $\vee$ $\pi^{\sqrt{2}}$ is irrational)

e. $(n > 200) \to ((n$ is prime$) \to (n > 210))$

g. (n is an integer) $\to$ (n is positive $\vee$ n is negative $\vee$ $n = 0$)

7. a.

Operator	*Precedence*
NOT	highest
*, /, DIV, MOD, AND	next highest
+, −, OR	next highest
<, <=, =, <>, >=, >	lowest

b. no; it will be interpreted as "A >= ((15 AND A) − B) >= 2"

c. yes; the parentheses cause (CURRENT <= LAST) to be evaluated first; then NOT FOUND is evaluated; then AND is applied

9. a. ASERVES AND AWINSVOLLEY **b.** (NOT ASERVES) AND (NOT AWINSVOLLEY)

c. (ASERVES AND (NOT AWINSVOLLEY)) OR ((NOT ASERVES) AND AWINSVOLLEY)
alternate solution: ASERVES <> AWINSVOLLEY **d.** ASERVES := NOT ASERVES

Section 1.3

1.

p	q	$\sim(p \wedge q)$	$\sim p \vee \sim q$
T	T	F	F
T	F	T	T
F	T	T	T
F	F	T	T

3.
$$\begin{aligned} q \to p &\equiv \sim q \vee p \\ &\equiv p \vee \sim q \\ &\equiv \sim(\sim p) \vee \sim q \\ &\equiv \sim p \to \sim q \end{aligned}$$

5. a. If a quadrilateral *ABCD* is not a rectangle, then the quadrilateral is not a parallelogram.
If a quadrilateral *ABCD* is a parallelogram, then the quadrilateral is a rectangle.
If a quadrilateral *ABCD* is not a parallelogram, then the quadrilateral is not a rectangle.

c. If quadrilateral *ABCD* is not a square, then either it is not a rectangle or it is not a rhombus.
If quadrilateral *ABCD* is both a rectangle and a rhombus, then it is a square.
If quadrilateral *ABCD* is not a rectangle or is not a rhombus, then it is not a square.

e. If a polygon *P* has the property that either it is both equiangular and not equilateral or it is both equilateral and not equiangular, then the polygon *P* is not a triangle.
If a polygon *P* is a triangle, then it has the property that it is equiangular if and only if it is equilateral.
If a polygon *P* is not a triangle, then it has the property that either it is both equiangular and not equilateral or it is both equilateral and not equiangular.

7. a. If both p implies q and $\sim q$ hold, then $\sim p$ holds. **b.** $((p \to q) \wedge \sim q) \to \sim p \equiv ((\sim p \vee q) \wedge \sim q) \to \sim p \equiv ((\sim p \wedge \sim q) \vee (q \wedge \sim q)) \to \sim p \equiv ((\sim p \wedge \sim q) \vee \text{F}) \to \sim p \equiv (\sim p \wedge \sim q) \to \sim p \equiv \sim(\sim p \wedge \sim q) \vee \sim p \equiv (p \vee q) \vee \sim p \equiv (p \vee \sim p) \vee q \equiv \text{T} \vee q \equiv \text{T}$

9. a.

p	q	r	$q \wedge r$	$p \vee q$	$p \vee r$	$p \vee (q \wedge r)$	$(p \vee q) \wedge (p \vee r)$
T	T	T	T	T	T	T	T
T	T	F	F	T	T	T	T
T	F	T	F	T	T	T	T
T	F	F	F	T	T	T	T
F	T	T	T	T	T	T	T
F	T	F	F	T	F	F	F
F	F	T	F	F	T	F	F
F	F	F	F	F	F	F	F

11. a. $(p \to q) \wedge (p \to \sim q) \equiv (\sim p \vee q) \wedge (\sim p \vee \sim q) \equiv \sim p \vee (q \wedge \sim q) \equiv \sim p \vee \text{F} \equiv \sim p$

c. $p \leftrightarrow q \equiv (p \to q) \wedge (q \to p) \equiv (q \to p) \wedge (p \to q) \equiv q \leftrightarrow p$

e. These are logically equivalent; use a truth table.

g. $(p \to q) \vee (p \to r) \equiv (\sim p \vee q) \vee (\sim p \vee r) \equiv \sim p \vee (q \vee r) \equiv p \to (q \vee r)$

Section 1.4

1. a. If p, then q.
$x = -2$ implies $x^3 = -8$.
$x = -2$ only if $x^3 = -8$.
$x = -2$ is sufficient for $x^3 = -8$.
$x^3 = -8$ is necessary for $x = -2$.

c. p is sufficient for q.
If one works hard, then one will pass this course.
Working hard implies passing this course.

One works hard only if one passes this course.
Passing this course is necessary for working hard.

e. q is necessary for p.
If triangle ABC is a right triangle, then the side lengths satisfy the Pythagorean theorem.
Triangle ABC being a right triangle implies that the side lengths satisfy the Pythagorean theorem.
Triangle ABC is a right triangle only if its side lengths satisfy the Pythagorean theorem.
Triangle ABC being a right triangle is sufficient for its side lengths to satisfy the Pythagorean theorem.

g. p is sufficient for q.
If a number's decimal expansion terminates, then the number is rational.
That a number's decimal expansion terminates implies that the number is rational.
A number's decimal expansion terminates only if the number is rational.
In order for a number's decimal expansion to terminate, it is necessary that the number be rational.

i. p only if q.
If one is to pass this course, then one must pass the final exam.
Passing this course implies passing the final exam.
Passing this course is sufficient for passing the final exam.
Passing the final exam is a necessary condition for passing this course.

Section 1.5

1. a. There exists an x such that x is prime.
c. There exists an x such that both x is prime and x is even.
e. There exists an x such that both x is prime and either x is even or $x \geq 3$.

3. a. $\forall n(n$ is an even integer $\rightarrow \exists m(m$ is an integer $\wedge\ n = 2m))$
c. $\exists T(T$ is a right triangle $\wedge\ T$ is isosceles$)$
e. $\exists x(x$ is an even integer $\wedge\ x$ is a prime$)$
g. $\exists x(x/2$ is an integer $\wedge\ \forall y(x/(2y)$ is not an integer$))$

5. a. $\forall x\ \forall y(x + y = y + x)$
$\exists x\ \exists y(x + y \neq y + x)$
There exist x and y such that $x + y \neq y + x$.
c. $\exists y\ \forall x((2x^2 + 1)/x^2 > y)$
$\forall y\ \exists x((2x^2 + 1)/x^2 \leq y)$
For every y there is an x such that $(2x^2 + 1)/x^2 \leq y$.
e. $\forall x\ \forall y\ \exists z(2z = x + y)$
$\exists x\ \exists y\ \forall z(2z \neq x + y)$
There exist x and y such that for all z, $2z \neq x + y$.

7. a. $\exists x(p(x) \vee q(x)$ is true$) \leftrightarrow p(x) \vee q(x)$ is true for some $x \leftrightarrow p(x)$ is true for some x or $q(x)$ is true for some $x \leftrightarrow \exists x\ p(x)$ is true $\vee\ \exists x\ q(x)$ is true
c. Use part a and the fact that $\exists x(p(x) \rightarrow q(x)) \equiv \exists x(\sim p(x) \vee q(x))$.

Section 1.6

1. a. $(\sim q \wedge (p \rightarrow q)) \rightarrow \sim p$; valid **c.** $((p \vee q) \wedge \sim q) \rightarrow p$; valid
e. $((p \rightarrow q) \wedge (p \rightarrow r) \wedge \sim r) \rightarrow \sim q$; invalid

3. Direct: If $x = y$, then $x \cdot x = y \cdot y$, so that $x^2 = y^2$.
Indirect: $x^2 \neq y^2 \rightarrow x^2 - y^2 \neq 0 \rightarrow (x + y)(x - y) \neq 0 \rightarrow x - y \neq 0 \rightarrow x \neq y$

5. Let M be the midpoint of side BC. Compare triangles AMB and AMC. These have side AM in common; sides MB and MC have the same length; and sides AB and AC have the same length by hypothesis. Hence, by side–side–side, triangles AMB and AMC are congruent. Therefore, angles ABM and ACM have the same measure.

7. a. If n is even, then $n = 2k$ for some integer k. Hence $n^2 = (2k)^2 = 4k^2 = 2(2k^2)$. Therefore, n^2 is even.

b. If n is odd (not even), then $n = 2k + 1$ for some integer k. Hence $n^2 = (2k + 1)^2 = 4k^2 + 4k + 1 = 2(2k^2 + 2k) + 1$. Therefore, n^2 is odd (not even).

9. Assume that x is real, that $x^3 + 4x = 0$, and that $x \neq 0$. Since $x^3 + 4x = 0$, by factoring, $x(x^2 + 4) = 0$. Thus either $x = 0$ or $x^2 + 4 = 0$; but $x \neq 0$, so $x^2 + 4 = 0$. This is a contradiction, since $x^2 + 4 > 0$ for all real numbers x.

11. Let $p = 11$.

Chapter 1 problems

1. a. Ralph reads The New York Times and watches the MacNeil–Lehrer Report implies Ralph jogs 3 miles.

c. If Ralph neither reads The New York Times nor watches the MacNeil–Lehrer Report, then Ralph jogs 3 miles.

e. If Ralph reads The New York Times, then either he watches the MacNeil–Lehrer Report or he jogs 3 miles.

g. Ralph reads The New York Times and he watches the MacNeil–Lehrer Report and he does not jog 3 miles.

3. a. Fred has taken CS 260 if and only if either Fred knows BASIC or Fred knows Pascal.

c. Fred has taken CS 260 and Fred does not know BASIC.

e. Either Fred knows both BASIC and Pascal or Fred has not taken CS 260.

5. a. $\sim p \vee \sim q \vee \sim r$ **c.** $p \wedge q \wedge \sim r$ **e.** $\sim(p \wedge q \wedge r)$ **g.** $\sim p \vee (q \wedge \sim r)$

7. Assume $(p \to q) \wedge (q \to r) \wedge (r \to p)$ holds. Then $((q \to r) \wedge (r \to p)) \to (q \to p)$, and $((r \to p) \wedge (p \to q)) \to (r \to q)$, and $((p \to q) \wedge (q \to r)) \to (p \to r)$. Therefore, $(p \leftrightarrow q) \wedge (q \leftrightarrow r) \wedge (r \leftrightarrow p)$ holds.

9. a.

p	q	$p \mid q$
T	T	F
T	F	T
F	T	T
F	F	T

b. $p \mid p \equiv \sim(p \wedge p) \equiv \sim p \vee \sim p \equiv \sim p$

c. $(p \mid q) \mid (p \mid q)$

d. $(p \mid p) \mid (q \mid q)$

11. a. If x is not odd, then x^2 is not odd.
If x^2 is odd, then x is odd.
If x^2 is not odd, then x is not odd.

c. If f is not differentiable at x, then f is not continuous at x.
If f is continuous at x, then f is differentiable at x.
If f is not continuous at x, then f is not differentiable at x.

e. If $F(x) \leq F(y)$, then $x \leq y$.
If $x > y$, then $F(x) > F(y)$.
If $x \leq y$, then $F(x) \leq F(y)$.

g. If f is not continuous at a, then f is not defined at a.
If f is defined at a, then f is continuous at a.
If f is not defined at a, then f is not continuous at a.

13. a. $(p \wedge q) \rightarrow r \equiv \sim(p \wedge q) \vee r \equiv (\sim p \vee \sim q) \vee r \equiv (\sim p \vee r) \vee \sim q \equiv \sim(p \wedge \sim r) \vee \sim q \equiv (p \wedge \sim r) \rightarrow \sim q$

b. If n is prime and n is even, then $n \leq 2$.
If $n > 2$ and n is even, then n is not prime.

c. If $f'(x) = 2x + 1$ and $f(x) \neq x^2 + x + 3$, then $f(0) \neq 3$.
If $f(0) = 3$ and $f(x) \neq x^2 + x + 3$, then $f'(x) \neq 2x + 1$.

15. a. $\sim(p \vee q)$ **b.** $\sim(p \leftrightarrow q)$ **c.** $\sim q$ **d.** $p \wedge \sim r$ **e.** $(\sim p \wedge q) \vee r$
f. $(p \vee q \vee r) \wedge \sim(p \wedge q \wedge r)$

17. a. $\forall y\ \exists x(xy < 3)$ **c.** $\exists x(x > 0 \wedge x^2 < x)$ **e.** $\exists x\ \forall y(x \geq y \vee f(x) < f(y))$
g. $\exists y(y > 0 \wedge \forall x(\log x \leq y))$

19. a. If the relation is not symmetric, then there exist x and y such that x is related to y and y is not related to x.

c. If the function f is not onto, then there exists a real number b such that, for all real numbers a, $f(a) \neq b$.

21. a. valid **c.** invalid **e.** valid

23. a. $x + \frac{1}{x} < 2 \rightarrow x + \frac{1}{x} - 2 < 0 \rightarrow \frac{x^2 - 2x + 1}{x} < 0 \rightarrow \frac{(x-1)^2}{x} < 0 \rightarrow$ x is not a positive real number

b. $\left(x > 0 \wedge x + \frac{1}{x} < 2\right) \rightarrow x^2 + 1 < 2x \rightarrow x^2 - 2x + 1 < 0 \rightarrow (x-1)^2 < 0$, which is a contradiction

25. If he wears white kid gloves, then he is not an opium eater.

CHAPTER 2

Section 2.1

1. a. $A = \{-3, -2, -1, 0, 1, 2, 3, 4\}$ **b.** $B = \{-1, 0, 2\}$

3. $A = B = D$ and $C = E$

Section 2.2

1. a. false **b.** true **c.** true **d.** false **e.** true **f.** false **g.** true **h.** true

3. a. ϕ **b.** $\phi, \{1\}$ **c.** $\phi, \{1\}, \{2\}, \{1, 2\}$
d. $\phi, \{1\}, \{2\}, \{3\}, \{1, 2\}, \{1, 3\}, \{2, 3\}, \{1, 2, 3\}$ **e.** $\phi, \{\phi\}, \{1\}, \{\phi, 1\}$
f. $\phi, \{\phi\}, \{1\}, \{\{1\}\}, \{\phi, 1\}, \{\phi, \{1\}\}, \{1, \{1\}\}, \{\phi, 1, \{1\}\}$

5. 2^n

7.

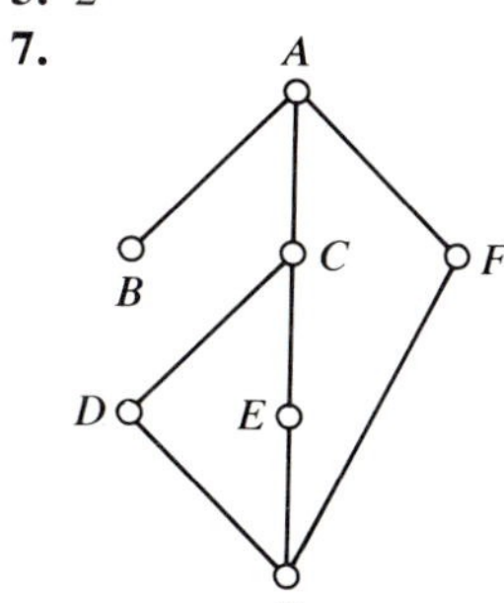

9. a. $A = \{1, 2\}, B = \{1\}, C = \phi$
b. $A = \{1, 2\}, B = \{1\}, C = \{2\}$
c. $A = \{1\}, B = \{2\}, C = \{3\}$
d. $A = \{1, 2\}, B = \{1\}, C = \{2\}, D = \phi$
e. $A = \{1, 2, 3\}, B = \{1\}, C = \{2\}, D = \{3\}, E = \phi$
f. $A = \{1, 2, 3\}, B = \{1, 2\}, C = \{1, 3\}, D = \{2\}, E = \{1\}, F = \phi$

Section 2.3

1.

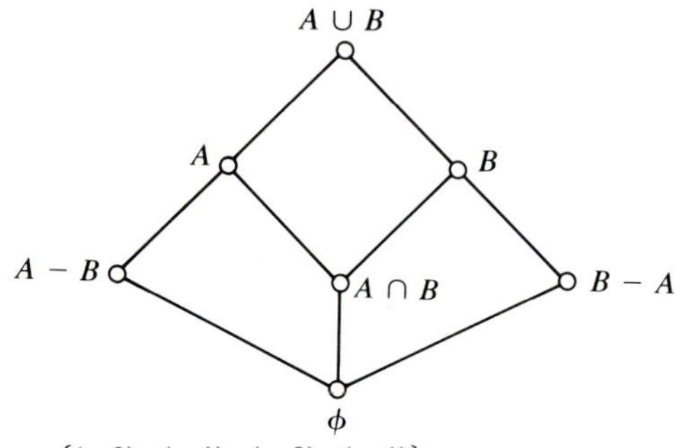

3. a. $\{\ldots, -12, -6, 0, 6, 12, \ldots\}$
c. $\{\ldots, -6, -2, 2, 6, \ldots\}$
e. ϕ
g. C

7. a. A **b.** B **c.** ϕ **d.** ϕ, ϕ
e. A, B **f.** $A = B$

9. a. $\{(x,0), (x,1), (y,0), (y,1)\}$
b. $\{(0,-1), (0,0), (0,1), (1,-1), (1,0), (1,1)\}$
c. $\{(x,0,-1), (x,0,0), (x,0,1), (x,1,-1), (x,1,0), (x,1,1), (y,0,-1), (y,0,0), (y,0,1), (y,1,-1), (y,1,0), (y,1,1)\}$
d. $\{((x,0),-1), ((x,0),0), ((x,0),1), ((x,1),-1), ((x,1),0), ((x,1),1), ((y,0),-1), ((y,0),0), ((y,0),1), ((y,1),-1), ((y,1),0), ((y,1),1)\}$
e. $\{(0,0,0,0), (0,0,0,1), (0,0,1,0), (0,0,1,1), (0,1,0,0), (0,1,0,1), (0,1,1,0), (0,1,1,1), (1,0,0,0), (1,0,0,1), (1,0,1,0), (1,0,1,1), (1,1,0,0), (1,1,0,1), (1,1,1,0), (1,1,1,1)\}$
f. $\{\phi, \{(0,0)\}, \{(0,1)\}, \{(1,0)\}, \{(1,1)\}, \{(0,0), (0,1)\}, \{(0,0), (1,0)\}, \{(0,0), (1,1)\}, \{(0,1), (1,0)\}, \{(0,1), (1,1)\}, \{(1,0), (1,1)\}, \{(0,0), (0,1), (1,0)\}, \{(0,0), (0,1), (1,1)\}, \{(0,0), (1,0), (1,1)\}, \{(0,1), (1,0), (1,1)\}, \{(0,0), (0,1), (1,0), (1,1)\}\}$

11. a. $A = B$ **b.** $A \cap B = \phi$

13. a. C := [2, 3, 5, 7, 11, 13, 17, 19, 23] **b.** C := [3..17] **c.** C := [1..25] − [7,14,21]
d. C := (A − B) + (B − A) or C := (A + B) − (A * B) **e.** C := C − (A + B)

15. a.
```
PROCEDURE FIND(WORD:STRING; VAR VOWS,CONS: SETTYPE);
  VAR LETTERS,VOWELS:SETTYPE; J:INTEGER;
  BEGIN
    LETTERS := ['A'..'Z'];  (*assume ASCII*)
    VOWELS := ['A','E','I','O','U'];
    VOWS := []; CONS := [];
    FOR J := 1 TO WORD.LENGTH DO
      IF WORD.VALUE[J] IN LETTERS THEN
        IF WORD.VALUE[J] IN VOWELS THEN VOWS := VOWS + [WORD.VALUE[J]]
        ELSE CONS := CONS + [WORD.VALUE[J]]
  END;
```
b. L1 := V1 + C1; L2 := C1 * C2; L3 := V1 − V2

17. a. `FOR J := MIN TO MAX DO C[J] := A[J] AND B[J]`
b. `FOR J := MIN TO MAX DO C[J] := A[J] OR B[J]`
c. `FOR J := MIN TO MAX DO C[J] := NOT A[J]`

Section 2.4

1. 3 **3. a.** 10^9 **b.** 9^9 **c.** $9^2 \cdot 10^7$
5. a. 41

b.

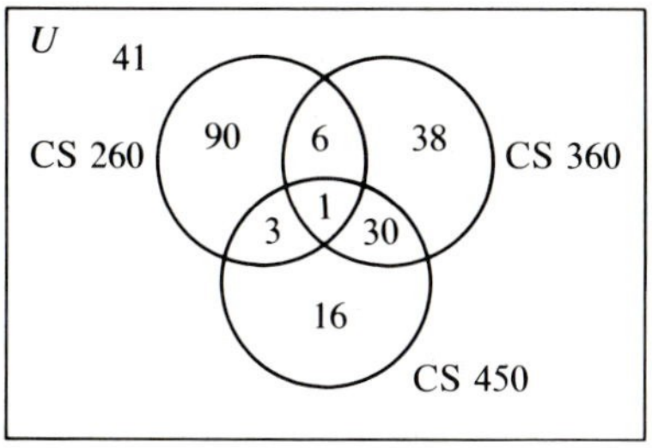

U 41
CS 260 90 6 38 CS 360
3 1 30
16
CS 450

7. a. 52^4 **b.** 39^4 **c.** 13^4 **d.** $4 \cdot 13^4$ **e.** $13 \cdot 48 \cdot 52^2$ **f.** $13 \cdot 52^3 + 48 \cdot 52^3 - 13 \cdot 48 \cdot 52^2$

9. a. 3^3

b.

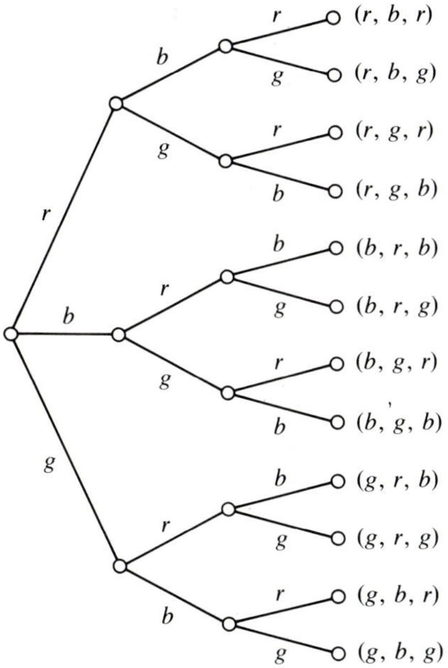

(r, b, r)
(r, b, g)
(r, g, r)
(r, g, b)
(b, r, b)
(b, r, g)
(b, g, r)
(b, g, b)
(g, r, b)
(g, r, g)
(g, b, r)
(g, b, g)

c.

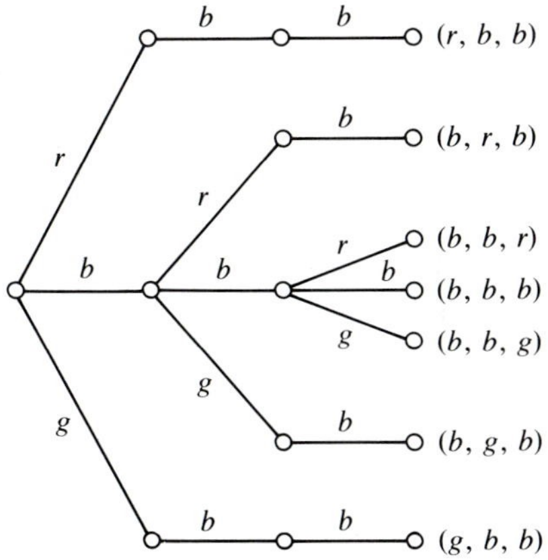

(r, b, b)
(b, r, b)
(b, b, r)
(b, b, b)
(b, b, g)
(b, g, b)
(g, b, b)

Section 2.5

1. $P(7, 3)$ **3. a.** $5!$ **b.** $P(5, 3) \cdot P(5, 2)$ **c.** $P(5, 3) \cdot 2^2$

5. a. $C(52, 13)$ **b.** $C(4, 2) \cdot C(48, 11)$ **c.** $C(4, 3) \cdot C(4, 2) \cdot C(4, 1) \cdot C(36, 7)$
d. $C(52, 13) - C(39, 13)$ **e.** $C(13, 4) \cdot C(13, 3)^3$ **f.** $C(4, 1) \cdot C(13, 4) \cdot C(13, 3)^3$
g. $C(52, 13) - C(39, 13) - C(39, 13) + C(26,13)$
h. $C(13, 6) \cdot C(39, 7) + C(13, 7) \cdot C(39, 6) - C(13, 7) \cdot C(13, 6)$

7. a. 6^4 **b.** $x = 1$: $C(6, 1)$; $x = 2$: $C(6,2) \cdot (2^4 - 2)$; $x = 3$: $C(6, 3) \cdot (3^4 - C(3, 2) \cdot 2^4 + C(3, 1))$;
$x = 4$: $C(6, 4) \cdot (4^4 - C(4, 3) \cdot 3^4 + C(4, 2) \cdot 2^4 - C(4, 1))$

9. $n \geq 10$

11. a. $C(15, 4) - C(11, 4)$
b. $C(4, 2) \cdot C(8, 1) \cdot C(3, 1) + C(4, 1) \cdot C(8, 2) \cdot C(3, 1) + C(4, 1) \cdot C(8, 1) \cdot C(3, 2)$
c. $C(3, 1) \cdot C(12, 3) + C(8, 2) \cdot C(7, 2) - C(3, 1) \cdot C(8, 2) \cdot C(4, 1)$

13. b.
```
FUNCTION COMBO(N,K:INTEGER):INTEGER;
  VAR RESULT,J:INTEGER;
  BEGIN
    RESULT := 1;
    FOR J := 1 TO K DO
      RESULT := (N - J + 1)*RESULT DIV J;
    COMBO := RESULT
  END;
```

Section 2.6

1. $\bigcap_{k=1}^{50} C_k = (\frac{-1}{50}, 2)$, $\bigcup_{k=1}^{50} C_k = (-1, 100)$

3. a. $\cap A_n = A_1$, $\cup A_n = \mathbb{Z}$ **b.** $\cap A_n = \{0\}$, $\cup A_n = (-1, 2)$ **c.** $\cap A_n = \phi$, $\cup A_n = \mathbb{N}$
d. $\cap A_n = \phi$, $\cup A_n = \{2m \mid m \in \mathbb{Z}\}$

5. a. $\cap A_i = \{1\}$; $\cup A_i = (0, \infty)$ **b.** $\cap A_i = \phi$; $\cup A_i = \mathbb{R} \times \mathbb{R}$
c. $\cap A_i = \{(1, 1, \ldots, 1)\}$; $\cup A_i = U - \{(0, 0, \ldots, 0)\}$
d. $\cap A_i =$ {identifiers declared in the main procedure, i.e., the "global" identifiers}; $\cup A_i = U$
e. $\cap A_i = B$; $\cup A_i = \mathcal{P}(B) - \{\phi\}$

Chapter 2 problems

3. a. $E' \cap D$ **b.** $E \cup B$ **c.** $E' \cap D \cap A$ **d.** $E \cap (C \cup D) \cap A'$

5. Hint: $A \times B \subseteq C \times D \leftrightarrow [(x,y) \in A \times B \rightarrow (x,y) \in C \times D] \leftrightarrow [(x \in A \wedge y \in B) \rightarrow (x \in C \wedge y \in D)]$

7. a. Hint: $(x,y) \in A \times (B \cap C) \leftrightarrow [x \in A \wedge y \in B \cap C] \leftrightarrow [x \in A \wedge y \in B \wedge y \in C]$

9. a. $X \in \mathcal{P}(A \cap B) \leftrightarrow X \subseteq A \cap B \leftrightarrow X \subseteq A \wedge X \subseteq B \leftrightarrow X \in \mathcal{P}(A) \wedge X \in \mathcal{P}(B) \leftrightarrow X \in \mathcal{P}(A) \cap \mathcal{P}(B)$
b. Let $A = \{1\}$, $B = \{2\}$. **c.** Let $A = \{1, 2\}$, $B = \{1\}$.

13. 19 students

15. a. $3!$ **b.** $C(5, 2) \cdot 3! = 5!/2!$ **c.** $5!/(2! \cdot 2!)$ **d.** $9!/(3! \cdot 2!)$ **e.** $\dfrac{n!}{n_1! \cdot n_2! \cdots n_m!}$

19. a. $C(13, 1) \cdot C(4, 2) \cdot C(12, 3) \cdot C(4, 1)^3$ **b.** $C(13, 1) \cdot C(4, 3) \cdot C(12, 2) \cdot C(4, 1)^2$
c. $C(13, 2) \cdot C(4, 2)^2 \cdot C(11, 1) \cdot C(4, 1)$ **d.** $C(13, 1) \cdot C(4, 3) \cdot C(12, 1) \cdot C(4, 2)$
e. $C(10, 1) \cdot C(4, 1)^5$ **f.** $C(4, 1) \cdot C(13, 5)$

21. a. $9!$ **b.** $5! \cdot 4!$ **c.** $4! \cdot 2^5$ **d.** $5! \cdot 5!$

23. a. $15!$ **b.** $3! \cdot 6! \cdot 5! \cdot 4!$

25. **a.** $C(4, 2) \cdot C(6, 1) \cdot C(8, 1)$ **b.** $C(3, 1) \cdot C(4, 1) \cdot C(5, 1) \cdot C(6, 1)$
c. $C(18, 4) - (1 + C(5, 4) + C(6, 4))$ **d.** $C(13, 4) + C(5, 1) \cdot C(13, 3) + C(5, 2) \cdot C(13, 2)$

29.
```
FUNCTION CARD(A:SETTYPE):INTEGER;
  VAR X:BASETYPE; COUNT:INTEGER;
  BEGIN
    COUNT := 0;
    FOR X := MIN TO MAX DO
      IF X IN A THEN COUNT := COUNT + 1;
    CARD := COUNT
  END;
```

31. **b.**
```
(*Assume the following declarations:
    CONST MIN = 1; MAX = 10; PMAX = 1024;
    TYPE  BASETYPE = MIN..MAX;
          SETTYPE = SET OF BASETYPE;
          POWERTYPE = ARRAY[1..PMAX] OF SETTYPE*)
   PROCEDURE POWERFIND(A:SETTYPE; VAR P:POWERTYPE; VAR T:INTEGER);
(*Procedure to find the power set P of the set A, where T is the cardinality of P*)
    VAR M:BASETYPE; J:INTEGER;
    BEGIN
      P[1] := []; T := 1;
      FOR M := MIN TO MAX DO
        IF M IN A THEN
          BEGIN
            FOR J := 1 TO T DO
              P[T + J] := P[J] + [M];
            T := 2 * T
          END
    END;
```

CHAPTER 3

Section 3.1

1. **a.** not well-ordered **b.** not well-ordered **c.** well-ordered **d.** well-ordered
e. well-ordered

3. **a.** 3 **b.** 2 **c.** 2 **d.** 1

Section 3.2

1. **a.** 27, 0 **b.** −7, 0 **c.** 9, 5 **d.** −12, 1 **e.** −7, 0 **f.** 6, 3

3. **a.** 0 **b.** 4 **c.** 1 **d.** 1 **e.** 1 **f.** 5

5. Hint: If m, $m + 1$, $m + 2$ are the integers, then $m = 3q + r$, where $r \in \{0, 1, 2\}$. Consider the cases for r.

Section 3.3

1. $d = 4$, $s = 25$, $t = -11$ **3.** $1 = (189)(-11) + (520)(4)$ **5.** $\gcd(a, b) = 1$

7. If m and $m + 1$ are the integers, then $1 = (m)(-1) + (m + 1)(1)$.

9. **a.** $\gcd(a, b) = \gcd(r_{i+1}, r_i) = 1$ or r_i
b. Check if $r_i \mid a$ and $r_i \mid b$.
c. 371 MOD 40 = 11, which is prime. Since $11 \nmid 40$, $\gcd(40, 371) = 1$.

11. a. Show that $11 \mid (n^2 + 1) \leftrightarrow n^2 \text{ MOD } 11 = 10$. Suppose $n \text{ MOD } 11 = m$, where $0 \le m \le 10$. Show that in each case $n^2 \text{ MOD } 11 \ne 10$.

b. $n = 13k + 5$ or $n = 13k + 8$, $k \in \mathbb{Z}$

13. Hint: First verify that $a \text{ MOD } m = b \text{ MOD } m \leftrightarrow m \mid (a - b)$. Then show $[n_1 \mid (a - b) \wedge n_2 \mid (a - b)] \to n \mid (a - b)$.

Section 3.4

1. a. $3^3 \cdot 5^2 \cdot 7$ **b.** $2 \cdot 3^2 \cdot 7^2 \cdot 11$ **c.** $5^4 \cdot 41$

3. a. $\gcd(a, b) = p_1^{\gamma_1} \cdot p_2^{\gamma_2} \cdots p_n^{\gamma_n}$, where $\gamma_i = \min\{\alpha_i, \beta_i\}$

b. $\text{lcm}(a, b) = p_1^{\delta_1} \cdot p_2^{\delta_2} \cdots p_n^{\delta_n}$, where $\delta_i = \max\{\alpha_i, \beta_i\}$

Section 3.5

1. Hint: Using the IHOP: $1 + 3 + \cdots + (2k - 1) = k^2$, we have that $1 + 3 + \cdots + (2k + 1) = k^2 + (2k + 1) = (k + 1)^2$.

3. Hint: The IHOP is that $6 \mid (k^3 + 5k)$. Then $(k + 1)^3 + 5(k + 1) = (k^3 + 5k) + 3k(k + 1) + 6$.

5. Hint: Using the IHOP: $a + ar + \cdots + ar^k = a(1 - r^{k+1})/(1 - r)$, we obtain $a + ar + \cdots + ar^{k+1} = a(1 - r^{k+1})/(1 - r) + ar^{k+1}$; now simplify.

9. Hint: Since $(2k + 3)^2 > 4(k + 1)(k + 2)$, it follows that $2\sqrt{k + 1} + 1/\sqrt{k + 1} > 2\sqrt{k + 2}$.

11. Hint: Use induction on b to prove that if execution reaches the top of the loop and the variables A, B, and P have the values a, b, and p, respectively, then the value $p + ab$ will be output.

Chapter 3 problems

1. $a \mid b \to at = b$ for some $t \in \mathbb{Z} \to a + at = c \to a(1 + t) = c \to a \mid c$.

3. $[m \mid (35n + 26) \wedge m \mid (7n + 3)] \to m \mid [(35n + 26) + (-5)(7n + 3)] \to m \mid 11$. Since $m \mid 11$ and $m > 1$, we must have $m = 11$.

5. $n \text{ DIV } d = q \to n = dq + r,\ 0 \le r < d \to (n - d) = d(q - 1) + r \to (n - d) \text{ DIV } d = q - 1$

7. Assume $n = m^2 = t^3$, where $m \text{ MOD } 7 = r$ and $t \text{ MOD } 7 = s$, $0 \le r, s < 7$. Then $m^2 \text{ MOD } 7 = r^2 \text{ MOD } 7 \in \{0, 1, 2, 4\}$ and $t^3 \text{ MOD } 7 = s^3 \text{ MOD } 7 \in \{0, 1, 6\}$. Thus $n \text{ MOD } 7 \in \{0, 1\}$.

9. Suppose $d = \gcd(a, b)$ and $e = \gcd(ca, cb)$. Then there exist $s, t, x, y \in \mathbb{Z}$ such that $d = as + bt$ and $e = cax + cby$. Now $(e \mid ca \wedge e \mid cb) \to e \mid (cas + cbt) \to e \mid cd$; also $(d \mid a \wedge d \mid b) \to d \mid (ax + by) \to cd \mid (cax + cby) \to cd \mid e$. Thus $e = cd$.

11. a. $17 = (357)(-7) + (629)(4)$ **b.** $1 = (1109)(-2353) + (4999)(522)$

13. $d = \gcd(a, b) \to$ there exist integers s, t such that $d = as + bt \to cd = acs + bct$. Then $(a \mid c \wedge b \mid c) \to (ab \mid bc \wedge ab \mid ac) \to ab \mid cd$.

15. $d = \gcd(a, b) \to d = as + bt$ for some $s, t \in \mathbb{Z} \to 1 = (a/d)s + (b/d)t \to 1 = \gcd(a/d, b/d)$

21. a. Hint: first show $A = \{0 \text{ MOD } 7, 3 \text{ MOD } 7, \ldots, 3^6 \text{ MOD } 7\}$ and then show $A = \{x \text{ MOD } 7, (x + 1) \text{ MOD } 7, \ldots, (x + 6) \text{ MOD } 7\}$.

b. Hint: show $A \ne \{0 \text{ MOD } 7, 2 \text{ MOD } 7, \ldots, 2^6 \text{ MOD } 7\}$.

23. a. Hint: $a^2 \text{ MOD } p = b^2 \text{ MOD } p \to p \mid (a^2 - b^2) \to p \mid (a + b)(a - b)$. Now apply Euclid's lemma.

CHAPTER 4

Section 4.1

1. b. **d.** **f.**

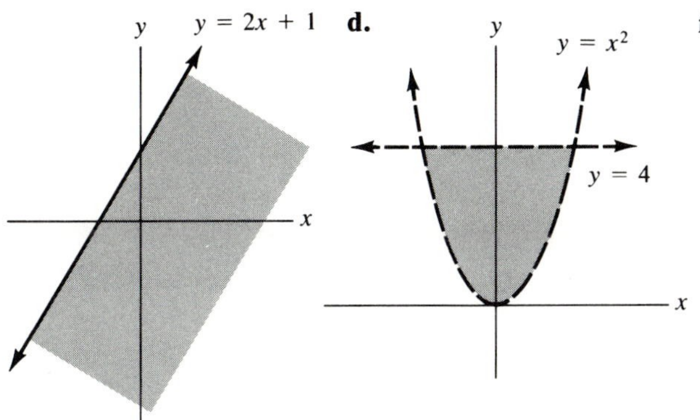

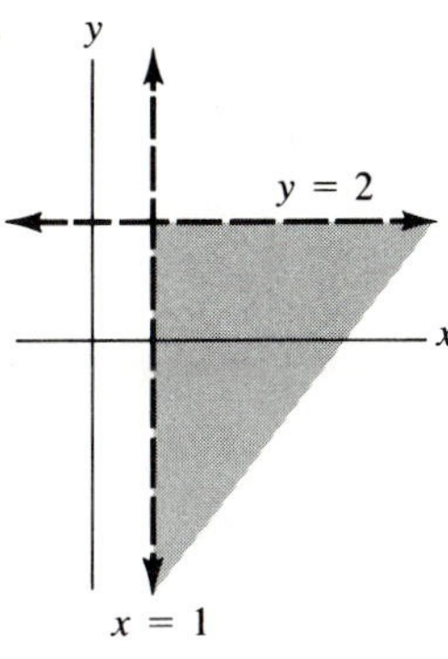

Section 4.2

1. a. reflexive, symmetric **b.** transitive, antisymmetric **c.** symmetric
d. reflexive, symmetric, transitive

3. a. symmetric **b.** reflexive, symmetric **c.** symmetric

5. a.

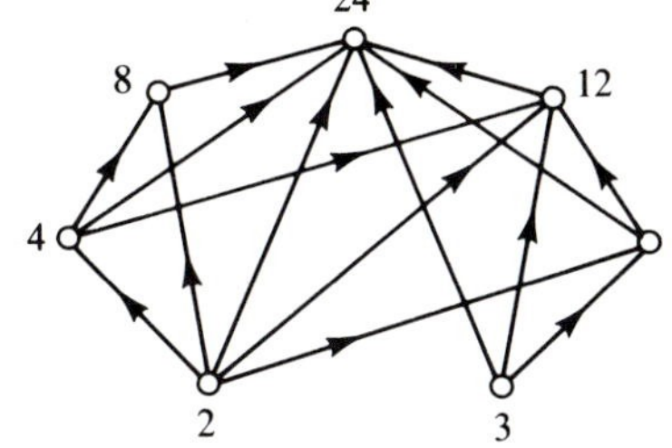

b.

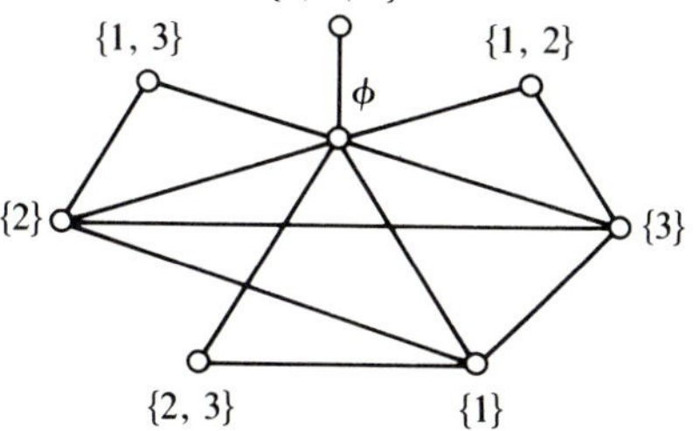

c.

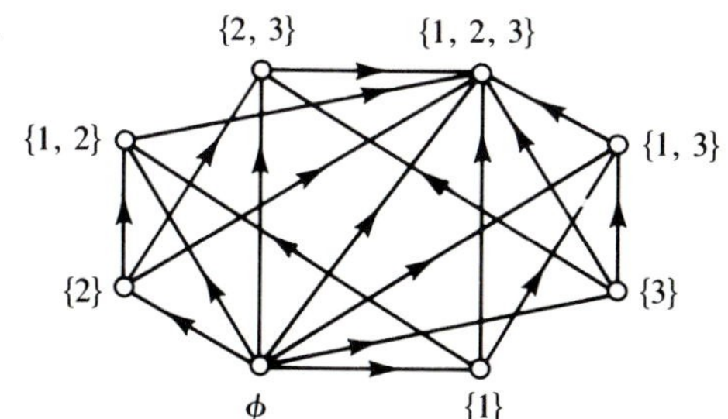

7. b. $[C] = \{\{2, 5\}, \{2, 3, 5\}, \{2, 4, 5\}, \{2, 3, 4, 5\}\}$

9. a. If $r_0^2 = x_0^2 + y_0^2$, then $(x_0, y_0) \in C_{r_0}$, so $\cup C_r = S$. Also $r \neq t \rightarrow C_r \cap C_t = \phi$.
b. $(2,4) \sim (4,-2)$; $(3,4) \sim (0,-5)$; $(2,3) \not\sim (1,4)$
c. $(a,b) \sim (c,d) \leftrightarrow a^2 + b^2 = c^2 + d^2$

Section 4.3

1. a. **b.**

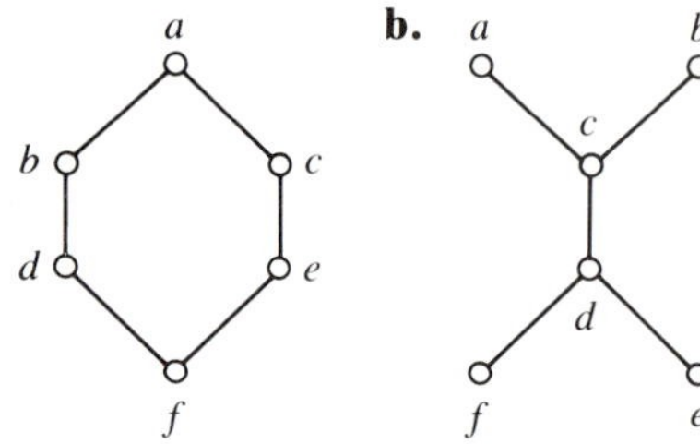

A lattice — Not a lattice

c. Not a lattice

d.

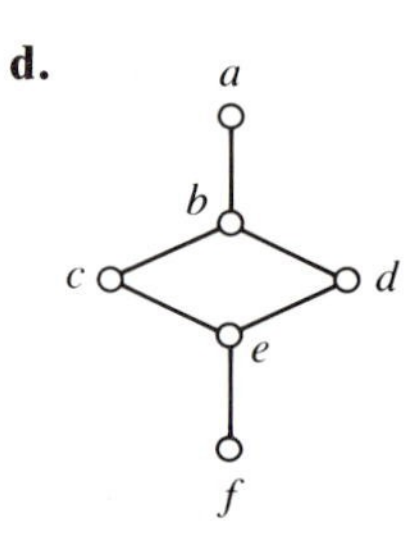

A lattice

3. b. $\text{lub}(A) = (3,4)$, $\text{glb}(A) = (-1,-3)$

c. $\text{lub}(A) = (u_1,u_2)$, where $u_1 = \max\{a_1, a_2, \ldots, a_n\}$ and $u_2 = \max\{b_1, b_2, \ldots, b_n\}$; $\text{glb}(A) = (v_1,v_2)$, where $v_1 = \min\{a_1, a_2, \ldots, a_n\}$ and $v_2 = \min\{b_1, b_2, \ldots, b_n\}$

d. yes

5.

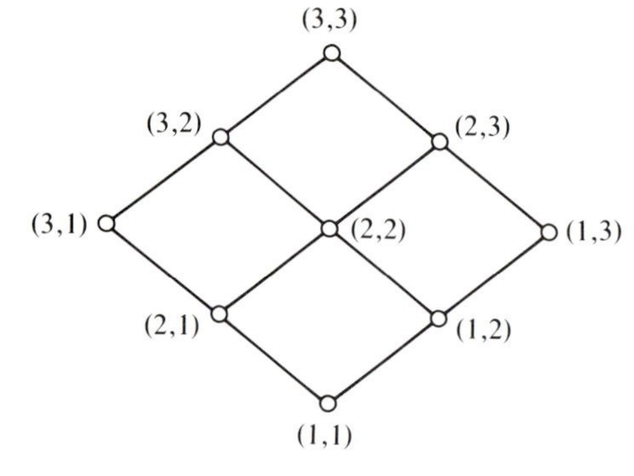

Section 4.4

1. a.

COURSEID	COURSENAME	TIME	LOCATION	INSTRUCTOR
CS105	Computing I	10	GEM 175	Moses
CS260	Data Structures	10	GEM 174	Lewis
MA331	Algebra	10	GEM 179	Straat

b.

COURSEID	TEXTAUTHOR
CS105	Allenton
CS260	Paston
CS340	Strong
MA122	Calvin
MA331	Seelo
MA350	Normal

c.

COURSEID	TEXTAUTHOR	PREREQ	CREDITS
CS105	Allenton	CS102	3
CS260	Paston	CS105	3
CS340	Strong	CS261	3
MA331	Seelo	MA231	3
MA350	Normal	MA231	3

d.

COURSEID	COURSENAME	TIME
CS105	Computing I	8
CS105	Computing I	9
CS105	Computing I	10
CS260	Data Structures	9
CS260	Data Structures	10

(*table continues on next page*)

COURSEID	COURSENAME	TIME
CS340	Software Design	2
MA122	Calculus I	11
MA122	Calculus I	1
MA331	Algebra	10
MA350	Statistics	1

(continued)

e.

COURSEID	COURSENAME	TIME	LOCATION	INSTRUCTOR
CS105	Computing I	10	GEM 175	Moses
CS260	Data Structures	9	GEM 175	Straat
MA331	Algebra	10	GEM 179	Straat
MA350	Statistics	1	GEM 174	Moses

3. a. `PROJECT R3 OVER TIME AND ENROLLMENT`

b. `SELECT FROM R3 WHERE COURSEID = 'CS105'`

c. `PROJECT (JOIN R1 and R3 OVER COURSEID AND TIME AND LOCATION) OVER COURSEID AND INSTRUCTOR AND ENROLLMENT`

d. `SELECT FROM (JOIN R1 AND R3 OVER COURSEID AND TIME AND LOCATION) WHERE INSTRUCTOR = 'LEWIS'`

Chapter 4 problems

1. yes; R is both symmetric and antisymmetric $\leftrightarrow ((x,y) \in R \to x = y)$

3. a. $\exists a \in A((a,a) \notin R)$ **b.** $\exists a, b \in A((a,b) \in R \wedge (b,a) \notin R)$

c. $\exists a, b \in A(a \neq b \wedge (a,b) \in R \wedge (b,a) \in R)$

d. $\exists a, b, c \in A((a,b) \in R \wedge (b,c) \in R \wedge (a,c) \notin R)$

5. a. equivalence relation; $[(a,b)]$ is the line through (a,b) with slope $= 1$

b. equivalence relation; $[(a,b)]$ is the circle with center $(1,0)$ and radius $\sqrt{(a-1)^2 + b^2}$

c. not an equivalence relation

d. equivalence relation; $[(a,b)]$ is the square with vertices $(r,0)$, $(0,r)$, $(-r,0)$, and $(0,-r)$, where $r = |a| + |b|$

e. equivalence relation; $[(0,0)]$ is the coordinate axes; if $ab \neq 0$, then $[(a,b)]$ is the hyperbola $xy = ab$

7. a. For transitivity, if $(a,b) \sim (c,d)$ and $(c,d) \sim (e,f)$, then $ad = bc$ and $cf = de$. Thus $acdf = bcde$, so $af = be$. Hence, $(a,b) \sim (e,f)$.

b. $[(a,b)] = \{(x,y) \mid x/y = a/b\}$

9. For antisymmetry, if $(a,b) <= (c,d)$ and $(c,d) <= (a,b)$, then $ad \leq bc$ and $bc \leq ad$, so $ad = bc$. Thus $a \mid bc$ and $c \mid ad$, which implies that $a \mid c$ and $c \mid a$, since $\gcd(a, b) = \gcd(c, d) = 1$. Hence $a = c$. Similarly, $b = d$.

11. a. reflexive, transitive **b.** symmetric **c.** symmetric **d.** reflexive, symmetric

e. symmetric

13. a. a **b.** i **c.** b **d.** h **e.** m **f.** a **g.** j **h.** m

15. a. $\{(a,b) \mid |a - b| = 1\}$ **b.** $\{(a,b) \mid a < b\}$

c. Suppose R is symmetric and transitive. Let $(a,b) \in R$. Since R is symmetric, $(b,a) \in R$. Then since R is transitive, $(a,b) \in R \wedge (b,a) \in R \to (a,a) \in R$. Therefore, R is not irreflexive.

17. b. $R = R^r \leftrightarrow R$ is reflexive

c. Let $(a,b) \in R^r$. If $a = b$, then $(a,b) \in R'$ since R' is reflexive. If $a \neq b$, then $(a,b) \in R^r \rightarrow (a,b) \in R \rightarrow (a,b) \in R'$. Thus, $R^r \subseteq R'$.
d. $\{(a,b) \mid a \leq b\}$

19. a. $R^t = \{(a,b) \mid a < b\}$
c. Let R be the relation from part a. Then $R^2 = \{(a,b) \mid a + 1 \leq b \leq a + 2\}$. So $R^2 \neq R^t$.

21. a. For each of the properties, if both R and S possess the property, then so does T_1. For each of properties (i), (ii), and (iii), if both R and S have the property, then so does T_2. However, having both R and S transitive is not sufficient for T_2 to be transitive.

23. a. The maximal elements are a and b; the minimal elements are i and j.
b. The maximal element is X; the minimal element is ϕ.
c. The primes.
d. Hint: to show that $A = \{a_1, a_2, \ldots, a_n\}$ contains a minimal element, use induction on n.
e. $(\mathbb{N}, |)$

25. a. $k < j < f < c < a,\quad k < j < g < c < a,\quad k < j < g < d < a,\quad k < j < g < d < b,$
$k < j < g < e < b,\quad k < j < h < d < a,\quad k < j < h < d < b,\quad k < j < h < e < b$
b. 4 **c.** $m + 1$
d. Hint: Prove the contrapositive: If a_1 is not minimal (a_n is not maximal), then the chain $a_1 < a_2 < \cdots < a_n$ is not a longest chain.
e. Hint: Let $a_1 < a_2 < \cdots < a_n$ be a longest chain in $(A, \preceq)$. Show that $a_1 < \cdots < a_{n-1}$ is a longest chain in $(A - M, \preceq)$.

27. a. See Theorem 2.6, part 3.
b. We must show that $\gcd(a, \operatorname{lcm}(b, c)) = \operatorname{lcm}(\gcd(a, b), \gcd(a, c))$ holds for all $a, b, c \in \mathbb{N}$. Express a, b, and c as follows: $a = p_1^{r_1}p_2^{r_2} \cdots p_n^{r_n}$, $b = p_1^{s_1}p_2^{s_2} \cdots p_n^{s_n}$, $c = p_1^{t_1}p_2^{t_2} \cdots p_n^{t_n}$, where $p_1, p_2, \ldots, p_n$ are distinct primes and the r_i, s_i, and t_i are nonnegative integers. Now apply Exercise 3 of Section 3.4.

CHAPTER 5

Section 5.1

1. a. a function **b.** a function **c.** not a function **d.** a function

3. a. a function; $\operatorname{im} f = [0, \infty)$ **b.** not a function **c.** a function; $\operatorname{im} h = P$ **d.** not a function
e. yes; $\operatorname{im} f = \{y \mid y$ is an eldest son and y's father has a grandchild$\}$

Section 5.2

1. a. neither one-to-one nor onto **b.** one-to-one and onto
c. onto but not one-to-one **d.** one-to-one but not onto
e. onto but not one-to-one **f.** one-to-one and onto

3. a. onto, not one-to-one **b.** one-to-one, not onto **c.** neither one-to-one nor onto
d. permutation **e.** permutation **f.** one-to-one, not onto **g.** one-to-one and onto

5. a. not a permutation; $f(C) = \{[2], [10]\}$; $f^{-1}(D) = \{[2], [4], [8], [10]\}$
b. not a permutation; $f(C) = \{[4], [8]\}$; $f^{-1}(D) = \{[1], [2], [4], [5], [7], [8], [10], [11]\}$
c. a permutation; $f(C) = C$; $f^{-1}(D) = D$

7. a. $f(C) = \{2, 3\}$; $f^{-1}(D) = \{1, 3, 4\}$ **c.** $f(C) = \{2\}$; $f^{-1}(D) = \{2, 4\}$ **e.** $f(C) = C$; $f^{-1}(D) = \{\ldots, -3, -2, -1\}$ **g.** $f(C) = \{0\}$; $f^{-1}(D) = \{2\}$

Section 5.3

1. a. $f^{-1}: \mathbb{Q} \rightarrow \mathbb{Q}$; $f^{-1}(x) = (x - 2)/4$
c. $f^{-1}: \mathbb{Z}_{12} \rightarrow \mathbb{Z}_{12}$; $f^{-1}([x]) = [5x]$, i.e., $f^{-1} = f$
e. $f^{-1}: \{1, 2, 3, 4\} \rightarrow \{1, 2, 3, 4\}$; $f^{-1}(1) = 2, f^{-1}(2) = 3, f^{-1}(3) = 4, f^{-1}(4) = 1$

3. a. $g \circ f: \mathbb{Z} \to \mathbb{Z}_5$; $(g \circ f)(n) = [-n]$

c. $g \circ f: \mathbb{R} - \{2\} \to \mathbb{R} - \{0\}$; $(g \circ f)(x) = x - 2$

e. $g \circ f: \{1, 2, 3, 4\} \to \{1, 2, 3, 4\}$; $(g \circ f)(1) = 2$, $(g \circ f)(2) = 3$, $(g \circ f)(3) = 4$, $(g \circ f)(4) = 1$

5. a. $f^{-1}(x) = x - 1$, $g^{-1}(x) = 2 - x$, $(f \circ g)(x) = 3 - x$, $(g \circ f)(x) = 1 - x$, $(f \circ g)^{-1}(x) = 3 - x = (g^{-1} \circ f^{-1})(x)$

c. $f^{-1}(1) = 2, f^{-1}(2) = 3, f^{-1}(3) = 4, f^{-1}(4) = 1$; $g^{-1}(1) = 3$, $g^{-1}(2) = 4$, $g^{-1}(3) = 1$, $g^{-1}(4) = 2$; $(f \circ g)(1) = 2$, $(f \circ g)(2) = 3$, $(f \circ g)(3) = 4$, $(f \circ g)(4) = 1$; $(g \circ f)(1) = 2$, $(g \circ f)(2) = 3$, $(g \circ f)(3) = 4$, $(g \circ f)(4) = 1$; $(f \circ g)^{-1}(1) = 4$, $(f \circ g)^{-1}(2) = 1$, $(f \circ g)^{-1}(3) = 2$, $(f \circ g)^{-1}(4) = 3$; $g^{-1} \circ f^{-1} = (f \circ g)^{-1}$

7. Hint: Let $c \in C$, and suppose $g^{-1}(c) = b$ and $f^{-1}(b) = a$. Then $(f^{-1} \circ g^{-1})(c) = a$. You must show that $(g \circ f)^{-1}(c) = a$.

Section 5.4

1. $k(m) = \dfrac{(n - m + 1)k(m - 1)}{m}$

3. a. $h_1(1) = 3$, $h_1(2) = 8$, $h_1(n) = 2(h_1(n - 1) + h_1(n - 2))$, $n \geq 3$

b. $h_2(1) = 3$, $h_2(2) = 7$, $h_2(n) = 2h_2(n - 1) + h_2(n - 2)$, $n \geq 3$

c. $h_3(1) = 3$, $h_3(2) = 9$, $h_3(3) = 26$, $h_3(n) = 2(h_3(n - 1) + h_3(n - 2) + h_3(n - 3))$, $n \geq 4$

d. $h_4(1) = 3$, $h_4(2) = 8$, $h_4(n) = 2h_4(n - 1) + h_4(n - 2) + \cdots + h_4(1) + 2$, $n \geq 3$

5. $t(1) = 1$, $t(n) = (2n - 1)t(n - 1)$, $n \geq 2$

Chapter 5 problems

1. a. $f(x) = x$ **b.** $f(x) = x/2$ **c.** $f(x) = (1 - 2x)^2$ **d.** $f(x) = 0$

3. Hint: For the necessity, suppose f is one-to-one, and let A and B be subsets of X. It must be shown that $f(A \cap B) \supseteq f(A) \cap f(B)$ (see Theorem 5.2). For the sufficiency, prove the contrapositive. Suppose f is not one-to-one. Then there exist $x_1, x_2 \in X$, with $x_1 \neq x_2$, such that $f(x_1) = f(x_2)$. Now let $A = \{x_1\}$ and $B = \{x_2\}$.

5. a. $(f \circ g)(x) = 9x^2 + 27x + 20$, $(g \circ f)(x) = 3x^2 + 3x + 4$

b. $(f \circ g)(x) = x/(x^2 + 1)$, $(g \circ f)(x) = (x^2 + 1)/x$

7. a. both one-to-one and onto, $f^{-1}(x) = \sqrt[3]{x - 1}$

c. onto, not one-to-one

e. both one-to-one and onto, $q^{-1}(x) = \sqrt{x}$

9. a. p_α, s_α, and r are one-to-one

b. $\operatorname{im} p_\alpha = \{y \in \mathcal{S}^+ \mid \text{the first character of } y \text{ is } \alpha\}$, $\operatorname{im} s_\alpha = \{y \in \mathcal{S}^+ \mid \text{the last character of } y \text{ is } \alpha\}$, $\operatorname{im} r = \mathcal{S} = \operatorname{im} t$

c. $r(x) = x \leftrightarrow x$ is a "palindrome"

d. $(r \circ p_\alpha)(x) = r(p_\alpha(x)) = r(\text{'}\alpha\text{'} \mathbin{\#} x) = r(x) \mathbin{\#} \text{'}\alpha\text{'} = s_\alpha(r(x)) = (s_\alpha \circ r)(x)$

11. a. $\chi_{B \cap C}(a) = 1 \leftrightarrow a \in B \cap C \leftrightarrow a \in B \wedge a \in C \leftrightarrow \chi_B(a) = 1 \wedge \chi_C(a) = 1 \leftrightarrow \chi_B(a) \cdot \chi_C(a) = 1$

b. $\chi_{B \cup C}(a) = 0 \leftrightarrow a \notin B \cup C \leftrightarrow a \notin B \wedge a \notin C \leftrightarrow \chi_B(a) = 0 \wedge \chi_C(a) = 0 \leftrightarrow \chi_B(a) + \chi_C(a) - \chi_B(a) \cdot \chi_C(a) = 0$

e. Use the fact that $B - C = B \cap (A - C)$, and the results of parts a and d.

13. Hint: consider the function f defined by $f(1) = 2$, $f(2) = 3$, $f(3) = 1$, and $f(n) = n$ for $n \neq 1, 2, 3$.

15. a. 2^{m^2} **b.** m^m **c.** $m!$

19. a. $f(n) = an + b$ **b.** $f(n) = an^2 + bn + c$ **c.** $f(n) = a^n + b$

CHAPTER 6

Section 6.1

1. a. commutative **b.** commutative, associative, the identity is 1, the inverse of m is $2 - m$

c. an operation **d.** an operation **e.** not an operation **f.** not an operation

3. **a.** $\exists a, b, c \in A((a * b) * c \neq a * (b * c))$ **b.** $\exists a, b \in A(a * b \neq b * a)$
c. $\exists b \in A(a * b \neq b \vee b * a \neq b)$ **d.** $a * b \neq e \vee b * a \neq e$

5. **b.**

⊕	[0]	[1]	[2]	[3]	[4]
[0]	[0]	[1]	[2]	[3]	[4]
[1]	[1]	[2]	[3]	[4]	[0]
[2]	[2]	[3]	[4]	[0]	[1]
[3]	[3]	[4]	[0]	[1]	[2]
[4]	[4]	[0]	[1]	[2]	[3]

d.

∪	ϕ	{1}	{2}	{3}	{1, 2}	{1, 3}	{2, 3}	{1, 2, 3}
ϕ	ϕ	{1}	{2}	{3}	{1,2}	{1,3}	{2,3}	{1,2,3}
{1}	{1}	{1}	{1,2}	{1,3}	{1,2}	{1,3}	{1,2,3}	{1,2,3}
{2}	{2}	{1,2}	{2}	{2,3}	{1,2}	{1,2,3}	{2,3}	{1,2,3}
{3}	{3}	{1,3}	{2,3}	{3}	{1,2,3}	{1,3}	{2,3}	{1,2,3}
{1, 2}	{1,2}	{1,2}	{1,2}	{1,2,3}	{1,2}	{1,2,3}	{1,2,3}	{1,2,3}
{1, 3}	{1,3}	{1,3}	{1,2,3}	{1,3}	{1,2,3}	{1,3}	{1,2,3}	{1,2,3}
{2, 3}	{2,3}	{1,2,3}	{2,3}	{2,3}	{1,2,3}	{1,2,3}	{2,3}	{1,2,3}
{1, 2, 3}	{1,2,3}	{1,2,3}	{1,2,3}	{1,2,3}	{1,2,3}	{1,2,3}	{1,2,3}	{1,2,3}

f.

⊙	[1]	[2]	[3]	[4]	[5]	[6]
[1]	[1]	[2]	[3]	[4]	[5]	[6]
[2]	[2]	[4]	[6]	[1]	[3]	[5]
[3]	[3]	[6]	[2]	[5]	[1]	[4]
[4]	[4]	[1]	[5]	[2]	[6]	[3]
[5]	[5]	[3]	[1]	[6]	[4]	[2]
[6]	[6]	[5]	[4]	[3]	[2]	[1]

7. **a.**

*	*a*	*b*	*c*	*d*
a	*a*	*b*	*c*	*d*
b	*b*	*c*	*d*	*a*
c	*c*	*d*	*a*	*b*
d	*d*	*a*	*b*	*c*

b.

*	*a*	*b*	*c*	*d*
a	*a*	*x*	*c*	*y*
b	*x*	*w*	*d*	*a*
c	*c*	*d*	*a*	*b*
d	*y*	*a*	*b*	*z*

c.

*	*a*	*b*	*c*	*d*
a	*a*	*b*	*c*	*d*
b	*b*	*w*	*d*	*a*
c	*c*	*d*	*a*	*b*
d	*d*	*a*	*b*	*z*

Hint for part a: $a * d = a * (c * b) = (a * c) * b = c * b = d$. For parts b and c, we have $x, y, z, w \in \{a, b, c, d\}$.

9. **a.** $\{2m \mid m \in \mathbb{N}\}$ **b.** $\{2m \mid m \in \mathbb{Z}\}$ **c.** $\{3^m \mid m \in \mathbb{Z}\}$

11.

∘	f_1	f_2	f_3	f_4	f_5	f_6
f_1	f_1	f_2	f_3	f_4	f_5	f_6
f_2	f_2	f_1	f_4	f_3	f_6	f_5
f_3	f_3	f_5	f_1	f_6	f_2	f_4
f_4	f_4	f_6	f_2	f_5	f_1	f_3
f_5	f_5	f_3	f_6	f_1	f_4	f_2
f_6	f_6	f_4	f_5	f_2	f_3	f_1

13. **a.** associative, commutative, *a* is the identity, each element has an inverse
c. not associative, not commutative, the identity is *b*, each element has an inverse

15.

$*$	a	b	c
a	a	b	c
b	b	a	a
c	c	a	a

Section 6.2

1.

$*$	x	y	z
x	x	y	z
y	y	z	x
z	z	x	y

3. **a.** $|[0]| = 1, |[1]| = 6, |[2]| = 3, |[3]| = 2, |[4]| = 3, |[5]| = 6$
c. $|a| = 1, |b| = |c| = |d| = 2$

5. **a.**

$*$	a	b	c
a	a	b	c
b	b	c	a
c	c	a	b

c.

$*$	a	b	c	d
a	a	b	c	d
b	b	a	d	c
c	c	d	a	b
d	d	c	b	a

7. **b.** (i) $r, t \in \mathbb{Q}^+ \rightarrow rt \in \mathbb{Q}^+$
(ii) $r, s, t \in \mathbb{Q}^+ \rightarrow (rs)t = r(st)$
(iii) 1 is the identity for $(\mathbb{Q}^+, \cdot)$
(iv) $1/r$ is the inverse of r
(v) $rt = tr$ for $r, t \in \mathbb{Q}^+$, so $(\mathbb{Q}^+, \cdot)$ is an abelian group

d. (i) $a, b \in E \rightarrow ab \in E$
(ii) $a, b, c \in E \rightarrow (ab)c = a(bc)$
(iii) there is no identity
(iv) $ab = ba$ for $a, b \in E$, so $\cdot$ is commutative

f. Let $a, b \in M_k$, where $a = kn$ and $b = km$ for some $m, n \in \mathbb{Z}$. Then
(i) $a + b = kn + km = k(n + m) \in M_k$
(ii) $a, b, c \in M_k \rightarrow (a + b) + c = a + (b + c)$
(iii) 0 is the identity for $(M_k, +)$
(iv) the inverse of a is $-a$
(v) $a + b = b + a$, so $(M_k, +)$ is an abelian group

9. **a.**

$\circ$	ε	α	β	γ	δ	ρ
ε	ε	α	β	γ	δ	ρ
α	α	ε	δ	ρ	β	γ
β	β	ρ	ε	δ	γ	α
γ	γ	δ	ρ	ε	α	β
δ	δ	γ	α	β	ρ	ε
ρ	ρ	β	γ	α	ε	δ

11. **a.** Only part b, with $m * n = m + n - 1$, yields a group.
b. This group is cyclic (and thus also abelian); it is generated by 2.

13. $x * y = y \rightarrow x * (y * y^{-1}) = e \rightarrow x * e = e \rightarrow x = e$

15. $x = y \leftrightarrow x^{-1} * x = x^{-1} * y \leftrightarrow e = x^{-1} * y \leftrightarrow e * y^{-1} = x^{-1} * (y * y^{-1}) \leftrightarrow y^{-1} = x^{-1} * e \leftrightarrow y^{-1} = x^{-1}$

17. Hint: $(a * b)^2 = a^2 * b^2 \leftrightarrow a^{-1} * (a * b)^2 * b^{-1} = a^{-1} * (a^2 * b^2) * b^{-1}$

19. a. $|[0]| = 1$, $|[1]| = 12$, $|[2]| = 6$, $|[3]| = 4$, $|[4]| = 3$, $|[5]| = 12$, $|[6]| = 2$, $|[7]| = 12$, $|[8]| = 3$, $|[9]| = 4$, $|[10]| = 6$, $|[11]| = 12$

c. $n/\gcd(m, n)$

21. a. $\{[0]\}$, $\mathbb{Z}_{12}$, $\langle[2]\rangle = \{[0], [2], [4], [6], [8], [10]\}$, $\langle[4]\rangle = \{[0], [4], [8]\}$

c. $\{[1]\}$, $\mathbb{Z}_7^{\#}$, $\langle[2]\rangle = \{[1], [2], [4]\}$, $\langle[6]\rangle = \{[1], [6]\}$

23. a.

$\circ$	e	r_1	r_2	r_3	h	v	d_1	d_2
e	e	r_1	r_2	r_3	h	v	d_1	d_2
r_1	r_1	r_2	r_3	e	d_1	d_2	v	h
r_2	r_2	r_3	e	r_1	v	h	d_2	d_1
r_3	r_3	e	r_1	r_2	d_2	d_1	h	v
h	h	d_2	v	d_1	e	r_3	r_2	r_1
v	v	d_1	h	d_2	r_2	e	r_1	r_3
d_1	d_1	h	d_2	v	r_3	r_1	e	r_2
d_2	d_2	v	d_1	h	r_1	r_2	r_3	e

b. $e^{-1} = e$, $|e| = 1$; $r_1^{-1} = r_3$, $|r_1| = 4$; $r_2^{-1} = r_2$, $|r_2| = 2$; $r_3^{-1} = r_1$, $|r_3| = 4$; $h^{-1} = h$, $|h| = 2$; $v^{-1} = v$, $|v| = 2$; $d_1^{-1} = d_1$, $|d_1| = 2$; $d_2^{-1} = d_2$, $|d_2| = 2$

c. $H = \{e, h, v, r_2\}$

25. a. $\langle a^{-1}\rangle = \{(a^{-1})^n \mid n \in \mathbb{Z}\} = \{a^{-n} \mid n \in \mathbb{Z}\} = \{a^m \mid m \in \mathbb{Z}\} = \langle a\rangle$

27. a. $(H, +)$ is a subgroup of $(\mathbb{Z}, +) \leftrightarrow H = \langle k\rangle$ for some fixed $k \in \mathbb{Z} \leftrightarrow$ $H = \{\ldots, -2k, -k, 0, k, 2k, \ldots\}$

c. $\{[1]\}$, $\langle[3]\rangle = \{[1], [3], [4], [5], [9]\}$, $\langle[10]\rangle = \{[1], [10]\}$, $\mathbb{Z}_{11}^{\#}$

29. b. $x^m = e \leftrightarrow (x^m)^{-1} = e \leftrightarrow (x^{-1})^m = e$. From this it follows that $|x| = |x^{-1}|$.

c. Hint: By the division algorithm, $m = nq + r$, $0 \le r < n$. Show that $r = 0$.

Section 6.3

1. a. not a ring; the distributive laws fail

b. a ring **c.** not a ring; the distributive laws fail

d. not a ring; $*$ is not associative

3. a. 1 **b.** 0 **c.** yes **d.** yes **e.** no; for example, 3 has no multiplicative inverse

5. yes

9. yes

11. a. $\gcd(a, m) = d > 1 \rightarrow [a] \odot \left[\frac{m}{d}\right] = \left[\frac{a}{d}m\right] = [0] \rightarrow$ a is a zero divisor; now apply Exercise 10a

b. From part a, it suffices to show sufficiency: $\gcd(a, m) = 1 \rightarrow 1 = as + mt$ for some $s, t \in \mathbb{Z} \rightarrow as = 1 - mt \rightarrow [a] \odot [s] = [1] \rightarrow [s] = [a]^{-1}$

c. If $m = st$, where $1 < s \le t < m$, then $[s] \odot [t] = [0]$. Thus $[s]$ is a zero divisor, so $(\mathbb{Z}_m, \oplus, \odot)$ is not an integral domain.

13. Hint: From previous work we know that $(E, +)$ is an abelian group. Thus, it remains to show (i) $*$ is an associative and commutative operation on E, (ii) $x, y, z \in E \rightarrow x * (y + z) = (x * y) + (x * z)$, (iii) $*$ has an identity, (iv) there are no zero divisors.

17. $(a \cdot x) + b = z \rightarrow a \cdot x = -b \rightarrow x = (a^{-1}) \cdot (-b)$

19. a. Let $r = a + b\sqrt{2}$ and $s = c + d\sqrt{2}$. Then $rs = (ac + 2bd) + (ad + bc)\sqrt{2}$, so $N(rs) = (ac + 2bd)^2 - 2(ad + bc)^2 = a^2c^2 - 2a^2d^2 - 2b^2c^2 + 4b^2d^2 = (a^2 - 2b^2)(c^2 - 2d^2) = N(r)N(s)$.

b. Suppose $s = r^{-1}$. Then $rs = 1 \rightarrow N(rs) = 1 \rightarrow N(r)N(s) = 1 \rightarrow N(r) = 1$ or $N(r) = -1$, since

$N(s)$ is an integer. Conversely, if $r = a + b\sqrt{2}$, then $N(r) = 1 \rightarrow r^{-1} = a - b\sqrt{2}$, whereas $N(r) = -1 \rightarrow r^{-1} = -a + b\sqrt{2}$.

c. Since $N(2) = 4$, from part b it follows that 2 has no multiplicative inverse.

Section 6.4

1. a. complemented: $a' = b$, $b' = a$, $c' = b$, $d' = a$, $0' = 1$, $1' = 0$; not distributive: $a \wedge (b \vee c) \neq (a \wedge b) \vee (a \wedge c)$; not a Boolean algebra

b. a Boolean algebra: $1' = 0$, $a' = f$, $b' = e$, $c' = d$, $e' = b$, $d' = c$, $f' = a$, $0' = 1$

c. not complemented; for example, b has no complement

3. a. **b.** **c.** See Exercise 5c, with $p = 2$, $q = 3$, and $r = 5$.

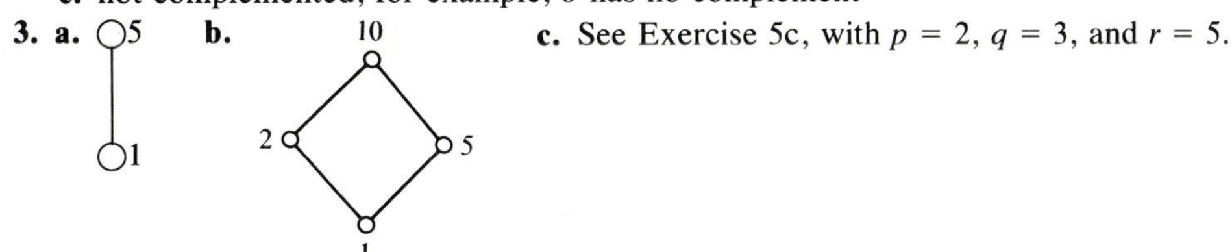

5. c.

$\vee$	n	pq	pr	qr	p	q	r	1
n	n	n	n	n	n	n	n	n
pq	n	pq	n	n	pq	pq	n	pq
pr	n	n	pr	n	pr	n	pr	pr
qr	n	n	n	qr	n	qr	qr	qr
p	n	pq	pr	n	p	pq	pr	p
q	n	pq	n	qr	pq	q	qr	q
r	n	n	pr	qr	pr	qr	r	r
1	n	pq	pr	qr	p	q	r	1

$\wedge$	n	pq	pr	qr	p	q	r	1
n	n	pq	pr	qr	p	q	r	1
pq	pq	pq	p	q	p	q	1	1
pr	pr	p	pr	r	p	1	r	1
qr	qr	q	r	qr	1	q	r	1
p	p	p	p	1	p	1	1	1
q	q	q	1	q	1	q	1	1
r	r	1	r	r	1	1	r	1
1	1	1	1	1	1	1	1	1

7. 2^m **9. a.** p **c.** r **e.** r **g.** qr

11. a.
$$\begin{aligned}(a \vee b) \wedge (a' \wedge b') &= (a \wedge (a' \wedge b')) \vee (b \wedge (a' \wedge b')) \\ &= ((a \wedge a') \wedge b') \vee (a' \wedge (b \wedge b')) \\ &= (0 \wedge b') \vee (a' \wedge 0) = 0 \vee 0 = 0 \\ (a \vee b) \vee (a' \wedge b') &= ((a \vee b) \vee a') \wedge ((a \vee b) \vee b') \\ &= ((a \vee a') \vee b) \wedge (a \vee (b \vee b')) \\ &= (1 \vee b) \wedge (a \vee 1) = 1 \wedge 1 = 1\end{aligned}$$

The complement of $a \vee b$ is unique, therefore $(a \vee b)' = a' \wedge b'$.

13. $a = a \vee (a \wedge c) = a \vee (b \wedge c) = (a \vee b) \wedge (a \vee c) = (b \vee a) \wedge (b \vee c) = b \vee (a \wedge c) = b \vee (b \wedge c) = b$

15. $a \wedge (b \vee c) = a \neq b = (a \wedge b) \vee (a \wedge c)$

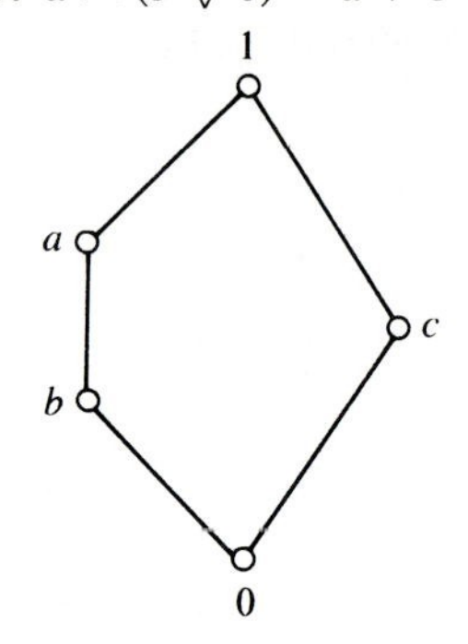

Section 6.5

1. a. **c.**

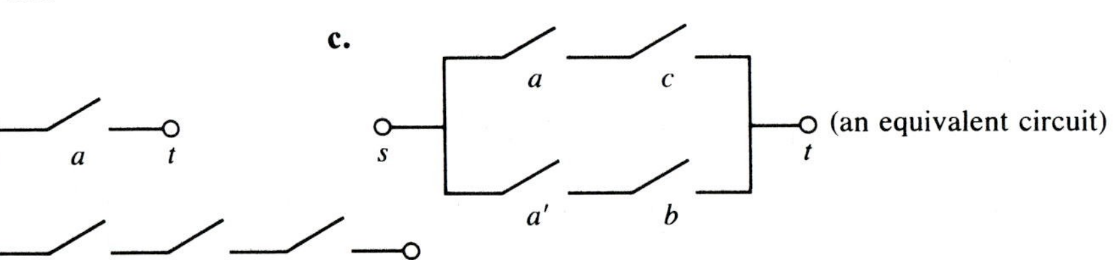

e. $s \quad a \quad b \quad c \quad t$

3. a. **b.**

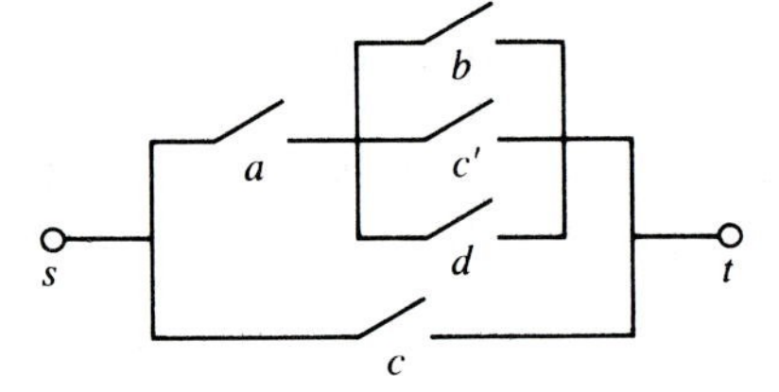

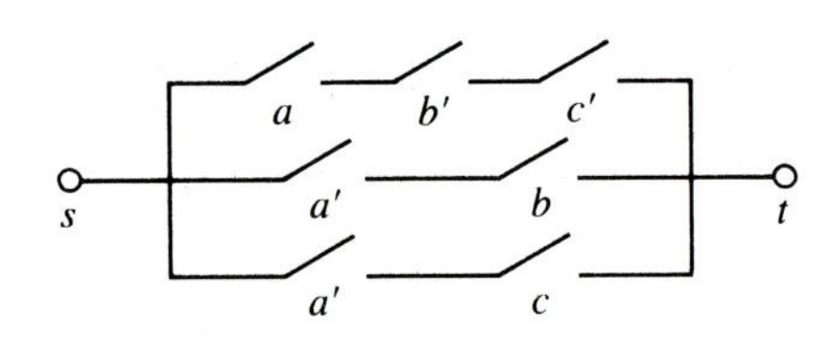

5. a. **b.**

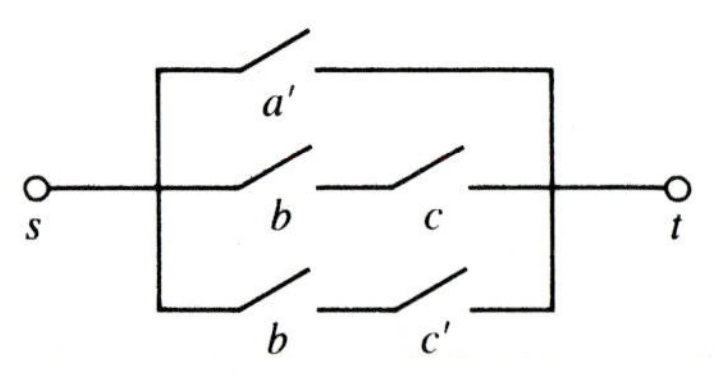

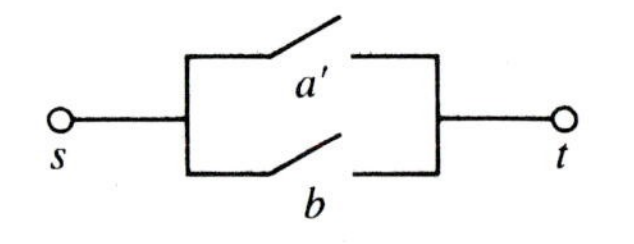

$$
\begin{aligned}
a' \vee (b \wedge c) \vee (b \wedge c') &= a' \vee (b \wedge (c \vee c')) \\
&= a' \vee (b \wedge 1) \\
&= a' \vee b
\end{aligned}
$$

7. $(B \wedge G \wedge H) \vee (B \wedge G' \wedge H') \vee (B' \wedge G \wedge H') \vee (B' \wedge G' \wedge H)$

9. a. $(A \wedge B \wedge C \wedge D) \vee (A \wedge B \wedge C) \vee (A \wedge B \wedge D) \vee (A \wedge C \wedge D) \vee (B \wedge C \wedge D)$

b. $(A \wedge B \wedge C \wedge D) \vee (A \wedge B \wedge C) \vee (A \wedge B \wedge D) \vee (A \wedge B \wedge E) \vee (A \wedge C \wedge D) \vee (A \wedge C \wedge E) \vee (A \wedge D \wedge E) \vee (B \wedge C \wedge D) \vee (B \wedge C \wedge E) \vee (B \wedge D \wedge E) \vee (C \wedge D \wedge E)$

Section 6.6

1. a. $f(e_G) = f(e_G * e_G) = f(e_G) \# f(e_G)$, therefore $f(e_G) = e_H$

b. $f(a^{-1}) \# f(a) = f(a^{-1} * a) = f(e_G) = e_H$. Similarly, $f(a) \# f(a^{-1}) = e_H$. Therefore $f(a^{-1}) = [f(a)]^{-1}$.

c. Hint: employ induction on n.

d.
$$
\begin{aligned}
f(a^{-n}) = f((a^{-1})^n) &= (f(a^{-1}))^n && \text{by part c} \\
&= [(f(a))^{-1}]^n && \text{by part b} \\
&= [f(a)]^{-n}
\end{aligned}
$$

3. a. Let n be an integer, $1 \le n \le m$; then $f([n]) = f([1] \oplus [1] \oplus \cdots \oplus [1]) = (f([1]))^n$.

5. a. $h([0]) = h([4]) = h([8]) = [0]$, $h([1]) = h([5]) = h([9]) = 2$, $h([2]) = h([6]) = h([10]) = [4]$, $h([3]) = h([7]) = h([11]) = [6]$; that is, $h([n]) = [2n]$

c. $h(m) = 2m$

7. a. $f([1]) = \delta, f([2]) = \rho, f([3]) = \varepsilon, f([4]) = \delta, f([5]) = \rho, f([0]) = \varepsilon$

b. No; if g is a homomorphism, then $g(\alpha) = g(\gamma \circ \delta) = g(\gamma) \oplus g(\delta) = [0] \oplus [2] = [2]$, but then $[0] = g(\varepsilon) = g(\alpha \circ \alpha) = g(\alpha) \oplus g(\alpha) = [2] \oplus [2] = [4]$, a contradiction.

c. yes; $h(\delta) = h(\rho) = h(\varepsilon) = [0]$, $h(\alpha) = h(\beta) = h(\gamma) = [3]$

9. Hint: define $f:\mathbb{Z}_6 \to \mathbb{Z}_7^{\#}$ by $f([1]) = [3]$.

11. Let $|x| = n$ and $|f(x)| = m$. Using Exercise 1c of this section, $(f(x))^n = f(x^n) = f(e_G) = e_H$. Thus, by Theorem 6.6, part 3, $m \mid n$.

13. a. The operation table for G must be as shown. Compare this with the operation table for $(\mathbb{Z}_3, \oplus)$.

$*$	e	a	b
e	e	a	b
a	a	b	e
b	b	e	a

b. $(\mathbb{Z}_4, \oplus)$ has two elements of order 4, whereas the Klein four-group has no elements of order 4. It follows from Exercise 12 that these two groups are nonisomorphic.

e. Hint: compare the orders of the elements and apply Exercise 12.

15. $n \mid md$, where $d = \gcd(a, n)$

17. a. f is an isomorphism $\leftrightarrow f([1]) = [3]$ or $f([1]) = [5]$

c. f is an isomorphism $\leftrightarrow f(1) = 2$ or $f(1) = -2$

19. a. Let $(G, *) = \langle a \rangle$ be cyclic of order n, and define $f:(\mathbb{Z}_n, \oplus) \to (G, *)$ by $f([1]) = a$. It is not difficult to show that f is one-to-one and well-defined. Furthermore, for integers s and t, $f([s] \oplus [t]) = f([s + t]) = a^{s+t} = a^s * a^t = f([s]) * f([t])$. Therefore, f is an isomorphism.

b. Hint: Let $(G, *) = \langle a \rangle$ be any infinite cyclic group, and define $f:(\mathbb{Z}, +) \to (G, *)$ by $f(1) = a$. Show that f is an isomorphism.

Chapter 6 problems

1. a. not associative, commutative, no identity

c. associative, commutative, identity is $\frac{1}{3}$, $x^{-1} = 1/(9x)$

e. not associative, not commutative, no identity

3. a. 56 **c.** 11 **e.** 3 8 * 2 7 * + −

5. a. n^{n^2} **b.** $n^{n(n+1)/2}$ **c.** n^{n^2-2n+2}

7. Let e denote the identity of $(G, *)$. If $(H, *)$ is a subgroup of $(G, *)$, then $e \in H$. Therefore (i) $H \neq \phi$. Also, $a,b \in H \to a,b^{-1} \in H \to a * b^{-1} \in H$. Thus (ii) $a,b \in H \to a * b^{-1} \in H$. Conversely, assume conditions (i) and (ii) hold. Since $H \neq \phi$, there is some $x \in G$ such that $x \in H$. Hence (iii) $x \in H \to x * x^{-1} = e \in H$, (iv) $e \in H \wedge x \in H \to e * x^{-1} = x^{-1} \in H$, and (v) $x,y \in H \to x,y^{-1} \in H \to x * (y^{-1})^{-1} = x * y \in H$. Also, $*$ is associative on H since $*$ is associative on G and $H \subseteq G$. Therefore, $(H, *)$ is a subgroup of $(G, *)$.

9. Hint: Let $e:\mathbb{R} \to \mathbb{R}$ be defined by $e(x) = 1$. Show that e is the identity of $(\mathcal{F}(\mathbb{R})^{\#}, \cdot)$. Also, given $f \in \mathcal{F}(\mathbb{R})^{\#}$, define $g:\mathbb{R} \to \mathbb{R}$ by $g(x) = 1/f(x)$. Show that $g \in \mathcal{F}(\mathbb{R})^{\#}$ and that g is the inverse of f.

11. a. $A_4 = \{\varepsilon, \alpha, \alpha^2, \beta, \beta^2, \alpha \circ \beta, \alpha \circ \beta^2, \alpha^2 \circ \beta, \beta \circ \alpha, \beta \circ \alpha^2, \beta^2 \circ \alpha, \alpha \circ \beta^2 \circ \alpha\}$

b. (i) $\{\varepsilon, \alpha \circ \beta\}$ (ii) $\{\varepsilon, \alpha, \alpha^2\}$ (iii) $\{\varepsilon, \alpha \circ \beta, \beta \circ \alpha, \alpha \circ \beta^2 \circ \alpha\}$

13. a. Since $e_G \in H \cap K$, it follows that $H \cap K \neq \phi$. Let $a, b \in H \cap K$. Then

$(a, b \in H \wedge a, b \in K) \rightarrow (a * b^{-1} \in H \wedge a * b^{-1} \in K) \rightarrow a * b^{-1} \in H \cap K$. Therefore, by the result of Problem 7, $(H \cap K, *)$ is a subgroup of $(G, *)$.

b. Let $H = \{\varepsilon, \alpha\}$ and $K = \{\varepsilon, \beta\}$ be subgroups of S_3. Then $H \cup K = \{\varepsilon, \alpha, \beta\}$ is not a subgroup of S_3.

15. a. yes **b.** no **c.** yes **d.** no **e.** yes

17. Hint: define $f:\mathbb{N} \rightarrow \mathbb{N}$ and $g:\mathbb{N} \rightarrow \mathbb{N}$ as follows:

$$f(n) = \begin{cases} n+1 & \text{if } n \text{ is odd} \\ n-1 & \text{if } n \text{ is even} \end{cases} \qquad g(n) = \begin{cases} 1 & \text{if } n = 1 \\ n+1 & \text{if } n \text{ is even} \\ n-1 & \text{if } n \text{ is odd, } n > 1 \end{cases}$$

19. Since the identity permutation $\varepsilon \in H_A$, it follows that $H_A \neq \phi$. Let $a \in A$ and let $f, g \in H_A$. Since $g(a) = a$, $g^{-1}(a) = a$, and so $(f \circ g^{-1})(a) = f(g^{-1}(a)) = f(a) = a$. Thus $f \circ g^{-1} \in H_A$. Therefore, by the result of Problem 7, H_A is a subgroup.

21. a. 0, 1 **b.** $\{f \in \mathcal{F}(\mathbb{R}) \mid \text{for each } x, \text{ either } f(x) = 0 \text{ or } f(x) = 1\}$

23. b. $U_8 = \{[1], [3], [5], [7]\}$; the Klein four-group

c. $U_{10} = \{[1], [3], [7], [9]\}$; $(\mathbb{Z}_4, \oplus)$

25. a. $(a/b)\cdot(c/d) = (a\cdot b^{-1})\cdot(c\cdot d^{-1}) = a\cdot(b^{-1}\cdot c)\cdot d^{-1} = a\cdot(c\cdot b^{-1})\cdot d^{-1} = (a\cdot c)\cdot(b^{-1}\cdot d^{-1}) = (a\cdot c)\cdot(b\cdot d)^{-1} = (a\cdot c)/(b\cdot d)$

27. a. $ns = n(e\cdot s) = (ne)\cdot s = z\cdot s = z$

b. Suppose, to the contrary, that n is composite, say $n = ab$, $1 < a \leq b < n$. Then $z = ne = (ab)e = (ab)(e\cdot e) = ae\cdot be$. This says that ae is a zero divisor, a contradiction.

29. a. $a = a \vee (c \wedge c') = (a \vee c) \wedge (a \vee c') = (b \vee c) \wedge (b \vee c') = b \vee (c \wedge c') = b$

33. Let $x, y \in G$ and let e_G denote the identity of G.

a. $f_a(x) = f_a(y) \rightarrow a * x = a * y \rightarrow a^{-1} * a * x = a^{-1} * a * y \rightarrow e_G * x = e_G * y \rightarrow x = y$, so f_a is one-to-one. Also, $f_a(a^{-1} * x) = x$, which shows that f_a is onto. If $a \neq e_G$, then $f_a(e_G) = a * e_G = a$; thus f_a is not an automorphism.

b. First, $g_a(x) = g_a(y) \rightarrow a * x * a^{-1} = a * y * a^{-1} \rightarrow x = y$, so g_a is one-to-one. Next, $g_a(a^{-1} * x * a) = x$, so g_a is onto. Last, $g_a(x * y) = a * (x * y) * a^{-1} = (a * x * a^{-1}) * (a * y * a^{-1}) = g_a(x) * g_a(y)$. Therefore, g_a is an automorphism.

37. Let $(L, \preceq)$ be a totally ordered set. Then it is easy to see that $(L, \preceq)$ is a lattice. We must show that the distributive laws are satisfied. Let $a, b, c \in L$. There are six cases to consider, depending on the order of a, b, and c. For example, if $a \preceq b \preceq c$, then $a \wedge (b \vee c) = a \wedge c = a = a \vee a = (a \wedge b) \vee (a \wedge c)$, and $a \vee (b \wedge c) = a \vee b = b = b \wedge c = (a \vee b) \wedge (a \vee c)$. The other five cases are similar.

CHAPTER 7

Section 7.1

1.

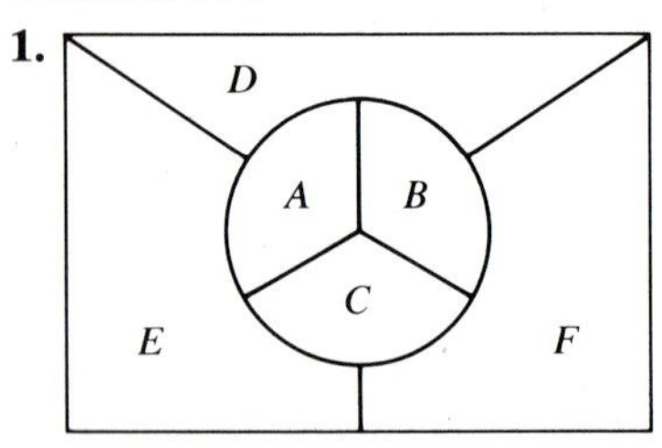

3. b. Suppose the map can be colored using red, blue and green. We may assume, without loss of generality, that J is colored red, K blue, and L green. Then D must be green, E blue, and F red. But then G is adjacent to a country of each color, a contradiction.

Section 7.2

1. Such a graph must have seven edges; an example of one is shown here.

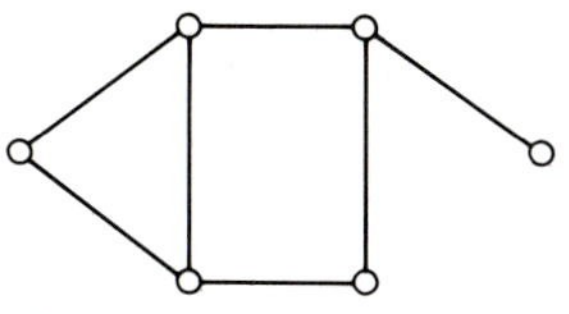

3. a.

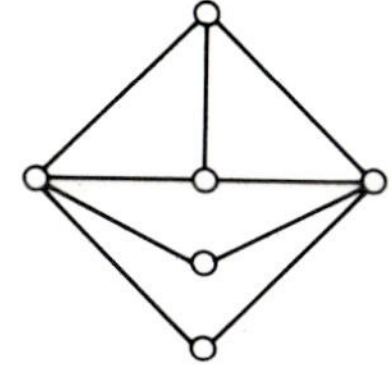

c. Having a vertex of degree 4 implies that the order is at least 5, a contradiction.

5. Hint: model the situation with a graph, representing each person by a vertex and making two vertices adjacent if the corresponding persons know one another.

9. a.

G_1

G_2

b.

G_1

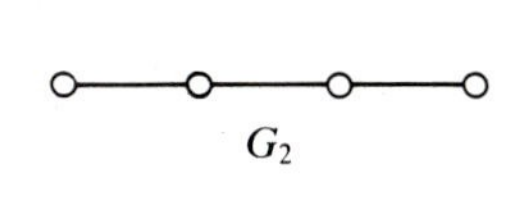

G_2

c.

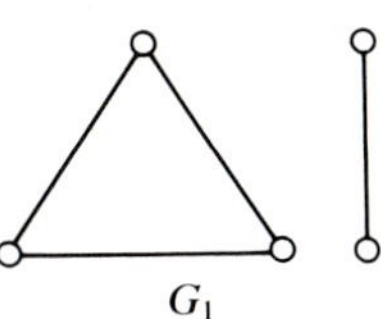

G_1

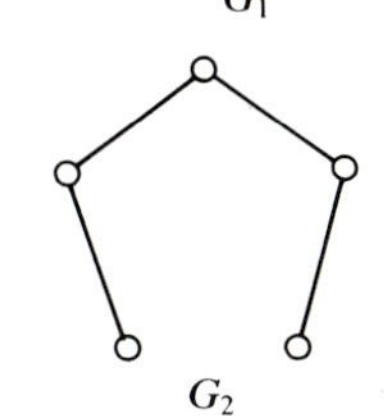

G_2

11. a. $\phi: V(G_1) \to V(G_2)$ defined by $\phi(u_1) = v_4$, $\phi(u_2) = v_2$, $\phi(u_3) = v_5$, $\phi(u_4) = v_1$, $\phi(u_5) = v_6$, $\phi(u_6) = v_3$ is an isomorphism; G_1 has size 7 and G_3 has size 6, so G_1 is not isomorphic to G_3.

c. $\phi: V(G_1) \to V(G_3)$ defined by $\phi(u_1) = x_2$, $\phi(u_2) = x_6$, $\phi(u_3) = x_4$, $\phi(u_4) = x_3$, $\phi(u_5) = x_1$, $\phi(u_6) = x_5$ is an isomorphism. Since G_2 contains three mutually adjacent vertices but G_1 does not, G_1 is not isomorphic to G_2.

13. a. $kp/2$ **b.** Since $kp/2$ must be an integer, k or p must be even.

15. Hint: count the number of edges in G two different ways: by summing the degrees of the vertices in V_1, and then by doing the same thing for V_2.

17. c. Let $V(G) = (V_1, V_2)$, where $V_1 = \{x_1, x_2, \ldots, x_{r+1}\}$ and $V_2 = \{y_1, y_2, \ldots, y_{r+1}\}$. Let $E(G) = \{x_i y_j \mid i \neq j\}$; that is, x_i is adjacent to each $y \in V_2$ except y_i.

19. a. $p = 6$ and $q = 9$; in fact, G is isomorphic to either $K_{3,3}$ or to the following graph:

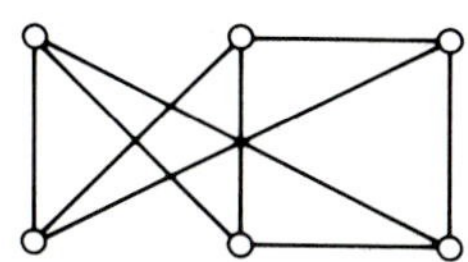

b. G is isomorphic to K_5.

21. a. 3-partite **c.** 4-partite

23. a.

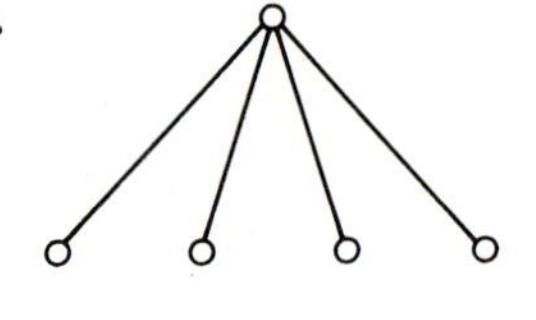

c.

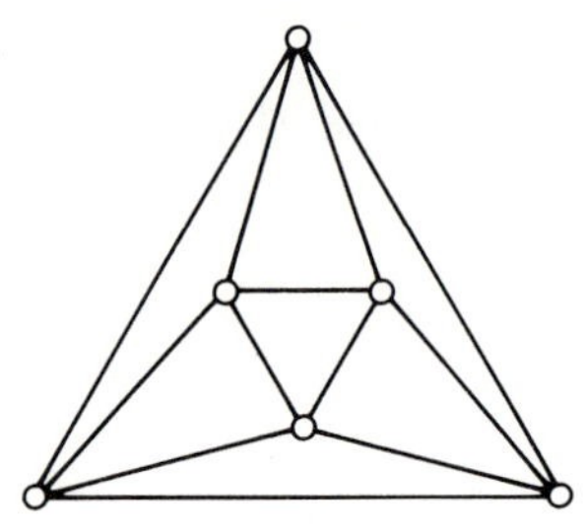

e.

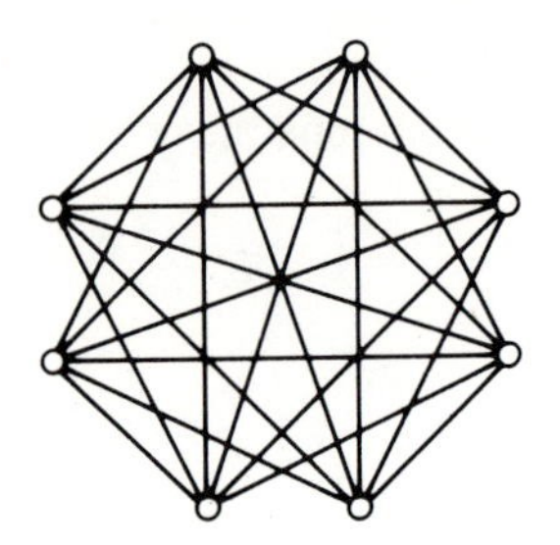

25. a. A possible schedule is given here.

Week: 1	2	3	4	5
A vs. *B*	*A* vs. *C*	*A* vs. *D*	*A* vs. *E*	*D* vs. *E*
C vs. *D*	*B* vs. *D*	*B* vs. *F*	*B* vs. *G*	*F* vs. *G*
E vs. *F*	*E* vs. *G*	*C* vs. *G*	*C* vs. *F*	bye for *A*, *B*, *C*
bye *G*	bye *F*	bye *E*	bye *D*	

Section 7.3

1. a.

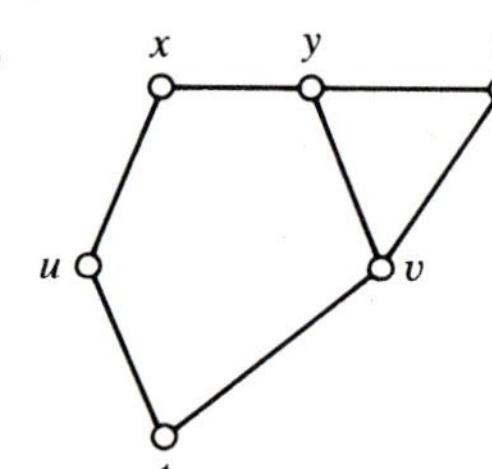

c.

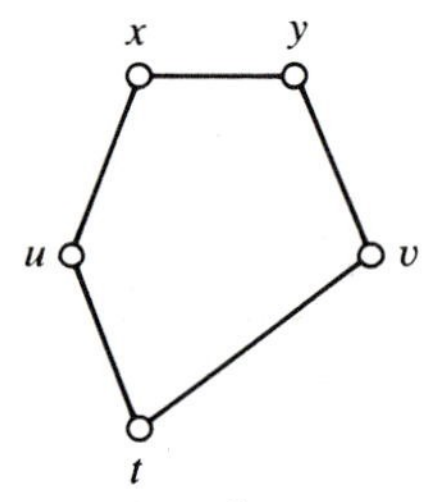

3. a. $H_1 = ((G - u) - t) - vz$ **c.** $E(H_3) = \{ux, xy, yz, wz, vw, tv, tu\}$

5.

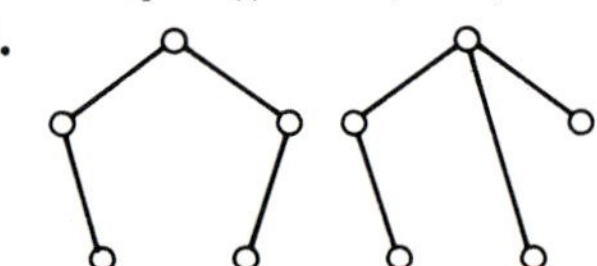

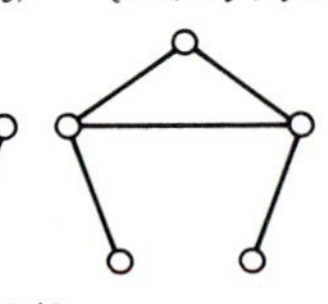

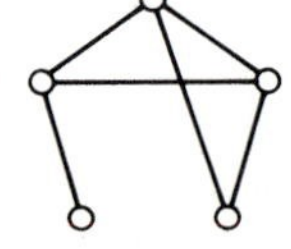

7. a. $q \geq p - 1$ **b.** $q \leq (p - 1)(p - 2)/2$

9. a. Consider the graph consisting of two disjoint copies of K_p.

b. Hint: prove the contrapositive: assume that G is disconnected, and let $C_1, C_2, \ldots, C_r$ $(r \geq 2)$ be the components of G.

13. a. u, x, v, z, u, x, y **b.** u, v, x, u, z, y **c.** u, x, y

d. u, z, v, x, y **e.** u, x, y, z, v, u **f.** u, z, v, x, u

15. a. x, y, z, w, t, v, w, x **c.** $y, x, u, t, w, x, t, v, y, z, w, v, z$

19. a. $A, C, E, G, H, F, E, B, A, D, H, B, C, F, G, B$

b. $A, C, E, G, H, B, C, F, G, B, E, F, H, D, A$

21. $A, D, H, G, B, E, F, C, A$

Section 7.4

1.

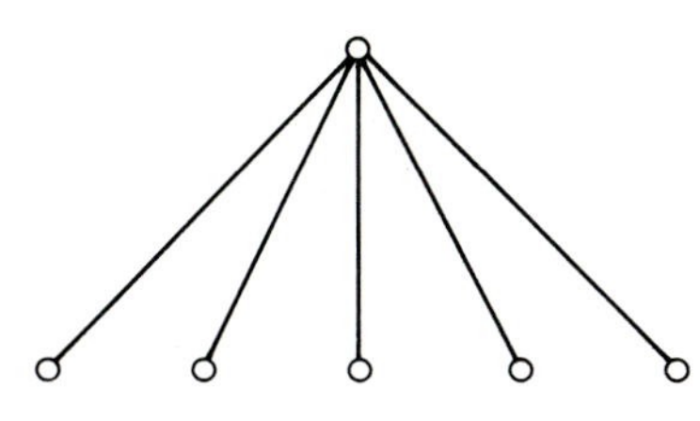

3. a. Hint for showing sufficiency: let P be a path in T of maximum length, and show $P = T$.

b. Let x_0 be the vertex of degree $n - 2$, and let $x_1, \ldots, x_{n-2}$ be the vertices of T adjacent to x_0. If x_{n-1} is the remaining vertex of T, then x_{n-1} is adjacent to some x_i, $1 \le i \le n - 2$. In all cases, the same graph (up to isomorphism) results.

5. Hint: there exist two such trees of order 6.

7. *Method 1:* Let T have order p and let ε denote the number of vertices of degree 1 in T. Summing degrees, we obtain $2p - 2 \ge \varepsilon + \Delta + 2(p - \varepsilon - 1)$. Use this inequality to get the desired result. *Method 2:* The result is true for $\Delta \le 2$ by Corollary 7.4.2, so assume $\Delta \ge 3$. Let $\deg u = \Delta$, where $v_1, v_2, \ldots, v_\Delta$ are adjacent to u in T. Now consider the components of $T - u$; argue that each component contributes a vertex of degree 1 to T.

9.

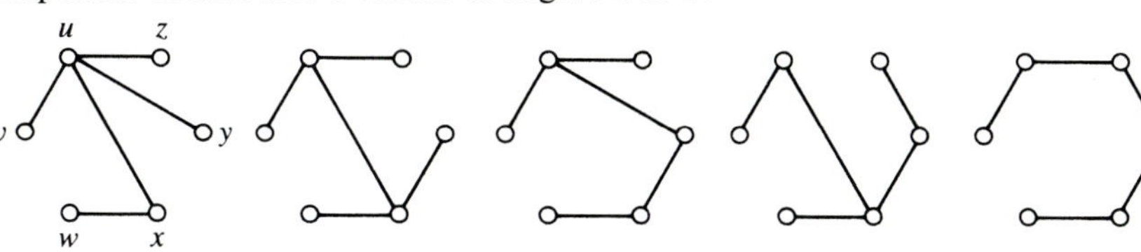

11. a.

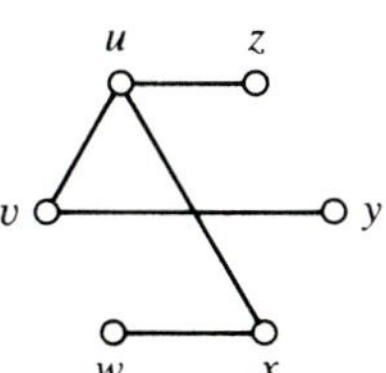

b. Hint: Assume G has order p and size q. Use induction on q. If $q = p - 1$, then $T = G$. If $q > p - 1$, let e be an edge of G that is not a bridge. Use the induction hypothesis to obtain a spanning tree T' of $G - e$ that is distance-preserving from v (relative to $G - e$). The graph H obtained by adding the edge e to T' contains a unique cycle. Let T be obtained from this cycle by removing the edge f that is "furthest" from v in H.

13. Hint: use induction on n.

15. a. P: x_1, x_2, x_3, x_4, x_5 **c.** $n - m \le 1$

e. Hint: Let P: $u_0, u_1, \ldots, u_k$ be a path of maximum length in G. If $k < p - 1$, show either $u_0u_k \in E(G)$ or, by the condition, $u_0u_t, u_{t-1}u_k \in E(G)$ for some t, $1 < t < k$. This yields a cycle C containing the vertices $u_0, u_1, \ldots, u_k$. Let x be a vertex of G not on C such that $xu_i \in E(G)$ for some i, $1 \le i \le k - 1$. Using this edge and all but one edge of C, construct a path P' of G of length $k + 1$, which contradicts the choice of P.

17. Hint: Let $e = uv$ be a bridge of G and assume, to the contrary, that every vertex of G has even degree. How many vertices of odd degree are there in the component G_u of $G - e$ containing the vertex u?

19.

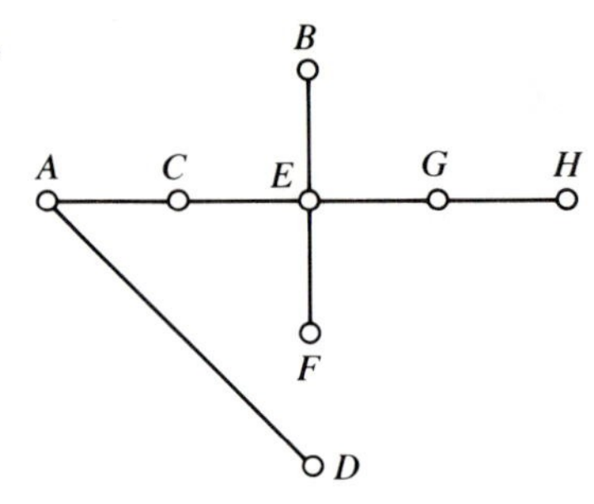

Section 7.5

1. a. x_2, x_5, x_4, x_3, x_2 **b.** $x_6, x_1, x_4, x_3, x_2, x_5$ **c.** $x_3, x_6, x_1, x_4, x_3, x_2, x_5, x_4$
d. $x_3, x_6, x_1, x_2, x_5, x_4, x_3, x_6, x_1, x_4, x_3, x_6, x_5$

5. a. Let $D_1 = (V_1, A_1)$ and $D_2 = (V_2, A_2)$ be digraphs. A mapping $\phi: V_1 \to V_2$ is called an *isomorphism* if (i) ϕ is one-to-one and onto, and (ii) $(u, v) \in A_1 \leftrightarrow (\phi(u), \phi(v)) \in A_2$.
b. $\phi: V(D_1) \to V(D_2)$ defined by $\phi(a) = y$, $\phi(b) = v$, $\phi(c) = x$, $\phi(d) = z$, $\phi(e) = u$ is an isomorphism.

7. See the digraph D_1 of Figure 7.26.

11. b. Hint: It's true: use induction on p. Let D' be an $(r - 1)$-regular digraph of order $p - 1$. Form an r-regular digraph D of order p by adding a new vertex and appropriate arcs to D'.

13. Hint: Let u and v be distinct vertices of D; we claim that v is reachable from u. If $(u, v) \in A(D)$, we're done, so assume $(u, v) \notin A(D)$. Now apply the condition to show there exists a vertex w such that both (u, w) and (w, v) are arcs of D.

17. a.

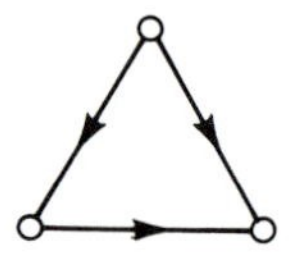

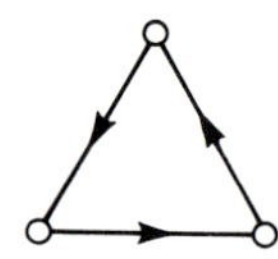

c. (See the definition of "score sequence" in Exercise 19 of this section.) There are tournaments of order 5 having the following score sequences:
0, 1, 2, 3, 4; 1, 1, 1, 3, 4; 0, 2, 2, 2, 4; 1, 1, 2, 2, 4; 0, 1, 3, 3, 3; 0, 2, 2, 3, 3; 1, 1, 2, 3, 3; 1, 2, 2, 2, 3; 2, 2, 2, 2, 2.

19. a. 0, 1, 2, 3; 1, 1, 1, 3; 0, 2, 2, 2; 1, 1, 2, 2
b. Hint: a possible score sequence is 1, 2, 2, 2, 3.
f. (i) 1, 2, 2, 2, 3 $\leftrightarrow$ 1, 1, 2, 2 $\leftrightarrow$ 1, 1, 1 $\leftrightarrow$ 0, 1. Therefore, yes.
(ii) 0, 1, 2, 3, 3, 5, 6, 7 $\leftrightarrow$ 0, 1, 2, 3, 3, 5, 6 $\leftrightarrow$ 0, 1, 2, 3, 3, 5 $\leftrightarrow$ 0, 1, 2, 3, 3 $\leftrightarrow$ 0, 1, 2, 2 $\leftrightarrow$ 0, 1, 1 $\leftrightarrow$ 0, 0. Thus, no.

21. a.

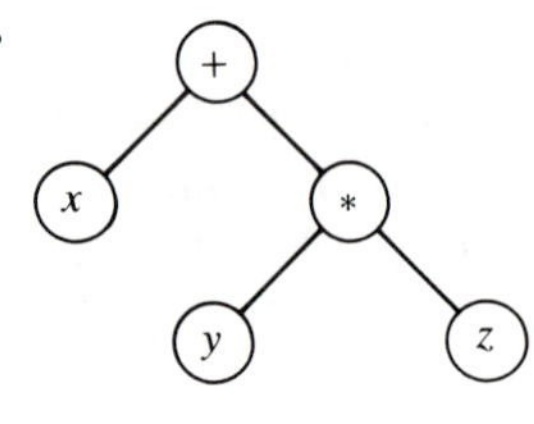

b.

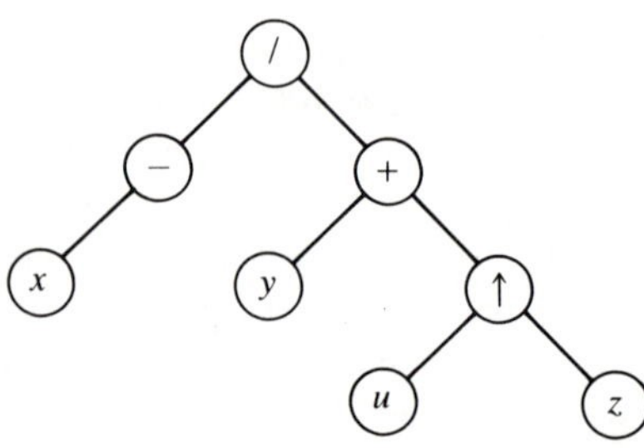

23. a. n **b.** 0 if $n = h$; $m + m^2 + \cdots + m^{h-n}$ if $n < h$

25. **a.** $a, b, e, k, f, c, g, m, b, d, h, i, j$
b. This is the prefix form of the expression $(a + b) * (c/\text{SQRT}(d))$.

Section 7.6

1. **a.** 5 **c.** 4
3. The chromatic number of the Petersen graph is 3.
5. If T is a nontrivial tree, then $\chi(T) = 2$. (This is easily proven by induction on the order of T. We can also apply Exercise 10 of Section 7.4.)
9. **a.** 3 **b.** The associated sets are pairwise disjoint.
c. No, since there is only one set having cardinality 2 and all the others have cardinality 3.
11. Let $\chi(G) = k$, and let $V_1, \ldots, V_k$ be the color classes of a k-coloring of G. Then for each i, $V_i \cap V(H)$ is either empty or an independent set in H. Hence H has an m-coloring for some $m \leq k$. Therefore, $\chi(H) \leq \chi(G)$.
13. **a.** Hint: It's true. Let v be a vertex on the unique odd cycle of G, and consider the graph $G - v$.
b. Assume $\chi(G) = n \geq 6$, and let $V_1, V_2, \ldots, V_n$ be the color classes of an n-coloring. Consider the subgraphs of G induced by $V_1 \cup V_2 \cup V_3$ and $V_4 \cup V_5 \cup V_6$. Each has chromatic number 3; thus each contains an odd cycle, and these odd cycles are vertex-disjoint.
17. If G is disconnected with components $C_1, C_2, \ldots, C_r$, then some C_i must have $\chi(C_i) = k$. Hence G is not k-critical. To show that G contains no cut-vertices, suppose to the contrary that v is a cut-vertex of G and that $H_1, \ldots, H_m$ are the components of $G - v$. Let $G_i = \langle V(H_i) \cup \{v\}\rangle$, $1 \leq i \leq m$. Then, since G is k-critical, $\chi(G_i) \leq k - 1$ for each i. Furthermore, we may assume that v receives the same color, say red, in each of the $\chi(G_i)$-colorings of the G_i. Since the G_i have no edges in common, it follows that $\chi(G) \leq k - 1$, a contradiction.
21. **a.** For K_5, $p = 5$ and $q = 10$, so $q > 9 = 3p - 6$.
b. For $K_{3,3}$, $p = 6$ and $q = 9$, and $K_{3,3}$ does not contain K_3 as a subgraph. So we see that $q = 9 > 8 = 2p - 4$.
23. **a.** planar **c.** nonplanar **e.** nonplanar

Section 7.7

1. (We give only the final labels.)
a. $L(u_0) = 0, L(u_1) = 2, L(u_2) = 3, L(u_3) = 5, L(u_4) = 4$
c. $L(u_0) = 0, L(u_1) = 5, L(u_2) = 2, L(u_3) = 7, L(u_4) = 6, L(u_5) = 17$
3. **b.** The edges that belong to some shortest path from u_0 induce a spanning tree of the graph. This spanning tree is distance-preserving from u_0 (see Exercise 11 of Section 7.4). For the graphs of Figure 7.50(a) and (c):

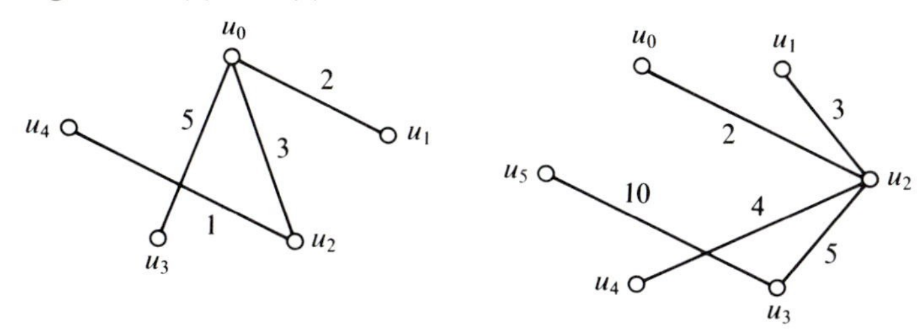

5. **a.** Change Step 1 as follows: For each $v \in A_i'$, let $L(u') = \min_{u \in A_i}\{L(u) + w(uv)\}$. If $L(u') + w(u'v) < L(v)$, then set $\text{Pred}(v) = u'$ and $L(v) = L(u') + w(u'v)$. Determine $\min_{v \in A_i'}\{(L(v)\}$. . . .
b. For Figure 7.50(a), $\text{Pred}(u_1) = \text{Pred}(u_2) = \text{Pred}(u_3) = u_0$, $\text{Pred}(u_4) = u_2$.
For Figure 7.50(c), $\text{Pred}(u_2) = u_0$, $\text{Pred}(u_1) = \text{Pred}(u_3) = \text{Pred}(u_4) = u_2$, $\text{Pred}(u_5) = u_3$.

Section 7.8

1. $E(T) = \{u_0u_4, u_4u_5, u_1u_3, u_1u_2, u_0u_3\}$

3. a. $E(T) = \{u_2u_4, u_0u_1, u_1u_2, u_3u_4\}$ **c.** $E(T) = \{u_0u_2, u_1u_2, u_2u_4, u_2u_3, u_3u_5\}$

5. True. Suppose that T_1 and T_2 are different minimum spanning trees of G. Choose the edge $e \in (E(T_1) - E(T_2)) \cup (E(T_2) - E(T_1))$ of smallest weight. Assume, without loss of generality, that $e \in E(T_2) - E(T_1)$. Then the graph $T_1 + e$ contains a unique cycle. Let f be an edge of this cycle such that $f \notin E(T_2)$. Then, by our assumption about e, $w(f) > w(e)$. Thus the spanning tree $(T + e) - f$ has weight less than the weight of T_1, a contradiction.

Section 7.9

1. a. **c.**

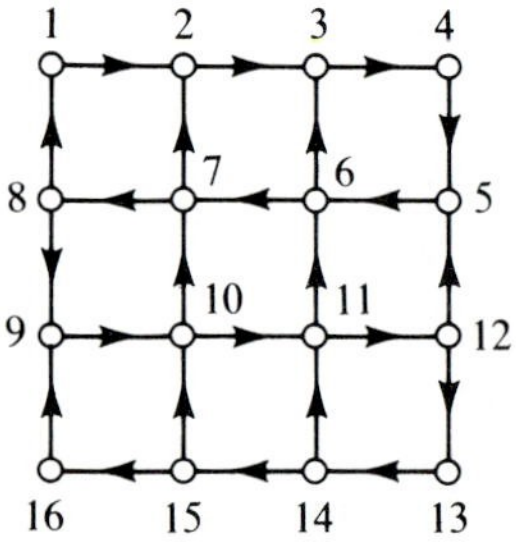

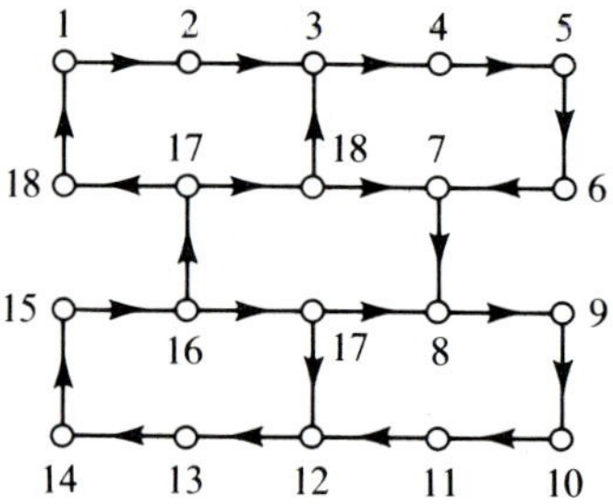

5. Suppose u and v are adjacent vertices of G and, during application of the algorithm, u was labelled before v. At the point when v was labelled, a vertex x adjacent to v of highest label was used. Thus $L(x) \geq L(u)$ and $L(v) = L(x) + 1$. Hence $L(v) > L(u)$.

7. b. G has exactly two strong orientations if and only if G is a cycle.

Chapter 7 problems

1. a. If G is regular, then let $H = G$. If G is not regular, let $f(G)$ denote the graph consisting of two copies of G, call them G and G', together with all edges xx', where $x \in V(G)$, $x' \in V(G')$, x and x' are corresponding vertices, and $\deg_G x < \Delta(G)$. Clearly G is an induced subgraph of $f(G)$. If $f(G)$ is regular, then $H = f(G)$; if not, then apply f again to obtain $f(f(G)) = f^2(G)$, and so on. Notice that if $\Delta(G) - \delta(G) = m$, then $H = f^m(G)$.

b. False; as a counterexample, let G be any graph of odd order with $\Delta(G)$ also odd.

3. a. Let G have size q. Then $\bar{G}$ has size q also. Thus $2q = p(p - 1)/2$, or $q = p(p - 1)/4$. Thus $p \equiv 0$ or $1 (\text{mod } 4)$.

b. We give the construction for $p \equiv 0 (\text{mod } 4)$. Start with two copies of the complete graph $K_{p/4}$ and a copy of the complete bipartite graph $K_{p/4,p/4}$. Add all edges joining the first $K_{p/4}$ with one partite set of $K_{p/4,p/4}$, and all edges joining the second $K_{p/4}$ with the other partite set of $K_{p/4,p/4}$. The resulting graph is a self-complementary graph of order p.

5. Hint for **b:** Let v_i be the vertex of degree d_i, $1 \leq i \leq n$, and let $H = \langle\{v_1, \ldots, v_n\}\rangle$. Then $d_1 + d_2 + \cdots + d_n \leq x + y$, where x is the sum of the degrees of the v_i in H and y is the number of edges of G that join a vertex of H with a vertex of $G - H$.

7. a. $e(s) = e(z) = e(u) = 4$, $e(t) = e(v) = e(x) = e(y) = 3$, $e(w) = 2$

b. $\text{rad } G = 2$, $\text{diam } G = 4$

c. Let $u, v, x \in V(G)$ be such that $d(u, v) = \text{diam } G$ and $e(x) = \text{rad } G$. Then $\text{diam } G = d(u, v) \leq d(u, x) + d(x, v) \leq e(x) + e(x) = 2 \text{ rad } G$.

e. Let u be a vertex of G, and let U_i denote the set of vertices at a distance i from u, $0 \leq i \leq e(u)$. First, to show that $\text{diam } G \leq 3$, assume to the contrary that $e(u) = 4$. Note that $\bar{G}$ contains all edges joining U_i with U_j if $|i - j| \geq 2$. This implies that $\text{diam } \bar{G} \leq 3$, which contradicts the assumption that diam $G \geq 4$, since diam G = diam $\bar{G}$. Next we must show rad $G = 2$; this is clearly true if diam $G = 2$, so assume diam $G = 3$, and let u be a vertex of G with eccentricity 3 in G. With U_3 as above, let $v \in U_3$. Then the eccentricity of v in $\bar{G}$ is easily seen to be 2, which shows that rad $\bar{G} = 2$. Thus rad $G = 2$.

9.

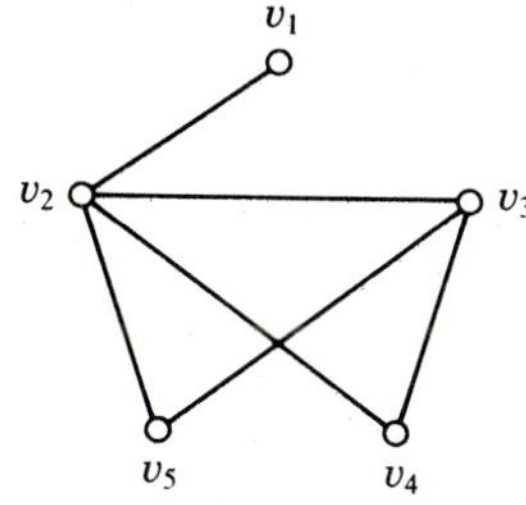

11. Proceed by induction on n. The result is obvious for $n = 1$. Assume, for some $k \geq 1$, that if G is any graph with $\delta(G) \geq k - 1$, and T is any tree of order k, then G contains a subgraph isomorphic to T. Let G be a graph with $\delta(G) \geq k$, and let T be any tree of order $k + 1$. Let v be an end vertex of T and consider the tree $T - v$. Since $\delta(G) \geq k > k - 1$ and T has order k, it follows from IHOP that G contains a subgraph T' isomorphic to $T - v$. Let u be the vertex of T adjacent to v, and let u' be the vertex of T' corresponding to u under the above isomorphism. Since $\deg_G u' \geq k$, and T' has order k, there is a vertex x of G such that $u'x \in E(G)$ and $x \notin V(T')$. Adding the vertex x and the edge $u'x$ to T' yields a tree isomorphic to T.

15. **a.** If G is a $(2,n)$-cage, then G must contain an n-cycle, say C. Since C is 2-regular and has girth n, it follows that $G = C$.

c. $K_{r,r}$

d. The Petersen graph (Figure 7.41) is a unique (3,5)-cage. The "Heawood graph," presented here, is the unique (3,6)-cage.

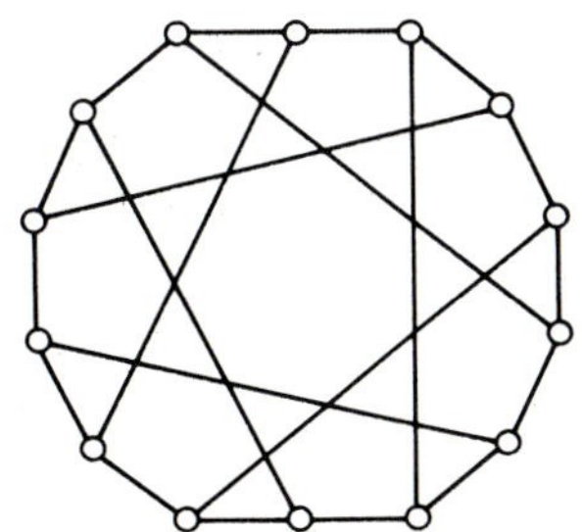

17. **a.** G^2 is as pictured here:

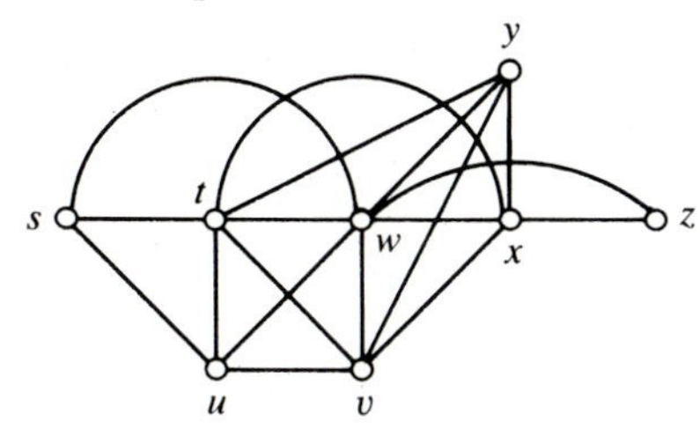

$E(G^3) = E(G^2) \cup \{sy, sv, sx, tz, uy, ux, vz\}$; $E(G^4) = E(G^3) \cup \{sz, uz\}$; $G^n \cong K_8$ for $n \geq 4$

19. **a.** 21

b. Consider the root of the tree. If it is an integer, then that integer is the value of the expression.

Otherwise, find the values of the expressions represented by the left and right subtrees of the root, and apply the operator at the root to these values to find the value of the expression.

c.

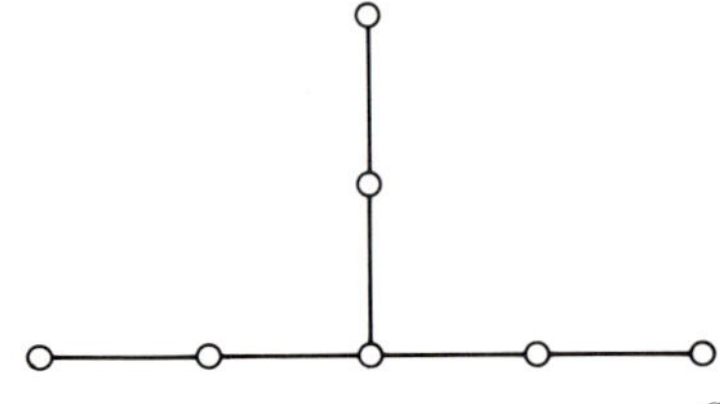

21. a.

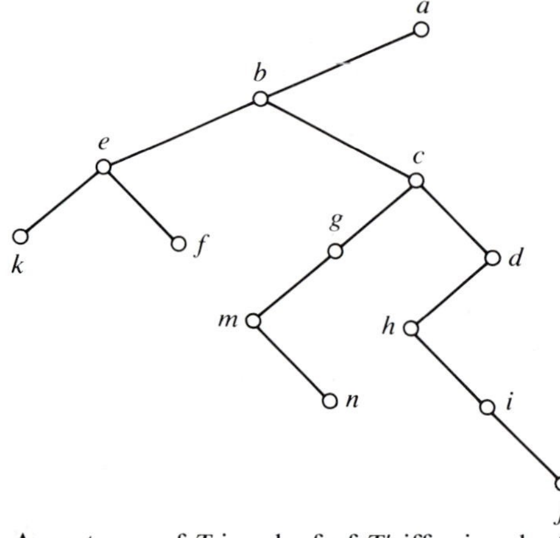

b. A vertex v of T is a leaf of T' iff v is a leaf of T and v is the rightmost child of its parent in T.

c. mh

d. yes

23. a. Let G consist of disjoint copies of K_k and $\bar{K}_{p-k}$.

b. Hint: Apply the result of Exercise 14 of Section 7.6. Let $k = \max(\delta(H))$, where the maximum is taken over all induced subgraphs H of G. Show that every induced subgraph of $\bar{G}$ has minimum degree at most $p - k - 1$.

27. For $n = 2$: $p_1 = 1$ and $p_2 \geq 1$, or $p_1 = 2$ and $p_2 \geq 2$. For $n = 3$: $p_1 = p_2 = 1$ and $p_3 \geq 1$, or $p_1 = 1$ and $p_2 = p_3 = 2$, or $p_1 = p_2 = p_3 = 2$. For $n = 4$: $p_1 = p_2 = p_3 = 1$ and $p_4 = 1$ or 2.

29. a. $\nu(K_5) \geq 1$ since K_5 is nonplanar; here is a drawing of K_5 with one crossing:

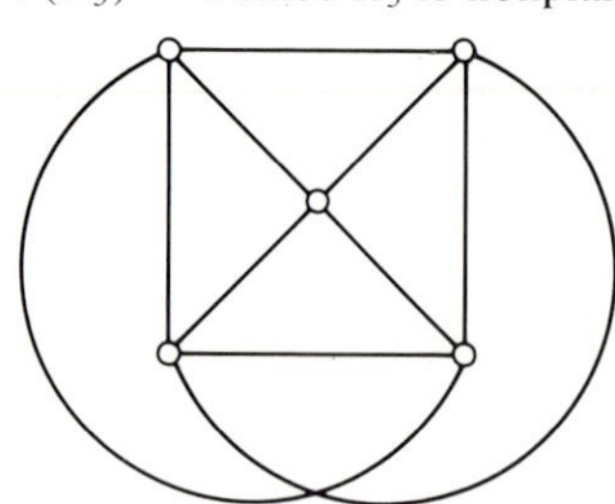

c. Here is a drawing of K_6 with three crossings:

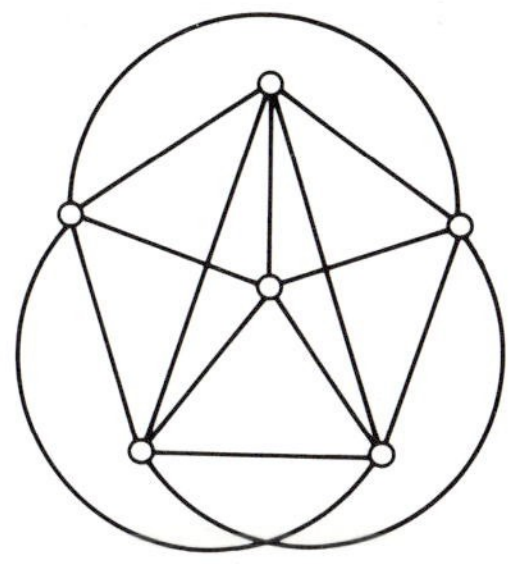

31. b. For the graph shown, $\{x, y\}$ is a minimal dominating set but is not an independent set.

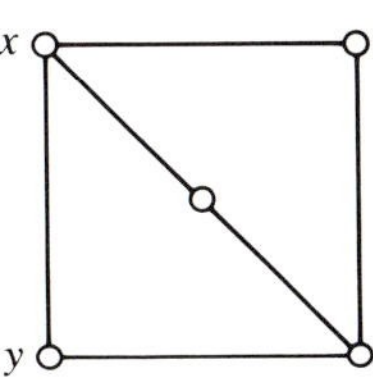

c. For $G = K_{m,m+n}$, we have $\beta(G) = m + n$ and $\sigma(G) = m$.

35. a. *Step 0.* Initially, choose an arbitrary vertex u, set $U = \{u\}$ and $V = V(G) - \{u\}$.

Step 1. Let $v \in V$. If v is adjacent to some vertex of U, replace V by $V - \{v\}$; else replace U by $U \cup \{v\}$ and V by $V - \{v\}$.

Step 2. If $V \neq \phi$, repeat Step 1; else U is a maximal independent set in G.

b. Let G be a path of order 4, and suppose the first U chosen consists of the end vertices of G. Then the "algorithm" will compute $\chi(G) = 3$, whereas the correct value is $\chi(G) = 2$.

c. It does produce an upper bound, namely n, on $\chi(G)$.

Index of Notation

Notation	Meaning	Page
Logic		
$p \vee q$	p or q	4
$p \wedge q$	p and q	4
$\sim p$	not p	4
$p \rightarrow q$	if p then q	9
$p \leftrightarrow q$	p if and only if q	10
$u \equiv v$	u is logically equivalent to v	15
$\forall$	for all	24
$\exists$	there exists	24
Sets of Numbers		
$\mathbb{N}$	the set of positive integers	47
$\mathbb{Z}$	the set of integers	47
$\mathbb{Q}$	the set of rational numbers	48
$\mathbb{R}$	the set of real numbers	48
(a, b)	the open interval from a to b: $\{x \in \mathbb{R} \mid a < x < b\}$	54
$[a, b]$	the closed interval from a to b: $\{x \in \mathbb{R} \mid a \leq x \leq b\}$	54
$[a, b)$	the half-open interval from a to b: $\{x \in \mathbb{R} \mid a \leq x < b\}$	54
$(a, b]$	the half-open interval from a to b: $\{x \in \mathbb{R} \mid a < x \leq b\}$	54
(a, ∞)	$\{x \in \mathbb{R} \mid x > a\}$	55
$[a, \infty)$	$\{x \in \mathbb{R} \mid x \geq a\}$	55
$(-\infty, b)$	$\{x \in \mathbb{R} \mid x < b\}$	55
$(-\infty, b]$	$\{x \in \mathbb{R} \mid x \leq b\}$	55
$(-\infty, \infty)$	$\mathbb{R}$	55
Sets		
$x \in A$	x is an element of A	48
$A = B$	A and B are equal	48
ϕ	the empty set	50

Cardinality

Divisibility

Relations and Functions

Notation	*Meaning*	*Page*
$x \wedge y$	the meet of x and y	150
$x \vee y$	the join of x and y	150
$f: A \to B$	f is a function with domain A and codomain B	165
$f(x)$	the image of the element x under f	166
$f(C)$	the image of the set C under f	174
$f^{-1}(D)$	the inverse image of the set D under f	174
f^{-1}	the inverse function of f	177
$g \circ f$	the composite function of f with g	178
I_A	the identity function on the set A	180
Algebra		
$(\mathcal{F}(A), \circ)$	the semigroup of functions on the set A under composition	194
$(\mathcal{S}(A), \circ)$	the group of permutations of the set A under composition	194
$(\mathbb{Z}_m, \oplus, \odot)$	the ring of integers modulo m	211
S_m	the symmetric group of degree m	202
a^{-1}	the inverse of an element a in a group	203
$\|a\|$	the order of an element a in a group	206
$\|G\|$	the order of a group G	206
$(G, *) = \langle a \rangle$	the group $(G, *)$ is cyclic with generator a	207
Q_8	the quaternion group	208
D_4	the group of symmetries of the square	210
z	the zero element in a ring	212
$-a$	the additive inverse of an element a in a ring	212
e	the (multiplicative) identity in a ring with identity	213
$(\mathbb{Z}[\sqrt{2}], +, \cdot)$	the ring of real numbers of the form $a + b\sqrt{2}$, where $a, b \in \mathbb{Z}$	212
$(\mathcal{F}(\mathbb{R}), +, \cdot)$	the ring of functions from $\mathbb{R}$ to $\mathbb{R}$ under pointwise addition and multiplication of functions	213
$S^{\#}$	$S - \{z\}$, where S is a ring and z is the zero element of S	216
$(\mathbb{Q}[\sqrt{2}], +, \cdot)$	the field of real numbers of the form $a + b\sqrt{2}$, where $a, b \in \mathbb{Q}$	219
1	unity for a lattice	224
0	zero for a lattice	224
a'	the complement of an element a in a lattice	224
$(\mathcal{P}(U), \subseteq)$	the Boolean algebra of subsets of a set U	225
$(G, *) \cong (H, \#)$	groups $(G, *)$ and $(H, \#)$ are isomorphic	237

Notation	*Meaning*	*Page*
$(U_n, \odot)$	the group of invertible elements in the ring $(\mathbb{Z}_n, \oplus, \odot)$	247
Graphs		
$G = (V, E)$	G is a graph with vertex set V and edge set E	251
uv	the edge incident with vertices u and v	252
$V(G)$	the vertex set of G	251
$E(G)$	the edge set of G	251
$G_1 = G_2$	graphs G_1 and G_2 are equal	252
$G_1 \cong G_2$	graphs G_1 and G_2 are isomorphic	254
K_p	the complete graph of order p	254
$K_{r,s}$	the complete bipartite graph with partite sets of cardinalities $r \leq s$	255
$\deg u$	the degree of the vertex u	256
$\Delta(G)$	the maximum degree of G	256
$\delta(G)$	the minimum degree of G	256
$H \subseteq G$	H is a subgraph of G	261
$G - v$	the graph obtained from G by deleting the vertex v and all edges of G incident with v	261
$G - e$	the graph obtained from G by deleting the edge e	262
$\langle U \rangle$	the subgraph of G induced by U	262
$d(u, v)$	the distance from u to v	265
P_n	the path of order n	270
$D = (V, A)$	D is a digraph with vertex set V and arc set A	278
(u, v)	the arc from u to v	278
$\text{id } v$	the indegree of vertex v	283
$\text{od } v$	the outdegree of vertex v	283
$\chi(G)$	the chromatic number of G	291
C_p	the cycle of order p	291
W_p	the wheel of order p	300
$G_1 + G_2$	the join of G_1 and G_2	299
$\bar{G}$	the complement of G	316

The Greek Alphabet

A, α	alpha	N, ν	nu
B, β	beta	Ξ, ξ	xi
Γ, γ	gamma	O, o	omicron
Δ, δ	delta	Π, π	pi
E, ε	epsilon	P, ρ	rho
Z, ζ	zeta	Σ, σ	sigma
H, η	eta	T, τ	tau
Θ, θ	theta	Υ, υ	upsilon
I, ι	iota	Φ, ϕ	phi
K, κ	kappa	X, χ	chi
Λ, λ	lambda	Ψ, ψ	psi
M, μ	mu	Ω, ω	omega

Index

D0880800